The Tools of Biochemistry

The Tools of Biochemistry

Terrance G. Cooper

University of Pittsburgh

A Wiley-Interscience Publication

JOHN WILEY & SONS, New York / London / Sydney / Toronto

Library of Congress Cataloging in Publication Data:

Cooper, Terrance G.   1942–
　The tools of biochemistry.

　"A Wiley-Interscience publication."
　Includes bibliographical references and index.
　1. Biological chemistry—Technique. I. Title.
QP519.7.C66 C.2 574.1'92'028   76-30910
ISBN 0-471-17116-6

Printed in the United States of America

10 9 8 7 6 5 4 3 2

To Carol, Mark, and Gregory

Preface

An essential condition for fruitful research is to have satisfactory techniques at one's disposal. Progress in biological sciences has largely depended on parallel advances in the technology of measuring and observing devices. Unfortunately, a lack of understanding of the tools used is commonly a weakness of many investigations. Careful evaluation is needed, not only of experimental data but also of the methods by which they were obtained.

The purpose of this work is to increase the number and quality of biochemical techniques available to the reader. Often students, during their undergraduate and graduate studies, become acquainted with only the techniques used in their immediate environments. They may find it difficult to apply entirely new approaches and techniques to a problem because they lack a general source of information that presents the potential and limitations of a particular method as well as an opportunity to use it under well characterized conditions. It is to this difficulty that I address this book. Although the text forms the basis of a senior level undergraduate course, it is intended also to serve as a ready source of useful information for graduate students and experienced investigators. Its purpose is to enlarge the sphere of the reader's experience, and its success will be measured by the degree to which this occurs.

The format used is to first present the theoretical basis and limitations of each technique. This section provides an understanding of the reasons for procedures that are subsequently performed but is not designed to be an absolutely comprehensive treatise; an experimental section follows in which a number of experiments using the method are described in detail. The particular experiments were selected because they are straightforward to execute and provide clear examples of data that can be obtained when the method is used properly. To permit application of this material to the greatest range of teaching and research requirements, each chapter is a self-contained unit. It is therefore possible to select those methods that are appropriate to the reader's needs and resources.

I wish to express my appreciation to Professor David Krogmann for providing the early stimulus and enthusiasm needed to undertake this work. Without his help it would still be just an idea. Many of my students and colleagues have read these chapters and offered advice for their improvement. This and the many hours spent by my wife Carol reading and rereading the text have avoided countless errors, confusing statements, and insults to the English language. I did not, however, follow their good advice on all occasions and am solely responsible for any problems that still remain. For technical preparation of the manuscript I am indebted to Mrs. Sandra Wight. The many photographs generously provided by independent investigators and by manufacturers of scientific equipment and supplies enhance this work and are greatly appreciated. To the staff at John Wiley and Sons I am grateful for unlimited patience and assistance during development and publication of this work.

T. G. COOPER

Pittsburgh, Pennsylvania
January 1977

Contents

The Tools of Biochemistry

Chapter 1

Potentiometric Techniques

Most of the chemical reactions that comprise a living organism are profoundly influenced by hydrogen ion concentration. So important is this characteristic that multicelled organisms have evolved a variety of sophisticated methods to maintain the solutions in which their cells are bathed within rigid limits of hydrogen ion concentration. The same care exercised by living organisms to maintain acceptable hydrogen ion concentrations must be duplicated in the laboratory if meaningful insights are to be gained into the functioning of organisms and their components. The subsequent discussion employs the Brønsted–Lowry definition of acids and bases: an acid is a compound that donates protons and a base is one that accepts protons. This definition may also be formulated as an acid dissociating into a base and a proton:

$$\text{acid} \rightleftharpoons \text{base} + H^+ \tag{1}$$

Therefore, HCl would be considered an acid and Cl^- would be its conjugate base.

Acid	Conjugate Base
HCl	Cl^-
CH_3COOH	CH_3COO^-
H_2CO_3	HCO_3^-
HCO_3^-	CO_3^{2-}
NH_4^+	NH_3

Acids and bases can be classified as strong or weak depending on the extent to which they ionize. A strong acid is one for which reaction 1

1

proceeds far to the right; that is, the acid is essentially totally ionized. For example, the hydrogen ion concentration of a $0.01M$ solution of the strong acid HCl is $0.01M$ because all of the acid is dissociated. On the other hand, a weak acid (such as acetic, boric, or carbonic) is one for which reaction 1 does not proceed significantly to the right.

pH CALCULATIONS

A Danish chemist, S. P. L. Sorensen, proposed a convenient notation for the hydrogen ion concentration of a solution. He defined the negative log of the hydrogen ion concentration as pH.

$$pH = -\log [H^+] \tag{2}$$

Therefore, for a $0.01M$ solution of HCl,

$$pH = -\log [10^{-2}]$$
$$= 2.0$$

The pH values most often encountered in biochemistry range from 4 to 11. Figure 1-1 depicts the relationship of pH to acidity and basicity or alkalinity.

It is clear that the pH of a $10^{-2}M$ solution of a strong acid is 2.0, but the

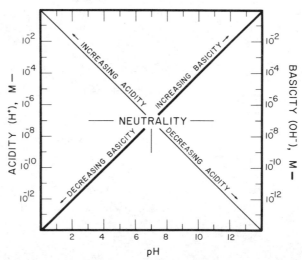

Figure 1-1. The relationship of acidity and basicity or alkalinity of a solution to its hydrogen ion and hydroxyl ion concentrations.

pH of a 0.01M solution of a strong base (reaction 3) is less obvious.

$$NaOH \rightleftharpoons Na^+ + OH^- \tag{3}$$

The hydroxyl ion concentration can be related to the hydrogen ion (H^+), or more accurately hydronium ion (H_3O^+), concentration by considering the dissociation of water:

$$H_2O \rightleftharpoons H^+ + OH^- \tag{4}$$

The equilibrium constant equation for this reaction is

$$K_{eq} = \frac{[H^+][OH^-]}{[H_2O]} \tag{5}$$

and thus

$$K_{eq}[H_2O] = [H^+][OH^-] \tag{6}$$

Since the concentration of water remains about constant throughout the ionization process this term can be combined with the equilibrium constant, generating a new constant, K_W:

$$K_W = [H^+][OH^-] \tag{7}$$

For pure water at 25°C both the hydronium and hydroxyl ion concentrations are equal to $1 \times 10^{-7}M$. Therefore,

$$K_W = (1 \times 10^{-7})(1 \times 10^{-7})$$
$$= 1 \times 10^{-14}$$

Since in aqueous solution, the product of the hydronium and hydroxyl ion concentrations must remain constant at $1 \times 10^{-14}M$, an increase in one term of equation 7 requires a corresponding decrease in the other term. Therefore, a 0.01M NaOH aqueous solution has a hydrogen ion concentration of

$$[H^+] = \frac{K_W}{[OH^-]} \tag{8}$$
$$= \frac{10^{-14}}{10^{-2}}$$
$$= 10^{-12}$$

and

$$pH = 12$$

Weak acids, by definition, are only partially ionized in aqueous solution.

$$HA \rightleftharpoons H^+ + A^- \tag{9}$$

The concentration of each species at equilibrium may be calculated from the dissociation constant of the acid as described by the equation

$$K_a = \frac{[H^+][A^-]}{[HA]} \qquad (10)$$

Rearrangement of equation 10 gives

$$[H^+] = \frac{[HA]K_a}{[A^-]} \qquad (11)$$

If the negative logarithms of both sides of equation 11 are taken,

$$-\log [H^+] = (-\log K_a) + \left(-\log \frac{[HA]}{[A^-]}\right) \qquad (12)$$

or

$$pH = pK_a + \log \frac{[A^-]}{[HA]} \qquad (13)$$

In more general terms,

$$pH = pK_a + \log \frac{[\text{conjugate base}]}{[\text{undissociated acid}]} \qquad (14)$$

This equation is known as the Henderson–Hasselbach equation. If

$$[A^-] = [HA]$$

then

$$pH = pK_a \qquad (15)$$

Many handbooks list dissociation constants (the equilibrium constant for the dissociation of acid into a proton and its conjugate base) as pK_a's.

Thus far it has been assumed that the molar concentration of any given ion is also its effective or active concentration. This, however, is true only at very low ion concentrations. As the number of ions in a given volume increases, the probability of ionic interactions also increases. These interactions tend to impede the movement of ions and hence decrease their effective concentration or activity. Activity is related to molar concentration by a normalization factor or activity coefficient

$$a_i = f_i[i] \qquad (16)$$

where a_i is the activity of ionic species i, and f_i is the activity coefficient. When ions are separated from one another (at low concentration) f_i approaches unity; f_i decreases as the concentration of i increases. The distinction between the activity and molar concentration of an ion is significant because all potentiometric measurements of hydrogen ion concentration yield hydrogen ion activity, not concentration.

Occasions arise when it is useful to calculate the pH of a solution resulting from addition of a known concentration of weak acid or base to water. The pH may be determined by considering the dissociation of the acid.

$$HA \rightleftharpoons H^+ + A^-$$

The dissociation is described by the equation

$$K_a = \frac{[H^+][A^-]}{[HA]}$$

In a system containing only a weak acid, it is reasonable to assume that $[H^+] = [A^-]$ and therefore

$$K_a = \frac{[H^+]^2}{[HA]} \qquad (17)$$

Equation 17 may be rewritten as

$$[H^+]^2 = K_a[HA] \qquad (18)$$

or

$$[H^+] = \sqrt{K_a[HA]} \qquad (19)$$

In logarithmic terms,

$$pH = \tfrac{1}{2}(pK_a - \log[HA])$$

or

$$pH = \frac{pK_a - \log[HA]}{2} \qquad (20)$$

The pH obtained on dissolving a weak base in water is arrived at in a similar manner except that K_b is used in place of K_a and the OH$^-$ concentration must be related to the H$^+$ concentration by equation 8.

The pH of solutions of the salts of weak acids or bases in water may be calculated using the relationships described above. However, here the hydrolysis of the salt must be considered. For example, NH$_4$Cl, the salt of the strong acid HCl and weak base NH$_4$OH, dissociates in water as follows:

$$NH_4Cl \rightleftharpoons NH_4^+ + Cl^- \qquad (21)$$

Since NH$_4$OH is a weak base, the free NH$_4^+$ ion reacts with water as follows:

$$NH_4^+ + Cl^- + HOH \rightleftharpoons NH_4OH + H^+ + Cl^- \qquad (22)$$

HCl, being a strong acid, remains totally ionized. Reaction 22 may be alternatively considered as

$$NH_4^+ + HOH \rightleftharpoons NH_4OH + H^+ \tag{23}$$

and described by the equation

$$K_{\text{hydrolysis}} = \frac{[NH_4OH][H^+]}{[NH_4^+][H_2O]} \tag{24}$$

Since the concentration of water is so much higher (approximately $55M$) than any other species, it may be assumed to remain constant. Hence

$$K_h = \frac{[NH_4OH][H^+]}{[NH_4^+]} \tag{25}$$

However, NH_4OH dissociates,

$$NH_4OH \rightleftharpoons NH_4^+ + OH^-$$

and

$$K_b = \frac{[NH_4^+][OH^-]}{[NH_4OH]} \tag{26}$$

If we assume that NH_4Cl is totally ionized, then $[NH_4Cl] = [NH_4^+]$. We also know that $[NH_4OH] = [H^+]$. These two facts can be used to rewrite equation 26:

$$K_b = \frac{[NH_4Cl][OH^-]}{[H^+]} \tag{27}$$

We indicated above that

$$K_w = [H^+][OH^-]$$

and

$$[OH^-] = \frac{K_w}{[H^+]}$$

Therefore

$$K_b = \frac{[NH_4Cl]K_w}{[H^+][H^+]} \tag{28}$$

or

$$[H^+]^2 = \frac{[NH_4Cl]K_w}{K_b} \tag{29}$$

From this point it may be easily shown that

$$pH = \frac{pK_w - pK_b - \log[NH_4Cl]}{2} \tag{30}$$

These considerations may be used to show that the pH obtained on dissolving the salt of a weak acid and strong base is

$$pH = \frac{pK_w + pK_a + \log[salt]}{2} \qquad (31)$$

and the salt of a weak acid and weak base is

$$pH = \frac{pK_w + pK_a - pK_b}{2} \qquad (32)$$

It should be noted that the pK_b of a base can be calculated from the pK_a of its conjugate acid and vice versa using the identity

$$pK_w = pK_a + pK_b \qquad (33)$$

It is left to the reader to verify this identity.

pH MEASUREMENT USING ORGANIC INDICATOR MOLECULES

In the preceding discussion, calculation of the pH for a number of simple solutions was demonstrated. In practice, however, such calculations are not always possible owing to the complexity and often unknown nature of the solution whose pH is sought. Here it is more practical to determine the pH by measurement. These measurements can be made at low resolution using colored indicator dyes or at high resolution using potentiometric methods. Historically, pH measurements were made using organic, chromophoric indicator molecules such as those listed in Table 1-1. These dyes function by changing from one color to another as the pH of their environment is altered. The indicators are themselves weak acids and, therefore, dissociate.

$$\text{H--indicator} \rightleftharpoons \text{H}^+ + \text{indicator}^- \qquad (34)$$

This equilibrium is described by the equation

$$K_a = \frac{[\text{H}^+][\text{indicator}^-]}{[\text{H--indicator}]} \qquad (35)$$

Rearranging gives

$$\frac{K_a}{[\text{H}^+]} = \frac{[\text{indicator}^-]}{[\text{H--indicator}]} \qquad (36)$$

At a pH value substantially below its pK_a, the indicator is largely protonated, and at pH values above its pK_a, it is largely dissociated. The protonation--deprotonation of the indicator usually involves the conversion of a benzenoid moiety to its quinoid form, and it is this structural

Table 1-1. Common Acid–Base Indicators

Indicator	pH Range	Acid Color	Base Color
Methyl violet	0.5–1.5	Yellow	Blue
Thymol blue	1.2–2.8	Red	Yellow
Methyl yellow	2.9–4.0	Red	Yellow
Methyl orange	3.1–4.4	Red	Yellow
Bromophenol blue	3.0–4.6	Yellow	Blue-violet
Bromocresol green	3.8–5.4	Yellow	Blue
Methyl red	4.2–6.3	Red	Yellow
Chlorophenol red	4.8–6.4	Yellow	Red
Bromothymol blue	6.0–7.6	Yellow	Blue
Paranitrophenol	6.2–7.5	Colorless	Yellow
Phenol red	6.4–8.0	Yellow	Red
Cresol red	7.2–8.8	Yellow	Red
Thymol blue	8.0–9.6	Yellow	Blue
Phenolphthalein	8.0–9.8	Colorless	Red
Thymolphthalein	9.3–10.5	Colorless	Blue
Alizarin yellow R	10.1–12.0	Yellow	Violet

change that brings about the observed color change. These changes are illustrated in Figure 1-2 for one of the simplest indicators, p-nitrophenol. At acidic pH the molecule is in a benzenoid form and hence is colorless. On dissociation of the phenolic hydroxyl group, the molecule is converted to a quinoid form, which possesses an intense yellow color.

Usually a 1 to 2 unit change in pH is required to convert the dye from one color to the other. In a two-color system, one color is discernible only if the species giving rise to that color is in ten-fold excess over the alternative species. Therefore,

$$K_a = \frac{[Ind^-][H^+]}{[HInd]}$$

$$= \frac{10[H^+]}{1}$$

and

$$\frac{K_a}{10} = [H^+]$$

or $pK_a + 1 = pH$ if the basic color is to be visible. If, on the other hand, the acidic color is to be visible,

$$K_a = \frac{1[H^+]}{10}$$

$$K_a \times 10 = [H^+]$$

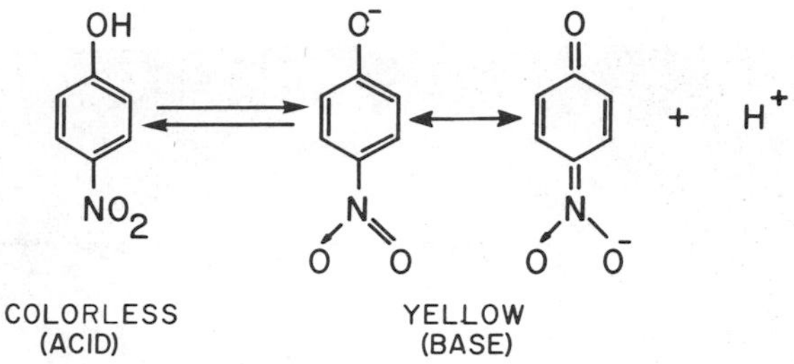

Figure 1-2. The structural changes associated with the ionization of *p*-nitrophenol.

or

$$pK_a - 1 = pH$$

The pH change resulting in shift from base to acid color would be

$$(pK_a + 1) - (pK_a - 1) = \Delta pH$$

or 2 pH units. As a result of this, single indicator systems are useful only if low resolution measurements are desired. However, the use of complex indicator systems composed of several indicators makes it possible to increase the resolution.

Today chromophoric pH indicators are used principally in pH indicator paper and as indicator media for the growth of microorganisms. In the production of pH paper, thin ribbons of paper are impregnated with an appropriate complex indicator system. If a drop of the solution to be tested is placed on a piece of this impregnated ribbon, the ribbon changes color in response to the pH of the solution. The pH is ascertained by comparing the color of the ribbon to a standard color key provided with the impregnated ribbon (Figure 1-3). Although this method is rather crude, it is very fast and is often used when only an approximate value is desired. The second widespread application of chromophoric pH indicators is in microbiological indicator media. These media are used to determine whether or not a microorganism possesses a given enzyme or set of enzymes. For example, lactose MacKonkey indicator medium can be used to determine whether or not a microorganism is capable of fermenting lactose. The medium is composed in such a way that all members of a given microbial population can grow, but only those organisms capable of fermenting lactose produce protons. These protons change the indicator, neutral red in the medium from pale pink to intense red. Thus wine red clones are those that are capable of fermenting lactose, and pale pink clones are incapable of this fermentation.

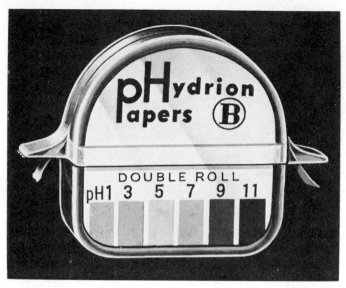

Figure 1-3. A dispenser of pH paper with its appropriate color key. (Courtesy of Micro Essential Laboratory, Inc., Brooklyn, N.Y.)

POTENTIOMETRIC MEASUREMENT OF pH

The most precise measurement of pH is performed potentiometrically with a pH meter (Figure 1-4). This instrument is composed of (1) a reference electrode, (2) a glass electrode whose potential depends on the pH of the solution surrounding it, and (3) an electrometer, which is a device capable of measuring very small potential differences in a circuit of extremely high resistance. The use of contemporary electrode design and solid state technology permits a resolution of 0.005 pH units by most high quality instruments.

Reference Electrode

The basic function of a reference electrode is to maintain a constant electrical potential against which deviations may be measured. The two most widely used reference electrodes are the calomel and silver–silver chloride electrodes. We discuss the latter since it is most often used in current instruments. As shown in Figure 1-5C, this electrode consists of a piece of metallic silver which is coated with silver chloride and immersed in a saturated solution of potassium chloride. The potential of this system

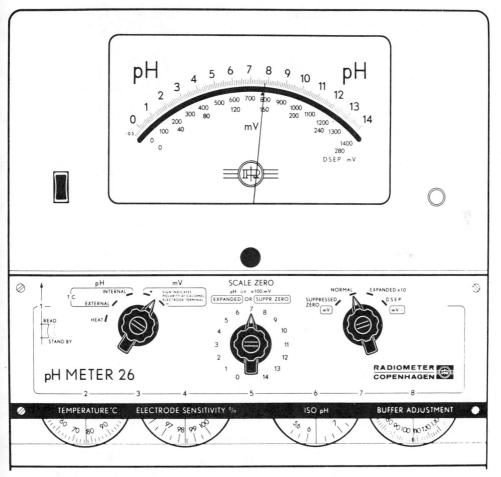

Figure 1-4. The front panel of a Radiometer PHM26 pH meter. (Courtesy of Radiometer A/S, Copenhagen.)

is derived from the reaction

$$AgCl + e^- \rightleftharpoons Ag^0 + Cl^- \tag{37}$$
$$\text{(solid)}$$

and may be calculated from the Nernst equation

$$E_{\text{ref}} = E^0 - \frac{RT}{\mathscr{F}} \ln a_{\text{Cl}^-}$$

$$= +0.2222 - 0.0592 \log a_{\text{Cl}^-} \qquad \text{at } 25°C \tag{38}$$

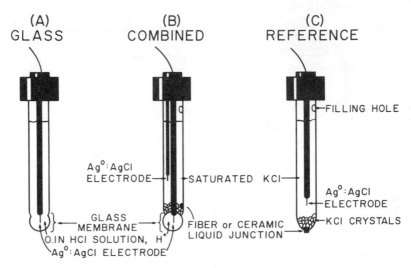

Figure 1-5. The structural features of glass, combined, and reference electrodes.

It is clear from equation 38 that the potential of the reference electrode is a function of the chloride ion concentration. In order to maintain a constant chloride ion concentration in varying conditions of humidity, a saturated solution is used. If the relative humidity decreases and evaporation of the reference electrode occurs, the excess chloride precipitates out of solution. Conversely, with high humidity the volume of solution increases slightly and additional potassium chloride dissolves. It is assumed, of course, that the temperature is constant. Electrical contact between the reference electrode and the solution being tested is maintained by means of a potassium chloride salt bridge. This junction is made through a fibrous or ceramic membrane (see Figure 1-5C, B) embedded in the bottom of the reference electrode or the side of a combination electrode.

Glass Electrode

The function of the glass electrode is to establish an electrical potential which responds to changes in the hydrogen ion activity of the solution being tested. The basic construction of a glass electrode is shown in Figure 1-5A. It consists of a high resistance glass tube with a thin, low resistance glass bulb at the bottom. Only the bulb responds to changes in pH; otherwise the potential of the electrode would be a function of the depth to which it is submerged. In the tube is a dilute aqueous solution of hydrochloric acid. In contact with this solution is a silver–silver chloride

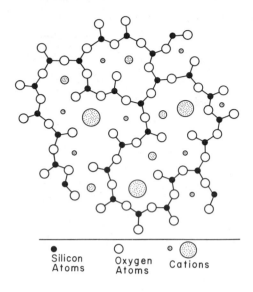

Silicon
Atoms

Oxygen
Atoms

Cations

Figure 1-6. Structure of a glass membrane as elucidated from X-ray diffraction studies. [Courtesy of the late George A. Perley and Leeds & Northrup Co., *Anal. Chem.*, **21**:395 (1949).]

electrode. The pH sensitive glass membrane was studied at the molecular level, using X-ray diffraction techniques, and was found to consist of a network of silicate and aluminate ions. This lattice structure is depicted in Figure 1-6. The holes in the lattice may be occupied by cations of differing size, but may not be occupied by anions owing to the strong repulsion of the oxygen-containing ions. It should be emphasized that although cations enter the glass matrix, they do not traverse it. In other words, appropriate ions are bound principally to the surfaces (inner and outer) of the glass membrane. By careful production of the bulb, it is possible to obtain a membrane whose matrix is available only to hydrogen ions. The precise mechanical details involved in the operation of pH sensitive membranes are not fully understood, but Figure 1-7 depicts one possible explanation of the events that occur when a glass electrode is placed in aqueous solution. Hydrogen ions from the concentrated inner side of the membrane are bound to the glass surface, resulting in concurrent release of an equal number of protons from the outer side of the membrane to preserve its electroneutrality (Figure 1-7B). If the electrode is placed in basic solution (Figure 1-7C) the hydrogen ion activity in the outer solution is very low and the "holes" created by release of protons into solution have a high probability of remaining empty. More of the released protons react with hydroxyl ions to yield water leaving an excess of positively charged cations. As long as the outside "holes" or "sites" are empty, the inner sites retain their bound protons. Since negative ions are excluded from the membrane, this condition results in an excess of negatively charged

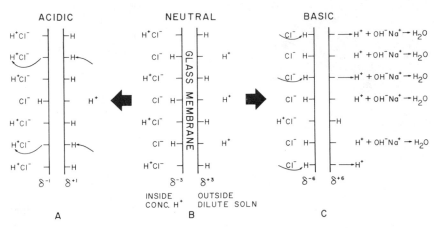

Figure 1-7. A conception of the molecular events responsible for the existence of the pH dependent response of a glass electrode.

chloride ions in the inner solution, whereas the outer solution has an excess of positively charged cations. This separation of charge is the source of the electrode potential. On the other hand, if the electrode is placed in acidic solution (Figure 1-7A), the hydrogen ion activity in the outer solution is high, resulting in protons occupying the vacant "holes" or "sites" on the outside of the membrane. In order to maintain electroneutrality of the glass an inner proton must be released, thereby decreasing the excess of chloride ions inside the bulb and decreasing the excess of protons outside. If a number of simplifying assumptions are made the potential (E) of the glass electrode is

$$E = E^0 + \frac{2.303RT}{\mathscr{F}}(\text{pH outside}) \tag{39}$$

where E^0 is the summation of a number of potentials that are not pH sensitive. These potentials include (1) the asymmetry potential, which is the potential that would exist across the membrane even if the inside and outside solutions possessed the same hydrogen ion activity; (2) the potential of the silver–silver chloride electrode within the glass electrode; and (3) the junction potential that arises between the solution being tested and the potassium chloride salt bridge (ceramic plug) of the reference electrode. Since E^0, R, and $\mathscr{F}$ of equation 39 are constants, the potential of the electrode is a function of pH and temperature. For this reason it is imperative for the temperature of the buffer used to standardize the meter and the temperature of the unknown sample to be the same. Recall that equation 38 also contains a temperature dependent term which is assumed to remain constant.

When a pH sensitive electrode and a reference electrode are placed in solution, a galvanic cell is set up. The complete cell is as follows:

	Glass membrane		Ceramic membrane	
Internal reference electrode	Inner electrolyte	Test Solution	Satd. KCl	Reference electrode
silver–silver chloride	(dil. HCl)			silver–silver chloride or calomel

pH Electrode   Reference Electrode

(single vertical lines indicate phase changes)

The potential of this cell is the algebraic sum of the potentials of the indicator (glass) and reference electrodes.

$$E_{cell} = E_{ref} + E_{glass} \qquad (40)$$

E_{cell} varies only as a function of the test solution pH if the temperature is constant. It is significant to point out that the potential of modern glass electrodes is a linear function of pH (equation 39). By using a test solution of known pH it is possible to relate the cell potential to hydrogen ion activity of a test solution. This "standardization" must be done each time a pH meter is used because of subtle changes in the various potentials owing to aging of the electrode. Therefore, the accuracy of a pH determination depends on the accuracy of the standard buffer. Table 1-2

Table 1-2. pH Values of National Bureau of Standards Primary Reference Buffer Solutions

Temperature (°C)	Potassium Acid Tartrate (Satd. at 25°C)	0.05M Potassium Acid Phthalate	0.025M KH$_2$PO$_4$, 0.025M Na$_2$HPO$_4$	0.0087M KH$_2$PO$_4$, 0.0302M Na$_2$HPO$_4$	0.01M Na$_2$B$_4$O$_7$
0	—	4.01	6.98	7.53	9.46
10	—	4.00	6.92	7.47	9.33
15	—	4.00	6.90	7.45	9.27
20	—	4.00	6.88	7.43	9.23
25	3.56	4.01	6.86	7.41	9.18
30	3.55	4.02	6.85	7.40	9.14
38	3.55	4.03	6.84	7.38	9.08
40	3.55	4.04	6.84	7.38	9.07
50	3.55	4.06	6.83	7.37	9.01

list five NBS primary standard buffers. Additional secondary standard buffers are available commercially.

Electrometer

Measurement of the potential generated in the galvanic cell of a pH meter is difficult because the potential is very small and the resistance of the circuit is very high (1–100 MΩ). Therefore the signal must be amplified greatly before it is sufficiently strong to activate a standard millivoltmeter or milliammeter. A variety of sophisticated circuits, many of them solid state, have been developed to perform this function, but a detailed discussion of the electronics behind their operation is beyond the scope of this work. Regardless of the circuitry that is used, the function of the electrometer is to multiply or amplify the galvanic cell potential many times (see Figure 1-8). Most of the amplifiers used for this purpose are

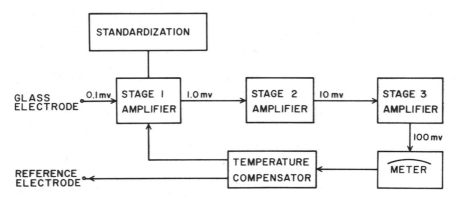

Figure 1-8. Block diagram of the electronic components of a negative feedback type pH meter. A full discussion of the functioning of this type of electrometer may be found in reference 6, pp. 596–598.

linear; that is, they accept a very small potential and, irrespective of its size, multiply it by a constant number. For example, if the amplification factor is 1000, input signals of 2 and 7 μV emerge from the amplifier as 2000 and 7000 μV signals, respectively. The amplified signal is channeled through a meter, which produces a deflection that reflects the strength of the signal activating it.

Equation 40 states that the pH sensitive, galvanic cell potential (E) to be amplified is

$$E_{cell} = E_{ref} + E_{glass} \qquad (40)$$

Substituting appropriately gives

$$E_{\text{cell}} = E_{\text{ref}} + E^0 + \frac{2.303RT}{\mathscr{F}}\text{pH} \tag{41}$$

Rearranging and collecting terms yield

$$E_{\text{cell}} = E_{\text{TOT}} + m\text{pH} \tag{42}$$

where $E_{\text{TOT}} = E_{\text{ref}} + E^0$ and m equals the constant $2.303RT/\mathscr{F}$. It should be noted that the glass electrode may be either positive or negative with respect to the reference electrode. This results in E_{cell} of equation 40 having both positive and negative values. This is the case in Figure 1-9. In this figure the cell potential has been plotted as a function of pH. At negative output values (high pH) the absolute potential of the glass electrode is greater than that of the reference electrode. The electrode potentials become equivalent at a pH near 6.5; below this pH the potential

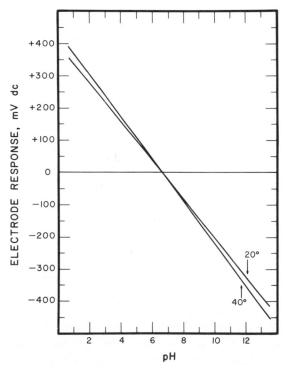

Figure 1-9. Effect of varying pH on the output potential of a glass electrode at 20 and 40°C. (Courtesy of Radiometer A/S, Copenhagen.)

of the glass electrode is smaller than that of the reference electrode, resulting in positive output potentials. To be of use, the meter of a pH electrometer must give a reading that is equivalent to the pH of the sample. Two factors must be considered in the signal amplification and meter readout processes of a pH meter: (1) the galvanic cell output potential for a given pH is different from the potential required to give a reading of that pH on the meter, and (2) the potential change of the cell when going from one pH (e.g., 5.0) to another (e.g., 9.0) (slope of the galvanic cell pH response curve) is different from the potential change needed by the meter to go from the first reading, pH 5.0, to the second, pH 9.0 (slope of the meter response curve). Therefore, it is necessary to standardize the meter to the electrode output. The first standardization is made by placing the electrodes in a standard buffer and adjusting the "calibration," "standardization," or "buffer adjust" control of the pH meter. These controls essentially correct the amplifiers for the E_{TOT} term of equation 42. From the earlier discussion it is known that this term is composed of both temperature dependent and independent terms; hence it is important that the temperatures of standardization and measurement be the same. The second correction is made by placing the electrodes in a second standard buffer (at least 2–3 pH units different from the first) and adjusting the temperature adjust or slope control until the meter reads the pH of the second buffer. This control corrects the amplifier circuits for the m term of equation 42. This control must be adjusted every time readings are made at a new temperature because m is a temperature dependent function (see Figure 1-9). Expensive instruments may have a temperature compensating system incorporated into their circuitry which automatically makes this correction once the instrument is initially calibrated. If this second temperature correction is necessary, the standardization of the instrument should be rechecked, because E_{TOT} of equation 42 also contains at least two temperature dependent terms.

Although the pH meter is a highly refined instrument, it is subject to a number of difficulties. When the pH of very alkaline solutions (above pH 10) is determined, the reading observed on the meter is lower than that of the test solution. This error, termed the alkali error, is the result of interactions between sodium ions and the glass membrane of the indicator electrode. Use of special glass in the production of expensive electrodes minimizes this error, but does not eliminate it completely. At the acidic end of the spectrum (usually less than pH 1) an error occurs in which the pH reading is higher than the true value. Another common source of error, found in instruments used to determine the pH of protein-containing solutions, is contamination of the electrode surface with protein. An electrode so contaminated gives spurious and unstable readings. This difficulty may be remedied by soaking the electrode in dilute HCl as described in steps 1-2 to 1-4 of the experimental section.

Buffers

The most common way of maintaining an acceptable pH in vitro is through the use of an appropriate buffer. The efficiency of using buffers to control pH is shown by the following example.

Example. (a) Calculate the pH of a phosphate buffer prepared by mixing 0.05 moles K_2HPO_4 and 0.05 moles KH_2PO_4 in 1 liter of solution. pK_{a_2} of phosphoric acid is 6.8. (b) Calculate the pH of a solution made by adding 5 ml $0.1M$ HCl to 95 ml water. (c) Calculate the pH of a solution made by adding 5 ml $0.1M$ HCl to 95 ml of the buffer prepared in part (a).

(a)
$$K_{a_2} = \frac{[H^+][HPO_4^{2-}]}{[H_2PO_4^-]}$$

$$[H^+] = \frac{K_{a_2}[H_2PO_4^-]}{[HPO_4^{2-}]}$$

$$pH = pK_{a_2} + \log\frac{[HPO_4^{2-}]}{[H_2PO_4^-]}$$

$$= 6.8 + \log\frac{0.05}{0.05}$$

$$= 6.8$$

(b)
$$(0.1M)(5 \times 10^{-3}\,l) = 0.5 \times 10^{-3}\text{ moles } H^+$$

$$pH = -\log[H^+]$$

$$= -\log\frac{0.5 \times 10^{-3}\text{ moles}}{0.11}$$

$$= -\log 5 \times 10^{-3}\frac{\text{moles}}{1}$$

$$= 2.30$$

(c)
$$H^+ + HPO_4^{2-} \longrightarrow H_2PO_4^-$$

$$[HPO_4^{2-}] = \frac{(\text{original moles } HPO_4^{2-}) - (\text{moles acid added})}{\text{total volume}}$$

$$= \frac{(0.095\text{ l})(0.05\ M) - (0.005\text{ l})(0.1\ M)}{0.11}$$

$$= \frac{(4.75 \times 10^{-3}\text{ moles}) - (0.5 \times 10^{-3}\text{ moles})}{0.11}$$

$$= 4.25 \times 10^{-2}\text{ moles/l}$$

$$[H_2PO_4^-] = \frac{\text{(original moles } H_2PO_4^-) + \text{(moles acid added)}}{\text{total volume}}$$

$$= \frac{(0.095 \text{ l})(0.05M) + (0.005 \text{ l})(0.1M)}{0.1 \text{ l}}$$

$$= \frac{(4.75 \times 10^{-3} \text{ moles}) + (0.5 \times 10^{-3} \text{ moles})}{0.1 \text{ l}}$$

$$= 5.25 \times 10^{-2} \text{ moles/l}$$

$$pH = pK_{a_2} + \log \frac{[HPO_4^{2-}]}{[HPO_4^-]}$$

$$= 6.8 + \log \frac{(4.25 \times 10^{-2})}{(5.25 \times 10^{-2})}$$

$$= 6.8 + \log 0.81$$

$$= 6.8 - 0.092$$

$$= 6.71$$

In Table 1-3 are tabulated the pK_a values of a variety of weak acids and bases. Selection of the most desirable buffer system depends on many considerations. The first consideration in selecting a buffer is its pK value. It is usually best to choose a buffer whose pK_a is within 0.5 to 1.0 pH units of the desired pH. If it is anticipated that only acidic materials will be produced or added to the buffer a somewhat lower pK_a is permissible, and if only basic compounds will be added a somewhat higher pK_a may be useful. In addition to the pK_a, the buffers' physical and chemical characteristics must be considered. For example, the potassium salts of phosphate buffers are usually preferred to the sodium salts, because the sodium phosphates are quite insoluble at low temperature. On the other hand, potassium phosphate buffers are useless if sodium dodecyl sulfate (SDS) is to be added to the buffer because potassium SDS is very insoluble. Occasions may arise when a volatile buffer is desired. Such a buffer is especially useful in the purification of compounds by ion exchange chromatography because they are easily removed from the preparation. A list of these buffers is presented in Table 1-4. This collection of buffers demonstrates that it is possible to combine a number of weak acids and bases to obtain a set of desired characteristics. In the above case it was desirable to achieve volatility. Thus, instead of selecting a buffer composed of a weak acid and a strong base (all the strong bases are nonvolatile), one composed of a volatile weak acid and a volatile weak base was selected. A second example of combining weak acids and bases to achieve a set of desired characteristics is the use of complex buffers to cover a wide pH range. Tris-phosphate buffer would be a good choice for

Table 1-3. Buffers

Compound	pK_a	Compound	pK_a	Compound	pK_a	Compound	pK_a
Diphenylamine	0.85	Benzoic acid	4.20	4- or 5-Hydroxymethyl-imidazole	6.40	Pyridine	8.85
Oxalic acid K_1	1.30	Oxalic acid K_2	4.26	Pyrophosphoric acid K_3	6.54	Diethanolamine	8.88
Maleic acid K_1	1.92	Tartaric acid K_2	4.37	Arsenic acid	6.60	Arginine K_2	9.04
Phosphoric acid K_1	1.96	Fumaric acid K_2	4.39	Phosphoric acid K_2	6.70	Boric acid	9.23
EDTA (versene) K_1	2.00	Acetic acid	4.73	Imidazole	6.95	Ammonium hydroxide	9.30
Glycine K_1	2.45	Citric acid K_2	4.75	2-Aminopurine	7.14	Ethanolamine	9.44
EDTA (versene) K_2	2.67	Malic acid K_2	5.05	Ethylenediamine K_1	7.30	Glycine K_2	9.60
Piperidine	2.80	Pyridine	5.19	2,4,6 Collidine	7.32	Trimethylamine	9.87
Malonic acid K_1	2.85	Phthalic acid K_2	5.40	4- or 5-Methylimidazole	7.52	Ethylenediamine K_2	10.11
Phthalic acid K_1	2.90	Succinic acid K_2	5.60	Triethanolamine	7.77	EDTA (versene) K_4	10.26
Tartaric acid K_1	2.96	Malonic acid K_2	5.66	Diethylbarbituric acid	7.98	Carbonic acid K_2	10.32
Fumaric acid K_1	3.02	Hydroxylamine	6.09	Tris-(hydroxymethyl)amino-methane	8.08	Ethylamine	10.67
Citric acid K_1	3.10	Histidine K_2	6.10			Methylamine	10.70
Glycylglycine K_1	3.15	Cacodylic acid	6.15	Glycylglycine K_2	8.13	Dimethylamine	10.70
β,β'-Dimethylglutaric acid K_1		EDTA (versene) K_3	6.16	2,4- or 2,5-Dimethyl-imidazole		Diethylamine	11.00
	3.66	β,β'-Dimethylglutaric acid K_2	6.20		8.36	Piperidine	11.12
Formic acid	3.75	Maleic acid K_2	6.22	Pyrophosphoric acid	8.44	Phosphoric acid K_3	12.32
Barbituric acid	3.79	Carbonic acid K_1	6.35	2-Amino-2 methyl-1,3-		Arginine K_3	12.50
Lactic acid	3.89	Citric acid K_3	6.40	propanediol	8.67		
Succinic acid K_1	4.18						

Table 1-4. Volatile Buffers

Buffer	pH Range
Ammonium acetate	4–6
Ammonium formate	3–5
Pyridinium formate	3–6
Pyridinium acetate	4–6
Ammonium carbonate	8–10

such an application, because it has a reasonable buffer capacity between pH 5.0 and 9.0. Use of such a complex buffer system permits determination of the effects of pH on a given enzymatic activity over a wide range of values using only one buffer. Chemical considerations may also bear on the choice, as exemplified by boric acid and Good's buffers. Borate buffers react with the cis-hydroxyl groups of carbohydrates to form stable complexes; if this complexation is unacceptable borate ion must be avoided. Good's buffers (Table 1-5) may not be used when the protein concentration of the solution is determined by the biuret or Lowry procedures, because all these buffers give extraordinarily high blanks. Finally, in enzymatic applications one must be cautious to select a buffer that does not react with or inhibit the enzyme under study. The buffers most notorious for such difficulties include phosphate, pyrophosphate, and arsenate. Good and his collaborators have recently synthesized a number of cyclic peptide buffers that appear to be quite inert as far as many enzymatic reactions are concerned. These buffers, listed along with their structures in Table 1-5, are quite expensive, but are most useful when handling delicate proteins or cellular organelles in vitro.

Some buffers display a large temperature coefficient. The most notorious of these is Tris. To a first approximation the pH of a Tris solution increases 0.03 pH units for every degree the temperature of the solution decreases (25 to 5°C). On the other hand, a 0.025 pH unit decrease is observed for every one degree rise from 25 to 37°C. In view of these considerations it is very hazardous to prepare Tris, or any buffer possessing a large temperature coefficient, at room temperature and then use it at 0 to 4°C. The only certain way to circumvent this problem is to measure the pH of the buffer at the temperature at which it will be used.

ION SPECIFIC ELECTRODES

Although pH determination is the most often used and well-established potentiometric method, electrodes capable of selectively responding to

Table 1-5. Physical Properties of Buffers

Proposed Name	pK_a at 20°C	Structure
MES	6.15	O ⬡ $\overset{+}{N}HCH_2CH_2SO_3^-$
ADA	6.6	$H_2NCOCH_2\overset{+}{\underset{H}{N}}\!\!\big\langle\!\!{}^{CH_2COO^-}_{CH_2COONa}$
PIPES	6.8	$NaO_3SCH_2CH_2\overset{+}{N}$ ⬡ $\overset{+}{N}HCH_2CH_2SO_3^-$
ACES	6.0	$H_2NCOCH_2\overset{+}{N}H_2CH_2CH_2SO_3^-$
Cholamine chloride	7.1	$(CH_3)_3\overset{+}{N}CH_2CH_2NH_2Cl^-$
BES	7.15	$(HOCH_2CH_2)_2\overset{+}{N}HCH_2CH_2SO_3^-$
TES	7.5	$(HOCH_2)_3\overset{+}{N}HCH_2CH_2SO_3^-$
HEPES	7.55	$HOCH_2CH_2\overset{+}{\underset{H}{N}}$ ⬡ $NCH_2CH_2SO_3^-$
Acetamidoglycine	7.7?	$H_2NCOCH_2\overset{+}{N}H_2CH_2COO^-$
Tricine	8.15	$(HOCH_2)_3\overset{+}{C}NH_2CH_2COO^-$
Glycinamide	8.2	$H_2NCOCH_2NH_2$
Tris	8.3	$(HOCH_2)_3CNH_2$
Bicine	8.35	$(HOCH_2CH_2)_2\overset{+}{N}HCH_2COO^-$
Glycylglycine	8.4	$H_3\overset{+}{N}CH_2CONHCH_2COO^-$

inorganic ions and gases have become available (see Table 1-6). As shown in this table, however, considerable care must be exercised when using these electrodes because they respond to many interfering ions. The operation of one electrode used to a significant extent in biochemistry is described here as an example. Principles underlying operation of other electrodes are outlined in Table 1-7.

The Clark oxygen electrode, depicted in Figure 1-10, is one of the most selective electrodes. Only hydrogen sulfide and sulfur dioxide interfere with its operation. Since these gases rarely occur in biological systems, their interference is of minimal consequence. As shown, the electrode assembly is composed of (1) the electrode elements molded into an epoxy plug and covered by a Teflon membrane, (2) a lucite block which fits snugly into a glass sample chamber, (3) the glass sample chamber, and (4) a disk shaped magnetic stirring bar that fits into the bottom of the sample chamber. The electrode itself possesses one platinum cathode and two

Table 1-6. Specific Ion Electrodes Available Commercially

Material Detected	Concentration Range (M)	Interfering Compounds and Ions
Ammonium	$10^{-6}-10^{0}$	CO_2 volatile amines
Bromide	$10^{-5}-10^{0}$	S^{2-}, I^-
Cadmium	$10^{-7}-10^{0}$	Ag^+, Hg^{2+}, Cu^{2+}
Calcium	$10^{-5}-10^{0}$	Zn^{2+}, Fe^{2+}, Pb^{2+}, Mg^{2+}
Carbon dioxide	$10^{-6}-10^{-2}$	SO_2, H_2S
Chloride	$10^{-5}-10^{0}$	I^-, NO_3^-, Br^-, HCO_3^-, SO_4^{2-}, F^-
Cupric	$10^{-7}-10^{0}$	S^{2-}, Ag^+, Hg^{2+}
Cyanide	$10^{-6}-10^{-2}$	S^{2-}, I^-
Fluoride	$10^{-6}-10^{0}$	OH^-
Iodide	$10^{-7}-10^{0}$	S^{2-}
Lead	$10^{-7}-10^{0}$	Ag^+, Hg^{2+}, Cu^{2+}
Nitrate	$10^{-5}-10^{-1}$	I^-, Br^-, NO_2^-
Nitrite	$10^{-6}-10^{-2}$	CO_2
Perchlorate	$10^{-5}-10^{-1}$	I^-
Potassium	$10^{-5}-10^{0}$	Cs^+, NH_4^+, H^+
Sulfide	$10^{-7}-10^{0}$	Hg^{2+}
Sodium	$10^{-6}-10^{0}$	Li^+, K^+
Sulfite	$10^{-6}-10^{-2}$	HF, HCl
Thiocyanate	$10^{-5}-10^{0}$	OH^-, Be^-, Cl^-

silver anodes. Note that the cathode element protrudes slightly from the bottom of the assembly. The three elements are bathed in a solution of half-saturated potassium chloride. A thin Teflon membrane is held tautly over the end of the electrode by a rubber "O" ring. It isolates the potassium chloride solution of the electrode from the sample and serves as the means of selectivity. Teflon is permeable to many gases, allowing

Table 1-7. Gas-sensing Electrode Systems

Measured Species	Diffusing Species	Equilibria in Electrolyte	Sensing Electrode
NH_3 or NH_4^+	NH_3	$NH_3 + H_2O \rightleftharpoons NH_4^+ + OH^-$	H^+
		$xNH_3 + M^{n+} \rightleftharpoons M(NH_3)_x^{n+}$	$M = Ag^+$, Cd^{2+}, Cu^{2+}
SO_2, H_2SO_3, or SO_3^{2-}	SO_2	$SO_2 + H_2O \rightleftharpoons H^+ + HSO_3^-$	H^+
NO_2^-, NO_2	$NO_2 + NO$	$2NO_2 + H_2O \rightleftharpoons NO_3^- + NO_2 + 2H^+$	H^+, NO_3^-
S^{2-}, HS^-, H_2S	H_2S	$H_2S + H_2O \rightleftharpoons HS^- + H^+$	S^{2-}
CN^-, HCN	HCN	$Ag(CN_2)^- \rightleftharpoons Ag^+ + 2CN^-$	Ag^+
F^-, HF	HF	$HF \rightleftharpoons H^+ + F^-$	F^-
		$FeF_x^{2-x} \rightleftharpoons FeF_y^{3-y} + (x-y)F^-$	Pt (redox)
$HOAc$, OAc^-	$HOAc$	$HOAc \rightleftharpoons H^+ + OAc^-$	H^+
Cl_2, OCl^-, Cl^-	Cl_2	$Cl_2 + H_2O \rightleftharpoons 2H^+ + ClO^- + Cl^-$	H^+, Cl^-
CO_2, H_2CO_3, HCO_3^-, CO_3^{2-}	CO_2	$CO_2 + H_2O \rightleftharpoons H^+ + HCO_3^-$	H^+
X_2, OX^-, X^-	X_2	$X_2 + H_2O \rightleftharpoons 2H^+ + XO^- + X^-$	$X = I^-$, Br^-

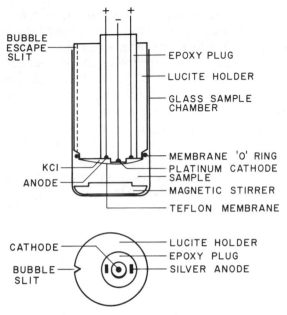

Figure 1-10. Clark oxygen electrode and sample chamber assembly.

them to pass through and come in contact with the platinum cathode. If a polarizing potential of 0.8 V is applied across the electrode elements, oxygen reacts at the cathode according to the following reaction:

$$O_2 + 2e^- + 2H_3O^+ \rightleftharpoons [H_2O_2] + 2H_2O$$
$$[H_2O_2] + 2e^- + 2H_3O^+ \rightleftharpoons 4H_2O$$

$$\text{Sum: } O_2 + 4H_3O^+ + 4e^- \rightleftharpoons 6H_2O$$

The circuit is completed by the following reaction occurring at the silver anode:

$$4Ag^0 + 4Cl^- \rightleftharpoons 4AgCl + 4e^-$$

$$\text{Total: } 4Ag^0 + 4Cl^- + 4H^+ + O_2 \rightleftharpoons 4AgCl + 2H_2O$$

Current flow is therefore directly proportional to the amount of oxygen diffusing across the Teflon membrane. This depends on the temperature of the Teflon membrane and the partial pressure of oxygen in the sample chamber.

Four principal precautions must be observed to obtain maximum accuracy from this instrument. Since the rate of oxygen diffusion depends on temperature (3–5% permeability variation/°C), the temperature of the sample chamber and electrode assembly must be carefully monitored and maintained constant. Fouling, creasing, or folding of the Teflon membrane is one of the most common causes of poor electrode response. Since the membrane can become clogged with biological materials from experimental samples, it is advisable to change it frequently and ensure that it is smooth and taut over the end of the electrode. Grease and body oils found on the skin of an investigator's fingers are also prime sources of fouling; hence the electrode membrane should never be touched. Though atmospheric concentrations of hydrogen sulfide are too low to interfere with most oxygen determinations, they are sufficiently high to cause a buildup of silver sulfide on the electrode anodes over extended periods. Such anode contamination results in spurious results. Therefore, care should be taken to ensure that the electrode elements are clean and bright. It is recommended that the electrodes be cleaned using a dilute solution of ammonium hydroxide. A serious and easily noticeable source of error is the presence of air bubbles, either within the electrode electrolyte solution (2–3 μl of half-saturated potassium chloride solution within the membrane) or within the sample chamber. Volume for volume, an air bubble contains about 20 times more oxygen than air-saturated water. Therefore, if a sample in which the uptake of oxygen is being monitored contains a bubble, oxygen will diffuse from the bubble into the sample solution as quickly as it is removed from solution by the experimental material. In commercially available instruments the sample is isolated from the atmosphere by the close tolerances between electrode holder and sample chamber wall. A slit running the length of the holder allows removal of trapped bubbles and permits additions to be made during a measurement. If, however, oxygen exchange occurring between the atmosphere and liquid in the slit is too great, instruments that are completely closed can be used (see reference 12).

EXPERIMENTAL

Preparation of a New or Unused Combination Electrode

1-1. Fill the electrode with the appropriate solutions as indicated in the manufacturer's instructions.

1-2. Carefully submerge the electrode bulb and salt bridge junction just below the surface of 0.1N HCl.

1-3. Leave the electrode in this position 4 to 6 hours.

1-4. Carefully remove the electrode from the solution and rinse it thoroughly. Do this by directing a stream of glass-distilled water from a wash bottle onto the sides of the electrode. Never wipe the rinsed electrode with tissue because this creates a static electric charge on the electrode and causes erroneous readings. After thorough rinsing, the last drop from the rinse may be left hanging on the end of the electrode. If it is necessary to remove this drop, do so by very carefully catching the drop with the edge of a beaker, as shown in Figure 1-11. Take extreme care when performing this operation so as not to damage the electrode. It is advisable to perform steps 1-1 to 1-4 on any combination electrode that has stood unused for an extended period or one that has been allowed to dry out.

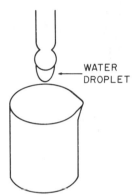

WATER
DROPLET

Figure 1-11. Position of a beaker lip used to remove the last water droplet from the glass membrane of a pH electrode.

1-5. Place the electrode in a standard buffer solution of pH 4.00 and stir the solution slowly. This ensures that a uniform solution is in contact with the electrode.

1-6. Read the pH of the solution following the manufacturer's instructions.

1-7. Set the pH meter so that it gives the same reading as that of the standard (pH 4.00) buffer solution.

1-8. Remove the electrode from this solution and repeat step 1-4.

1-9. Repeat steps 1-5 and 1-6 using a standard pH 7.00 buffer solution. If the meter is functioning properly, it should read 7.0. Otherwise consult the instrument manual for appropriate adjustment of the instrument.

1-10. Repeat step 1-4.

1-11. The electrode and pH meter are now ready to begin the titration procedures. If the electrode is not going to be used immediately, leave it with the electrode bulb and salt bridge submerged in glass-distilled water.

Titration of an Amino Acid

1-12. Prepare 50 ml 0.1M solution of each compound to be titrated (H_3PO_4, cysteine, histidine, glycine, and arginine).

1-13. Prepare 200 ml 0.2M KOH and 200 ml 0.2M HCl for each compound to be titrated.

1-14. Place 20 ml water, 20 ml 0.1M solution of the substance to be titrated, and a small stirring bar in a 250 ml beaker.

1-15. Arrange the beaker on a stirring motor as shown in Figure 1-12 and submerge the pH electrode in the solution. Take care that both the bulb of

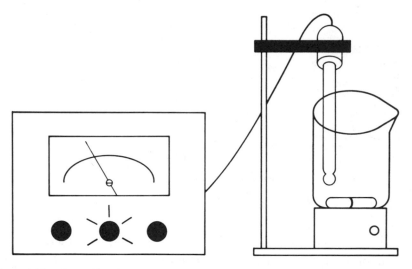

Figure 1-12. Assembly of a pH meter electrode and stirring apparatus used to measure pH during the manual titration of an amino acid.

the electrode and the salt bridge membrane are below the surface of the solution and that the spinning bar will not touch the electrode. Failure to do this will result in the destruction of the electrode when the stirring motor is turned on.

1-16. Turn the adjustable control of the stirring motor to its lowest position.

1-17. Turn on the stirring motor.

1-18. Slowly increase the speed of the motor until a small vortex appears at the surface of the solution.

1-19. Read the pH of the solution.

1-20. Carefully add 2.0 ml 0.2M HCl to the beaker.

1-21. Wait 5 seconds for the resulting solution to become homogeneous before measuring and recording the pH. The pH should be recorded to the nearest 0.05 pH unit.

1-22. Repeat steps 1-20 and 1-21 until the pH of the solution reaches 1.1.

1-23. During the course of the above additions, raise the pH electrode so that the end of the electrode and the salt bridge are just below the surface of the solution. Also, increase the speed of the stirring motor from time to time to ensure proper mixing. Make these adjustments following the completion of every fourth or fifth addition of acid or base.

1-24. Discard the solution from step 1-23, rinse the electrode (step 1-4), and repeat steps 1-14 to 1-19 inclusively.

1-25. Carefully add 2.0 ml 0.2M KOH to the beaker.

1-26. Repeat step 1-21.

1-27. Repeat steps 1-25 and 1-21 until the pH of the solution reaches 12.0.

1-28. After each titration is completed, repeat steps 1-4 and 1-11.

1-29. The data resulting from the titration of H$_3$PO$_4$, cysteine, histidine, and arginine are shown in Figures 1-13 and 1-14.

Titration of an Amino Acid in the Presence of Formaldehyde

1-30. Into a 250 ml beaker, place 20 ml water, 20 ml 0.1M glycine, and a small stirring bar.

1-31. Repeat steps 1-15 to 1-19 inclusively.

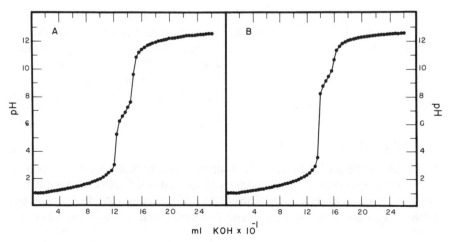

Figure 1-13.   Titration curves for H$_3$PO$_4$ (A) and arginine (B).

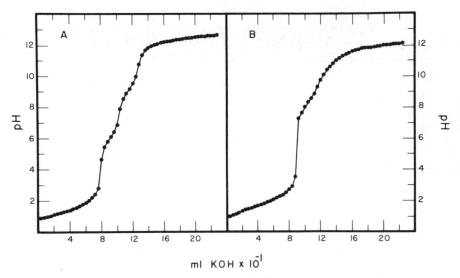

Figure 1-14. Titration curves for histidine (*A*) and cysteine (*B*).

1-32. Add 8.0 ml 37% formaldehyde.

1-33. Repeat step 1-19.

1-34. Repeat steps 1-20 to 1-27 inclusively.

1-35. Repeat steps 1-14 to 1-27 inclusively using 20 ml 0.1*M* glycine as the compound to be titrated.

1-36. Repeat steps 1-4 and 1-11.

1-37. Plot the data obtained in steps 1-34 and 1-35 and compare the curves that are obtained. This comparison is displayed in Figure 1-15. The set of reactions depicted in Figure 1-16 explains why the pK_a of the amino group of glycine is shifted to a lower pH in the presence of excess formaldehyde. This compound reacts with the unprotonated amine and thereby serves to drive the dissociation of the fully protonated amino group.

Calibration of the Clark Electrode

1-38. Assemble a Clark electrode (with a fresh membrane, if necessary), constant temperature bath, constant voltage source, and recorder following the manufacturer's instructions. In the subsequent discussion the Yellow Springs Instrument's Clark electrode and sample chamber are used.

1-39. Since passage of oxygen through the Teflon membrane depends

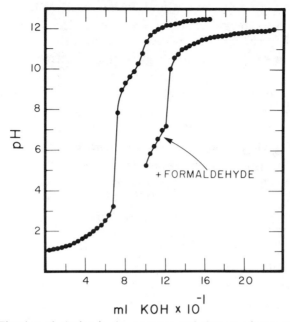

Figure 1-15. Titration of glycine in the presence and absence of excess formaldehyde.

greatly on temperature, carefully regulate the circulating water bath to 30°C and check periodically throughout the experiment to see that this temperature is maintained.

1-40. Place 5 ml of a freshly prepared $10mM$ sodium dithionate solution in the sample chamber.

1-41. Assemble the electrode.

1-42. Check to be sure that there are no bubbles in the sample chamber. Small bubbles often adhere to the wall of the sample chamber if it is not clean. If this occurs wash the glass chamber thoroughly with soap and water.

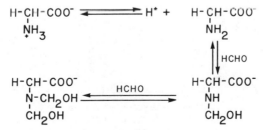

Figure 1-16. Reactions of glycine and formaldehyde.

1-43. Return the electrode assembly to the constant temperature bath.

1-44. Turn on the magnetic stirrer and recheck the sample chamber for bubbles. If any are present, remove them before proceeding.

1-45. Apply 0.8 V polarizing voltage to the electrode and activate the recorder.

1-46. No current should be flowing through the electrode because all of the oxygen has been reduced. Therefore, adjust the recorder to zero using the recorder zero adjustment.

1-47. Disassemble the sample chamber and wash it and the electrode and its housing thoroughly.

1-48. Bubble air through 400 ml water (placed in a 500 ml Erlenmeyer flask) for 1 hour or more to ensure full saturation. If a medium other than water is to be used, saturate it with oxygen in a similar manner. Be sure that this operation is performed at 30°C also.

1-49. Transfer a 5 ml sample of the saturated solution to the Clark electrode sample chamber.

1-50. Repeat steps 1-41 to 1-45 inclusively.

1-51. The current now flowing through the electrode represents that derived from an oxygen saturated solution. Table 1-8 lists the nanomolar amounts of oxygen per milliliter of water at various temperatures. In the present case, the current flow corresponds to 237 nmoles/ml.

Table 1-8. Oxygen Content of Air-Saturated Water at Various Temperatures

Temperature	Oxygen (nmoles/ml H_2O)
5	397
10	351
15	314
20	284
25	258
28	244
30	237
35	222
37	217

Derived from data in *Handbook of Chemistry and Physics* (1959). Atmospheric pressure, 760 mm Hg.

1-52. After the recorder stabilizes to a perfectly horizontal trace, adjust it to a full scale (100%) reading using the variable output voltage divider which is part of the constant voltage source. Full scale on the recorder now corresponds to an oxygen concentration of 237 nmoles/ml.

1-53. A more precise method of calibration, especially for complex solutions, is described in reference 14. This method involves complete oxidation of a known amount of NADH (spectrophotometrically determined) by an oxygen generating enzyme system such as heart mitochondria electron transport particles or diaphorase.

Oxygen Uptake by *Saccharomyces cerevisiae*

1-54. Grow a diploid strain of *Saccharomyces cerevisiae* to a cell density of 10^7 cells/ml (50 Klett units) in a glucose (0.1%) ammonia minimal medium (reference 15 describes one usable medium). Aerate the cells vigorously during growth either by shaking or with an asparger.

1-55. Calibrate the instrument as described in steps 1-38 to 1-52.

1-56. Repeat step 1-47.

1-57. Place 5 ml of the yeast suspension in the sample chamber and repeat steps 1-41 to 1-45 inclusively.

1-58. Allow the yeast to utilize about half the available oxygen. If for any reason stirring of the sample stops the trace will immediately drop towards zero, owing to oxygen depletion in the immediate vicinity of the electrode. If this occurs, simply restart the stirring process.

1-59. When half the available oxygen has been used, add to the chamber 0.1 ml 50mM potassium cyanide. If addition is made by inserting a Hamilton syringe down the bubble escape groove, it can be accomplished without interrupting the recording.

1-60. Data from this experiment are shown in Figure 1-17.

Oxygen Uptake by Castor Bean Mitochondria

1-61. Prepare castor bean mitochondria as described in steps 9-1 to 9-16.

1-62. Repeat step 1-48 using the mitochondrial resuspension medium in place of water. This medium (adjusted to pH 7.4 with HCl) is composed of 45 g mannitol, 6.05 g Tris, and 0.37 g EDTA per liter of solution.

1-63. Calibrate the electrode and recorder assembly (steps 1-38 to 1-52 inclusively).

1-64. Repeat step 1-47.

1-65. Place 5 ml of the oxygen-saturated solution (prepared in step 1-62) in the sample chamber.

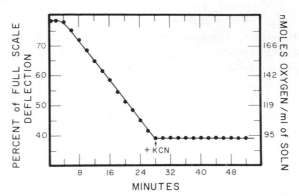

Figure 1-17. Oxygen consumption of *Saccharomyces cerevisiae* in the presence and absence of cyanide.

1-66. Repeat steps 1-41 to 1-45. Add 0.05 to 0.20 ml of a suspension of castor bean mitochondria to the sample chamber.

1-67. Record the oxygen utilization for 8 to 12 minutes.

1-68. Add 20 μl 20mM ADP to the sample chamber using a Hamilton syringe.

1-69. Add 50 μl 0.5M sodium succinate to the sample chamber using a Hamilton syringe.

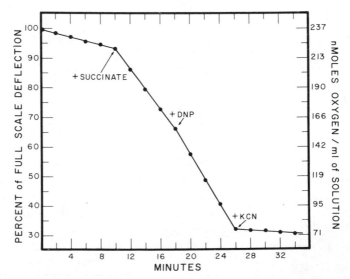

Figure 1-18. Oxygen consumption of castor bean mitochondria in the presence and absence of metabolic inhibitors.

1-70. Repeat step 1-67.

1-71. Add 20 μl 4mM dinitrophenol to the sample chamber using a Hamilton syringe. At this point the quality of the mitochondrial preparation can be ascertained. Mitochondria in good condition give a substantially higher rate of oxygen consumption when they are uncoupled with dinitrophenol. Explain this observation.

1-72. Repeat step 1-67.

1-73. Add 50 μl 50mM potassium cyanide to the sample chamber using a Hamilton syringe.

1-74. Repeat step 1-67.

1-75. Data from this type of experiment are shown in Figure 1-18.

1-76. Repeat this experiment using other metabolic inhibitors and/or citric acid cycle intermediates.

REFERENCES

1. Harold F. Walton, *Principles and Methods of Chemical Analysis*, 2nd ed., Prentice-Hall, Englewood Cliffs, N.J., 1964.

2. M. Dole, *The Glass Electrode*, Wiley, New York, 1940.

3. G. A. Perley, *Anal. Chem.*, **21**:391–393 (1949). Composition of pH-Responsive Glasses.

4. G. A. Perley, *Anal. Chem.*, **21**:394–401 (1949). Glasses for Measurement of pH.

5. G. A. Perley, *Anal. Chem.*, **21**:559–562 (1949). pH Response of Glass Electrodes.

6. H. H. Willard, L. L. Merritt, and J. A. Deon, *Instrumental Methods of Analysis*, 4th ed., Van Nostrand, Princeton, N.J., 1967.

7. J. S. Fritz and G. H. Schenk, *Quantitative Analytical Chemistry*, 2nd ed., Allyn and Bacon, Boston, 1969.

8. R. M. Bates, *Determination of pH: Theory and Practice*, 2nd ed., Wiley, New York, 1964.

9. G. Eisenmann, Ed., *Glass Electrodes for Hydrogen and Other Cations: Principles and Practice*, Marcel Dekker, New York, 1967.

10. G. A. Rechnitz, *Chem. Eng. News*, 146–158 (June 12, 1967). Ion Selective Electrodes.

11. C. E. Meloan, *Instrumental Analysis Using Physical Properties*, Lea and Febiger, Philadelphia, 1968.

12. W. D. Brown and L. B. Mebine, *J. Biol. Chem.*, **244**:6696–6701 (1969). Autooxidation of Oxymyoglobins.

13. I. M. Kolthoff and J. J. Lingane, *Polarography*, 2nd ed., Interscience, New York, 1952.

14. R. W. Estabrook, *Methods Enzymol.*, **10**:41–47 (1967). Mitochondrial Respiratory Control and the Polarographic Measurement of ADP:O Ratios.

15. L. J. Wickerham, *J. Bacteriol.*, **52**:293–301 (1946). A critical evaluation of the nitrogen assimilation tests commonly used in the classification of yeasts.

Chapter 2

Spectrophotometry

One of the earliest studied characteristics of chemical compounds was their colors. Color intensity is the basis of the most widely used set of biochemical assay procedures. Things are colored because of their ability to absorb or remove certain components of light that impinges upon them. For example, if one looks at white light through a glass of wine, the wine appears red. It appears red because all the blue and yellow components of white light were removed as it passed through the wine, leaving only a red component to be detected by the eye. As shown in Figure 2-1, light visible to the human eye occupies only a very small portion of the electromagnetic spectrum (400–800 nm). The area of the spectrum to be discussed here includes visible and ultraviolet light, which lie between 200 and 800 nm.

To discuss the measurement of light properly it is necessary to define operational terms that describe light. If light is assumed to be a wave phenomenon (Figure 2-2) the distance between two peaks or troughs is the wavelength (λ) and the number of waves passing a given point per unit time is the frequency (ν). The relationship between light wavelength and frequency is

$$\lambda = \frac{c}{\nu} \tag{1}$$

where c is the speed of light. Another useful parameter is the wave number of light ($\bar{\nu}$); this is the number of waves per unit distance (usually in centimeters). Wave numbers are related to wavelength and frequency by the equation

$$\bar{\nu} = \frac{1}{\lambda} = \frac{\nu}{c} \tag{2}$$

Energy and frequency are related by the equation

$$E = h\nu \tag{3}$$

36

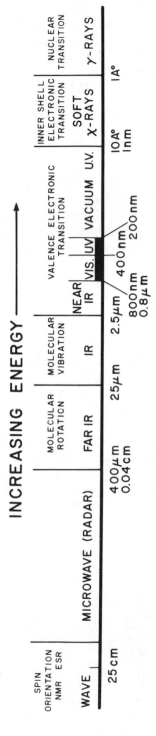

Figure 2-1. Diagrammatic representation of the electromagnetic spectrum.

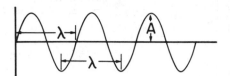

Figure 2-2. A sine wave. λ indicates the wavelength and A the amplitude.

or
$$h\nu = h\bar{\nu}c$$

where h is Planck's constant.

As light passes through a substance certain quantized energies may be transferred to the sample, raising its electrons to higher energy states. The possible types of molecular transitions and their relative energy changes are shown in Figure 2-3. The wavelengths of light participating in these transitions may be calculated using the equation

$$\Delta E = h\,\Delta\bar{\nu} \tag{4}$$

together with equation 2. Regions of the electromagnetic spectrum associated with these transitions are depicted in Figure 2-1. Visible and ultraviolet portions of the spectrum are associated only with transitions of energetic valence electrons. In biochemically significant molecules these transitions involve π electrons. Excitation of a single π electron absorbs very little visible light, but absorption increases as the number of π electrons increases. Participation of five or more π electrons of a benzenoid ring system or seven π electrons of a linear conjugated system yields significant absorption of visible light.

The various quantities of energy (wavelengths of light) absorbed by a given molecule can be used to detect and identify it quantitatively and qualitatively. This was recognized as early as 1729 by Bouguer, who performed the experiment summarized in Figure 2-4. He placed a series of identical containers full of an absorbing material in a row and measured

MOLECULAR TRANSITIONS

ELECTRONIC   VIBRATIONAL   ROTATIONAL

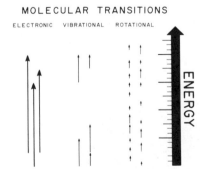

Figure 2-3. Quantitative relationships of the transition energies for three types of molecular transitions.

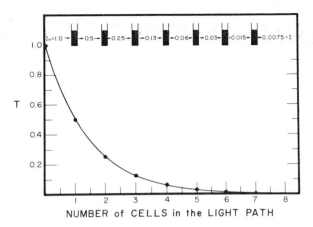

Figure 2-4. The Bouguer experiment.

the amount of light impinging on each cell and the amount that passed through it or was transmitted by it. He assumed radiation incident upon the first cell had a value of 1.0; only 50% was transmitted by it. In like manner only 25% of the original light could be detected after passing through two cells. To describe these observations quantitatively, transmittance (T) was defined as

$$T = \frac{I}{I_0} \tag{5}$$

where I is the amount of radiant energy transmitted by the cell and I_0 is the amount of radiant energy incident upon the cell. The data collected from the experiment were then described mathematically as

$$T = e^{-\alpha b} \tag{6}$$

where e is the base of natural logarithms, α is a constant (specific to each absorbing material) termed the absorption coefficient, and b is the path length of light as it passes through the absorbing material. Much later Beer performed analogous experiments but described them in terms of concentration as follows:

$$T = 10^{-abc} \tag{7}$$

where a is the absorptivity constant analogous to α in equation 6, b is the light path length, and c is the concentration of absorbing material. Although transmittance was initially used as the predominant measure of absorption, it has been largely replaced by absorbance or the less used term optical density which are denoted as A and O.D., respectively.

$$A = -\log T = -\log(10^{-abc}) = abc \qquad (8)$$

Also, absorptivity, a, has been largely replaced by use of the molar extinction coefficient, ϵ, which is the absorbance of a $1M$ solution of a pure compound under standard conditions of solvent, temperature, and wavelength. Therefore, in its most used form the Bouguer–Beer law is

$$A = \epsilon bc \qquad (9)$$

Use of this law to determine the extinction coefficient of a given material is illustrated in the following example.

Example.   Calculate the extinction coefficient of NADH if a $1.37 \times 10^{-4}M$ solution exhibits an absorbance of 0.85 at a wavelength of 340 nm in a 1 cm cell.

$$A = \epsilon bc$$

or

$$\epsilon = \frac{A}{bc}$$

Substitution of the above experimental values gives

$$\epsilon = \frac{0.85}{(1.0 \text{ cm})(1.37 \times 10^{-4} \text{ moles/l})}$$
$$= \frac{(6.20 \times 10^{6})(10^{3} \text{ cm})}{\text{cm} \cdot \text{moles}}$$
$$= 6.20 \times 10^{6} \text{ cm}^{2}/\text{mole}$$

Absorption characteristics may at times be used as one means of identifying a specific molecule. This is particularly true of highly colored molecules such as carotenoids and cytochromes. The identification process begins by constructing an absorbance spectrum of the sample under study. This is done by measuring the absorbance of the sample at all wavelengths of visible light. With present instrumentation, this can be accomplished automatically using a scanning spectrophotometer. Spectra of three carotenoids, shown in Figure 2-5, illustrate the unique absorbance characteristics of each compound. The red carotenoid astacene has a single maximum at 480 nm, whereas the golden yellow xanthophyll ester possesses three maxima at 481, 449, and 422 nm, respectively. This illustration raises an interesting question. What would the absorbance spectrum of a mixture look like? It can be shown that it is simply the sum of the single spectra. Mathematically formulated,

$$A_{\text{TOT}} = A_1 + A_2 + A_3 + \cdots + A_n \qquad (10)$$

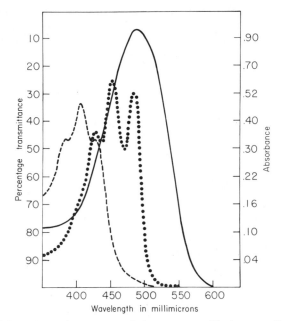

Figure 2-5. Visible absorption spectra of three carotenoid pigments. Taken from D. B. Meyer, in *The Structure of the Eye* (J. W. Rohen, Ed.), F. K. Schattauer-Verlag, Stuttgart, 1964, pp. 521–533.

This permits analysis of a mixed solution, provided each component of the mixture is known and can be obtained in pure form. Analysis is begun on a mixture of n components by determining the extinction coefficient of each separate, pure component at n different wavelengths. Next the absorbance of the mixture is determined at these same wavelengths. These data are finally used to construct and solve n simultaneous equations of the form used in equation 10.

$$A_1 = \epsilon_{11}bc_1 + \epsilon_{12}bc_2 + \cdots + \epsilon_{1n}bc_n$$
$$A_2 = \epsilon_{21}bc_1 + \epsilon_{22}bc_2 + \cdots + \epsilon_{2n}bc_n$$
$$\vdots$$
$$A_n = \epsilon_{n1}bc_1 + \epsilon_{n2}bc_2 + \cdots + \epsilon_{nn}bc_n$$

where A_n is the absorbance of the mixture at each wavelength n, ϵ_{nn} is the extinction coefficient of the nth pure compound at wavelength n, and c_n is the concentration of the nth component. $c_1, c_2, \ldots, c_n$ are constant for all the equations; b is known and can be ignored if a 1 cm cell is used in the measurements. $A_1, A_2, \ldots, A_n$ and all the extinction coefficients are measured directly. Although this approach may be used to resolve mixed solutions, it must be emphasized that its routine application is severely restricted by (1) the requirement of knowing with certainty the composition of the mixture and (2) the availability of each component in pure form.

Although the above relationships were developed and discussed for visible light, it must be emphasized that they are not restricted to this narrow range of energies. On the contrary, they may be applied broadly across the electromagnetic spectrum.

SPECTROPHOTOMETER

Widespread application of spectral assay procedures has resulted from the development of the spectrophotometer. This instrument is capable of producing monochromatic light (light of only one wavelength) and then accurately measuring the amount of that light absorbed by a given sample. As shown in Figure 2-6 most spectrophotometers are composed of five major components: a light source, a monochromator and associated band pass filters, a sample chamber, a detector, and an electronic unit for signal processing and data display. Each component is considered separately and is approached in terms of its mode of operation and factors affecting the quality of its performance.

Light Source

Characteristics of a good light source (*A* and *a* in Figure 2-7) include (1) high intensity but small surface area; (2) wide spectral range; (3) lack of sharp emission lines breaking up its spectral range; (4) stable output; and

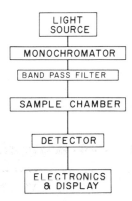

Figure 2-6. Block diagram of the components of a visible and ultraviolet spectrophotometer.

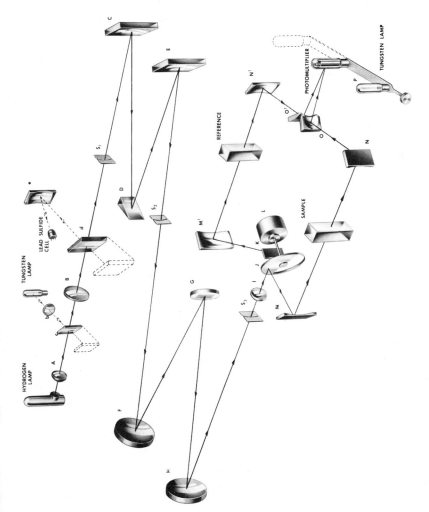

Figure 2-7. Optical diagram of a Cary 14 visible and ultraviolet spectrophotometer. (Courtesy of Varian Instrument Division, Palo Alto, Calif.)

(5) long life at reasonable cost. Three sources conform to these requirements and are found in most high quality instruments. For work in the
ultraviolet region a high pressure hydrogen or deuterium lamp is employed. The emission spectra of the two lamps are nearly identical but a
deuterium lamp has threefold greater intensity. These lamps possess
continuous emission spectra from 375 nm to the limit of transmission of
their envelopes (250–350 nm for glass or 160 nm for fused high purity
silica). Above 375 nm most instruments use a low voltage, high intensity
tungsten bulb. Xenon arc sources are being increasingly used for
wavelengths between 280 and 800 nm because they are less expensive and
more intense and have a greater life span than deuterium lamps. The only
disadvantage of this source is instability of the arc caused by its tendency
to "wander" from one point to another on the electrodes.

Monochromator

Light from either a tungsten or deuterium bulb is selected by means of a
movable mirror assembly (C in Fig. 2-7), and concentrated by a lens
system (B in Figure 2-7). The concentrated beam is then collimated by
passing it through either an optical collimating system or a narrow slit (S_1
in Figure 2-7). To be of use the collimated beam of polychromatic (white)
light must be separated into its colored components. This is accomplished
by the monochromator (prism D and grating G in Figure 2-7). The
characteristics desired of a monochromator include (1) minimum absorption of light as it passes through the system; (2) high degree of accuracy in
wavelength selection; and (3) high spectral purity over a broad spectral
range. The heart of a monochromator is a prism (D), a grating (G), or in
more expensive instruments a combination of both (as in Figure 2-7).
Figure 2-8 depicts how monochromatic light is produced using a prism.
Polychromatic light passing through the prism is broken into its component wavelengths. By rotating the prism in one direction or the other, a
particular, small area of the spectrum passes through the monochromator
exit slit (S_2 and S_3 in Figure 2-7) to the sample. Other spectral areas strike
the nonreflecting interior walls of the monochromator and hence are lost.
In automatically scanning instruments, used for production of continuous
spectra such as those shown in Figure 2-5, the prism is slowly rotated by
an electric motor. Although prisms remain an important element in
current instruments, they are deficient in several ways. The degree of
dispersion is limited and the angle of deviation between wavelengths does
not follow a geometric law; that is, the longer wavelengths are spaced
closer together than the shorter wavelengths. Prisms are also sensitive to
variations in temperature. These disadvantages have been largely over-

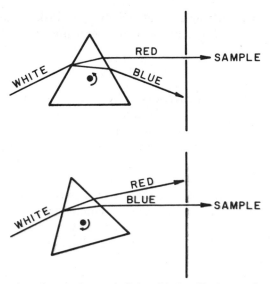

Figure 2-8. Generation of monochromatic light with the aid of a rotating prism. An arrow indicates the direction of rotation.

come through use of a diffraction grating as the light dispersing element. A diffraction grating consists of a front surface mirror into which is cut a series of parallel grooves (15,000/in.). Gratings have the advantage of giving a wide linear dispersion of light. As shown in Figure 2-9, each groove serves as a source of light. However, the reflected radiation from one groove overlaps that from neighboring grooves, allowing individual waves to interfere with one another. When the distance between the grooves is a whole number of wavelengths, the waves are in phase and light is reflected. On the other hand, if the distance is not an integral number of wavelengths, the waves are out of phase and cancel one another. Hence no light is reflected. The particular wavelengths reflected

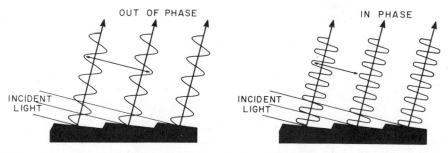

Figure 2-9. Results of reflecting different wavelengths of light off a diffraction grating.

are determined by the angle, θ, at which light strikes the grating. The relationship between θ, the distance between the grooves (d), and the wavelength is

$$n\lambda = 2d \sin \theta \qquad (11)$$

where n is the integer cited above. Unfortunately, diffraction gratings produce a whole series of overlapping arrays of monochromatic light, which are designated as first, second, and third order, corresponding to values of n equal to 1, 2, and 3, respectively (Figure 2-10). For any given angle θ radiation at wavelengths of λ, $\lambda/2$, $\lambda/3$, etc., is reflected. Therefore, the angles of reflection for light of 750, 375, and 250 nm are all the same. Though appropriate grating design results in 75% of the reflected light being first order, the remaining higher order light also reflected would cause large errors in absorbance measurements. This problem has been circumvented by placing a band pass filter in the output

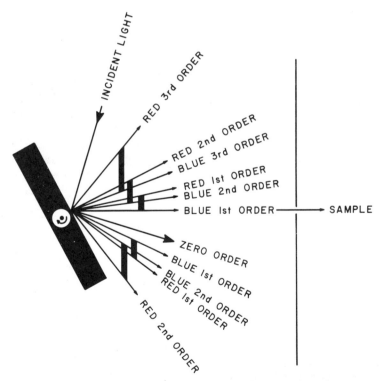

Figure 2-10. Spatial orientation of first and second order light obtained from a diffraction grating. Heavy lines connect extremes of each spectrum. An arrow indicates the direction of rotation of the grating. Also indicated is the external slit of the monochromator.

beam of the monochromator (usually after lens I in Figure 2-7). These filters, which are usually set in an automatically controlled disk, transmit light over relatively large spectral areas. If, for example, light of 750 nm is desired, a band pass filter transmitting light between 500 and 850 nm would likely be used. In this way the unwanted higher orders of diffracted light (250 and 375 nm) are prevented from reaching the sample. A second means of eliminating higher orders of light is to place a prism in tandem with the diffraction grating. A restricted area of the spectrum emerging from the prism (500 to 800 nm for example) is directed onto the diffraction grating. In this configuration a beam of 750 nm light can be reflected from the grating in the absence of 375 nm light which was removed from the incident beam. This technique also permits much greater dispersion of the light beam.

As depicted in Figures 2-8 and 2-10, light reaching the sample is that which passes through the exit slit of the monochromator (S_3 in Figure 2-7). It is the width of this slit, usually denoted as the slit width, along with the dispersing ability of the monochromator, that determines the spectral purity or resolution of the system. Figure 2-11 shows a plot of the intensity of light reaching the sample through a minimum usable slit width as a function of its wavelength. As can be seen, a small wavelength interval with a maximum at the desired wavelength exits through the slit. The greater the physical dispersion of wavelengths from one another, the closer the light passing through the slit is to monochromatic. Alternatively, decreasing the slit width decreases the size of the exiting wavelength interval. The width of this wavelength interval is designated as the spectral slit width and the interval at half maximum intensity (see

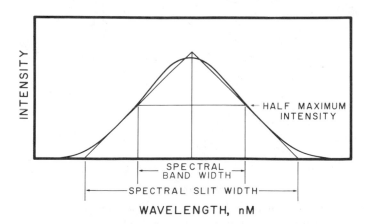

Figure 2-11. Spectrum of light passing through a minimum usable slit width.

Figure 2-11) as the spectral bandwidth. The spectral bandwidth provides a way of quantitating the degree to which a monochromator produces light of high spectral purity; the smaller the spectral bandwidth, the higher the spectral purity. But how do these parameters affect a spectrophotometric measurement in practice? Depicted in Figure 2-12 are two spectra of the same compound taken at a large and small bandwidth. At the larger bandwidth the center peak can no longer be resolved as a distinct entity. This should be rationalized by the reader in terms of the information provided in Figure 2-11. If decreasing the slit width increases spectral purity it is reasonable to inquire how far the slit width may be decreased. The degree of decrease is limited because as the slit width decreases so does the amount of light reaching the sample and detector. Although a light source of high intensity permits quite small slit widths, it is predominantly the detector that determines just how far it may be decreased. As the amount of light reaching the detector is decreased, the

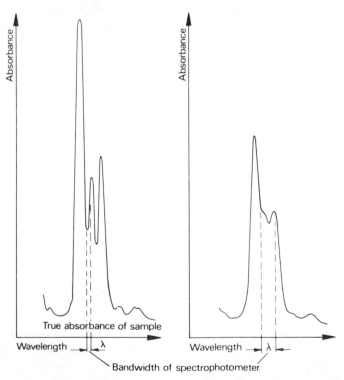

Figure 2-12. Effect of slit width on the resolution of a recording spectrophotometer. (From R. L. Manning, Ed., *Introduction to Spectroscopy*, Pye Unicam Ltd., Cambridge, England, 1969.)

detector's sensitivity must be correspondingly increased if a measurement is to be made. If this is done, the detector also becomes more sensitive to erroneous electronic background "noise" generated in the photomultiplier tube and circuits (discussed in greater detail in Chapter 3). As soon as the "noise" signal level reaches a few hundredths of a percent of the light generated signal, quality of the spectral trace is compromised.

Not only must a monochromator produce a beam of small bandwidth, it must also produce a beam with a wavelength that is accurately known. Wavelength accuracy may be determined in practice either by locating specific emission lines produced by the light source or by determining the spectrum of a standard holmium oxide or didymium glass filter. As shown in Figure 2-13 such standards possess a large number of reasonably sharp absorbance maxima at known wavelengths in the visible portion of the spectrum.

Sample Chamber

The sample compartment of most contemporary instruments accept a large variety of cells or cuvettes. These may be very small with an internal volume of only 0.05 ml or sufficiently large to contain an entire 20 cm

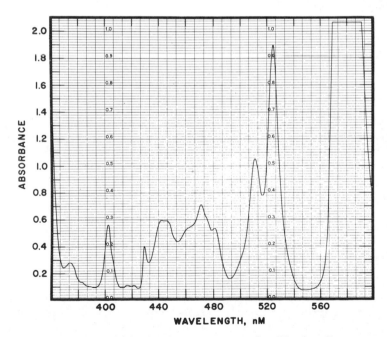

Figure 2-13.   Visible absorbance spectrum of a didymium filter.

polyacrylamide gel. Also accepted, usually with some modification, are continuous flow assemblies used to continuously measure the absorbance of a stream of liquid. Cuvettes and sample chambers of high quality instruments are constructed for maintenance of constant temperature conditions. This is accomplished by circulating water or an ethylene glycol solution through the walls of the sample chamber, cuvette holder, and even the cuvette itself.

Cuvettes are one of the most important parts of the spectrophotometric system because no matter how high the quality of the instrument, poor cuvettes yield poor measurements or spectra. For measurements in the visible region, cuvettes of optical glass are sufficient. However, below 350 nm standard glass absorbs light and more expensive quartz or fused silica cuvettes must be used. Careful use and storage of cuvettes are necessary if their quality is to be maintained. A few simple precautions follow. (1) Never allow abrasive materials to come in contact with a cuvette because a scratch on the optical surface reflects light and results in erroneous measurements. (2) Never leave materials in cuvettes when measurements are completed. This is especially true of protein or nucleic acid solutions, which stick tenaciously to the glass. If this occurs wash the cuvettes carefully with a very mild detergent and rinse them thoroughly before drying. (3) Use lens paper rather than facial tissues to wipe the cuvettes because extended use of facial tissues will scratch the optical surfaces.

A final problem affecting the quality of a measurement is stray light. This is light of unwanted wavelengths that most often arises in the monochromator, but may also result from a light "leak" in the chamber. For this reason, the interior walls of the monochromator and sample chamber are painted with black nonreflecting paint. Stray light arising in the monochromator is removed by one of two mechanisms. In expensive instruments use of two light dispersing elements, prisms and/or diffraction gratings, in tandem decreases stray light, usually by several orders of magnitude (see Figure 2-7). Other instruments use a band pass filter assembly between the monochromator and sample chamber to decrease the amount of stray light reaching the sample.

Detector

The detection system of a spectrophotometer is a light sensitive photomultiplier tube. Operation of this device is described in detail in Chapter 3 and is not, therefore, discussed here. In instruments with a broad spectral range it is usually necessary to use two different photomultipliers, one most sensitive between 200 and 600 nm and a second sensitive to wavelengths between 600 and 1000 nm (*P* and photomultiplier in Figure 2-7).

Although detailed discussion of the electronic circuits used in spectrophotometers is beyond our purpose, a discussion of two methods for measuring absorbance is useful. In the discussion thus far, only one sample has been used. This of course implies that for each measurement performed a corresponding "blank" measurement was also performed to ascertain contributions to observed absorbances from the solvent, cuvette, and other undesirable materials that might be present. For example, if the absorbance of a sample was found to be 0.80 and the absorbance of the blank (containing all materials present in the original sample with the exception of the material whose absorbance was to be determined) was found to be 0.40, then steps should be taken to improve the assay. The blank value should never account for greater than 10% of the observed absorbance. Though this procedure is acceptable for single readings, it is unacceptably cumbersome in the case of spectra. Therefore, the dual beam spectrophotometer was developed. A light path diagram for such an instrument is depicted in Figure 2-7. Construction of a dual beam instrument is identical to that of a single beam except that there is a light beam splitter inserted just prior to the sample (components *J*, *K*, *M*, and M_1 in Figure 2-7). Beam splitting may be accomplished with a prism or more often with a rotating disk (*J* in Figure 2-7), which directs the beam one-half of the time through the sample and the other half through a reference standard. Electronically, the signals emanating from the two samples are recombined (in time) and the reference signal is subtracted from the sample signal. This permits measurement of a continuous spectrum for a compound and at the same time provides for display of only the pertinent absorbances.

EXPERIMENTAL

Biuret Protein Determination

The biuret reaction for protein determination was one of the first colorimetric protein assay methods developed (1) and still enjoys wide use today. It is most often used in applications requiring a fast, but not highly accurate, determination. Failure of ammonium sulfate to interfere with color formation makes it advantageous for determinations during the early steps of purifying a protein. The nature of the colored complex, though unknown in detail, is thought to involve complexation of copper in alkaline solution with the peptide linkage of proteins and also with tyrosine residues.

Substances interfering with this assay include buffers which are amino acid or peptide in nature, such as Tris and Good's buffers, because they give positive color formation. The Cu^{2+} ions are also susceptible to reduction. This can be detected in practice by appearance of a red precipitate in the reaction mixture.

Preparation of Biuret Reagent

2-1. Place 1.50 g $CuSO_4 \cdot 5H_2O$, 6.00 g sodium potassium tartrate $(NaKC_4O_6 \cdot 4H_2O)_4$, and a stirring bar in a 1 liter volumetric flask.

2-2. Add 500 ml glass-distilled water to the flask and dissolve the above solids.

2-3. While stirring the contents of the flask vigorously, add 300 ml 10% (w/v) NaOH.

2-4. Remove the stirring bar from the flask and bring the volume of liquid to 1 liter with glass-distilled water.

2-5. Mix the contents thoroughly and transfer them to a plastic bottle.

2-6. The biuret reagent just prepared should be a deep royal blue. This solution may be stored indefinitely. If a black precipitate is observed in the storage container, however, discard the solution and prepare a fresh solution.

Determination of Protein by the Biuret Reaction

2-7. Turn on the spectrophotometer or colorimeter to be used and allow it to warm up as directed by the manufacturer.

2-8. Number 10 16×150 mm test tubes and place them in a test tube rack. In each tube carefully pipette one of the following volumes of a 10 mg/ml solution of bovine serum albumin: 0, 0.1, 0.2, 0.3, 0.4, 0.5, 0.6, 0.7, 0.8, and 1.0 ml.

2-9. Bring the total volume of liquid in each tube to 1.0 ml by adding an appropriate amount of glass-distilled water.

2-10. Add 4.0 ml biuret reagent to each tube and vortex the mixture for a few seconds to effect thorough mixing of the solutions.

2-11. Incubate the tubes for 20 minutes at 37°C.

2-12. Determine the absorbance of each sample at a wavelength of 540 nm. Color observed at the conclusion of the incubation period is stable for a short time (1–2 hours), but slowly increases over a period of several hours. Therefore, read the samples as soon as possible.

2-13. Plot the data obtained at step 2-12 as shown in Figure 2-14. Assay unknown samples in the same way and determine their protein content from the standard curve just prepared.

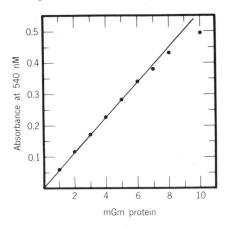

Figure 2-14. Absorbances observed with increasing amounts of protein assayed by means of biuret reagent.

Lowry Protein Determination

The Lowry reaction for protein determination is an extension of the biuret procedure (2, 3). The first step involves formation of a copper–protein complex in alkaline solution. This complex then reduces a phosphomolybdic–phosphotungstate reagent to yield an intense blue color. This assay procedure is much more sensitive than the biuret method but is also more time-consuming. It is susceptible to the same interfering ions as the biuret reaction but is far more greatly affected by them. In addition the presence of mercaptans and a variety of other compounds produce erroneous results. The only precaution to be observed when performing this assay concerns addition of the Folin's reagent. This reagent is stable only at acidic pH; however, the reduction indicated above occurs only at pH 10. Therefore, when Folin's reagent is added to the alkaline copper–protein solution, mixing must occur immediately so that the reduction can occur before the phosphomolybdic–phosphotungstate (Folin's) reagent breaks down.

2-14. Prepare reagent A by dissolving 100 g Na_2CO_3 in 1 liter (final volume) 0.5 N NaOH.

2-15. Prepare reagent B by dissolving 1 g $CuSO_4 \cdot 5H_2O$ in 100 ml (final volume) glass-distilled water.

2-16. Prepare reagent C by dissolving 2 g potassium tartrate in 100 ml (final volume) glass-distilled water. The reagents prepared in steps 2-14 to 2-16 may be stored indefinitely.

2-17. Turn on the spectrophotometer or colorimeter to be used and allow it to warm up as directed by the manufacturer.

2-18. Number 10 16 × 150 mm test tubes and place them in a test tube rack. In each tube carefully pipette one of the following volumes of a 0.3 mg/ml solution of bovine serum albumin: 0, 0.1, 0.2, 0.3, 0.4, 0.5, 0.6, 0.7, 0.8, and 1.0 ml.

2-19. Bring the total volume of liquid in each tube to 1.0 ml by adding an appropriate amount of glass-distilled water.

2-20. Mix thoroughly 15 ml reagent A, 0.75 ml reagent B, and 0.75 ml reagent C in a 50 ml Erlenmeyer flask.

2-21. Add 1 ml of the solution made in step 2-20 to each of the tubes prepared in steps 2-18 and 2-19. Vortex the tubes to mix them thoroughly.

2-22. Incubate the tubes for 15 minutes at room temperature.

2-23. While the tubes are being incubated, add 5.0 ml 2N Folin–phenol reagent to 50 ml distilled water in a 125 ml Erlenmeyer flask. Mix the solution thoroughly.

2-24. At the conclusion of the incubation period (step 2-22), forcibly pipette 3.0 ml of the solution made in step 2-23 into each tube. Vortex the resulting solution immediately. It is very important that the additional mixing be accomplished in as short a time as possible. The addition to and mixing of each tube should be completed before proceeding to the next.

2-25. Incubate the sample at room temperature for 45 minutes.

2-26. Determine the absorbance of each sample at a wavelength of 540 nm. In addition to 540 nm, the absorbance of the samples may be

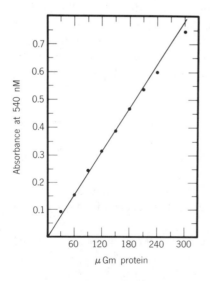

Figure 2-15. Absorbances observed with increasing amounts of protein assayed by the Lowry procedure.

determined at 660 or 750 nm. At these higher wavelengths the samples exhibit a much greater absorbance. The absorbances should be determined as soon as possible after the conclusion of incubation.

2-27. The color of the samples remains stable for about 45 minutes to 1 hour after the conclusion of the incubation period.

2-28. Plot the data as shown in Figure 2-15 to obtain a standard curve. Test samples should always be assayed using procedures identical to those employed for the standard curves.

Determination of Inorganic Phosphate

This procedure for phosphate determination (4) is a modification of the early methods of Fiske and Subbarow (5). It depends on the formation of a phosphomolybdate complex with ammonium molybdate and then reduction of this complex with ascorbic acid. It differs from the Fiske–Subbarow reaction in the nature of the reducing agent. The latter method employed a combination of sodium sulfite and aminonaphthosulfonic acid for this purpose. The extremely high sensitivity of this reaction requires use of acid-washed glassware, because glassware washed with phosphate-containing soaps usually contains sufficient residue to turn the reagents black in the absence of added phosphate.

2-29. Dissolve 10.0 g ascorbic acid in 100 ml glass-distilled water. This reagent may be stored up to 7 weeks at 2 to 4°C.

2-30. Dissolve 2.5 g ammonium molybdate, $(NH_4)_6Mo_7O_{24} \cdot 4H_2O$, in glass-distilled water. Final volume of the solution should be 100 ml.

2-31. Prepare a $6N$ solution of sulfuric acid by slowly adding 18 ml concentrated acid to 90 ml water.

2-32. Prepare reagent A by mixing thoroughly 1 volume (10 ml) $6N$ sulfuric acid with 2 volumes (20 ml) distilled water, 1 volume (10 ml) ammonium molybdate solution, and finally 1 volume (10 ml) ascorbic acid solution. This reagent should be prepared fresh each day because it is quite unstable.

2-33. Number 10 16×150 mm test tubes and place them in a test tube rack. Into each of the tubes carefully pipette one of the following volumes of a 34.5 mg/1000 ml solution of potassium dibasic phosphate (K_2HPO_4): 0, 0.1, 0.2, 0.3, 0.4, 0.5, 0.6, 0.7, 0.8, and 1.0 ml.

2-34. Add an appropriate amount of distilled water to each tube to give a final volume of 1.0 ml.

2-35. Add 4.0 ml reagent A (step 2-32) to each tube and mix thoroughly by vortexing.

2-36. Cover each tube (Parafilm is suitable) and incubate at 37°C for 2 hours.

2-37. Cool the tubes to room temperature and determine the absorbance of each tube at a wavelength of 820 nm.

2-38. Plot the data yielded from step 2-37 as shown in Figure 2-16. What is the extinction coefficient of the complex produced in this procedure? Unknown samples are assayed in an identical manner and their phosphate content determined from the standard curve prepared above.

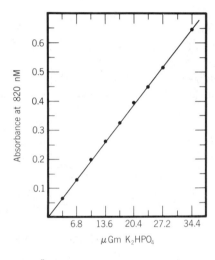

Figure 2-16. Absorbances observed with increasing amounts of inorganic phosphate assayed by a modification of the Fiske–Subbarow procedure.

Determination of Nucleic Acids by the Orcinol Reaction

The orcinol reaction (6) is used to assay nucleic acids, hexuronic acids, pentoses, and some aldopentoses. Although DNA reacts positively under the conditions of this assay, it exhibits an extinction of only 25% of that observed with an equivalent amount of RNA owing to the poorer reactivity of deoxypentose. Also, RNA has been observed to yield only 40% of the extinction observed for an equivalent amount of adenylic acid. This emphasizes that the material used to prepare the standard curve should be as close as possible to the actual experimental material being assayed. The color forming reaction involves (1) production of pentoses or deoxypentoses from nucleic acids; (2) conversion of these to furfural by heating in the presence of strong acid; and (3) reaction of furfural with orcinol to yield a mixture of blue-green condensation products (Figure 2-17).

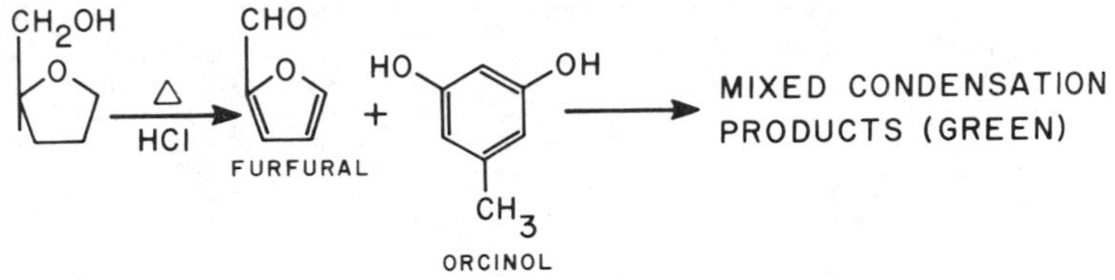

Figure 2-17. Reactions involved in the assay of nucleic acids by the orcinol procedure.

Determination of RNA

2-39. Dissolve 13.5 g ferric ammonium sulfate and 20.0 g orcinol in 500 ml glass-distilled water. This reagent may be stored at 4°C for short periods. If the reagent begins to darken, it should be discarded.

2-40. Add 5.0 ml of the reagent prepared in step 2-39 to a solution composed of 85 ml concentrated HCl and 10 ml glass-distilled water (final total volume should be 100 ml).

2-41. Number 10 16 × 150 mm test tubes and place them in a test tube rack. Into each of the tubes carefully pipette one of the following volumes of a 10 mg/100 ml solution of RNA: 0, 0.1, 0.2, 0.3, 0.4, 0.5, 0.6, 0.7, 0.8, and 1.0 ml.

2-42. Add an appropriate amount of distilled water to each tube to give a final volume of 1.0 ml.

2-43. Add 3.0 ml of the solution prepared in step 2-40 to each tube and mix the solution thoroughly by vortexing.

2-44. Cover each tube with a marble.

2-45. Place the covered tubes in a rack and place the rack in a pan of vigorously boiling water for 20 minutes. This should be done as soon as possible after step 2-43.

2-46. Remove the tubes from the bath and cool to room temperature.

2-47. Determine the absorbance of each sample at a wavelength of 670 nm.

2-48. Plot the data obtained in step 2-47 as shown in Figure 2-18. Test samples are handled in an identical manner.

Park and Johnson Method for a Reducing Sugar

This method for determining reducing sugars (7) is based on the reduction of ferricyanide ions in alkaline solution by a reducing sugar. The ferrocyanide produced can then react with a second mole of ferricyanide producing the ferric–ferrocyanide (Prussian blue) complex. Potassium

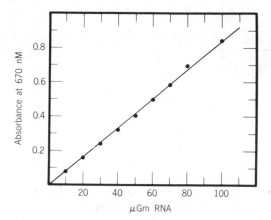

Figure 2-18. Absorbances observed with increasing amounts of RNA assayed by the orcinol reaction.

cyanide and sodium dodecyl sulfate are included in the formulation to speed the rate of reduction and maintain the Prussian blue complex in suspension, respectively. This reaction works well with any reducing sugar but is inhibited by acid, ammonia, and high ionic strength.

2-49. Prepare reagent A by dissolving 0.5 g potassium ferricyanide in 1000 ml distilled water. This solution should be stored in a brown bottle.

2-50. Prepare reagent B by dissolving 5.3 g sodium carbonate and 0.65 g potassium cyanide in 1000 ml distilled water.

2-51. Prepare reagent C by dissolving 1.5 g ferric ammonium sulfate and 1.0 g sodium dodecyl sulfate in 1000 ml 0.05N sulfuric acid (1.4 ml concentrated sulfuric acid per liter).

2-52. Number 10 16×150 mm test tubes and placed them in a test tube rack. Into each of the tubes carefully pipette one of the following volumes of a 28 mg/1000 ml solution of glucose: 0, 0.1, 0.2, 0.3, 0.4, 0.5, 0.6, 0.7, 0.8, and 1.0 ml.

2-53. Add an appropriate amount of distilled water to each tube to give a final volume of 1.0 ml.

2-54. Add 1.0 ml solution B and then 1.0 ml solution A to the sample tubes. DO NOT MOUTH-PIPETTE THESE SOLUTIONS.

2-55. Mix the contents of the sample tubes thoroughly by vortexing.

2-56. Cover each tube with a marble.

2-57. Place the samples in a boiling water bath for 15 minutes. Remove the tubes from the bath and let them cool to room temperature.

2-58. Add 5.0 ml reagent C to each sample and mix thoroughly. DO NOT MOUTH-PIPETTE THIS SOLUTION.

2-59. Allow the sample to incubate at room temperature for 15 minutes.

2-60. Determine the absorbance of each sample at a wavelength of 690 nm.

2-61. Plot the data obtained from step 2-60 as shown in Figure 2-19. Unknown samples should be processed in an identical manner and their sugar content determined using the standard curve prepared above.

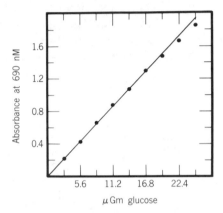

Figure 2-19. Absorbances observed with increasing amounts of glucose assayed by the Park and Johnson procedures.

Determination of the pK_a for Bromophenol Blue

2-62. Prepare 100 ml 0.1M citrate buffer by dissolving 2.94 g trisodium citrate in 70 ml glass-distilled water and adding sufficient concentrated HCl to produce a pH of 5.20. The pH adjustment should be made using a pH meter and should be as accurate as possible.

2-63. Add sufficient water to the above solution to give a final volume of 100 ml and recheck the pH.

2-64. Repeat steps 2-62 and 2-63 eight times, adjusting one of the solutions to each of the following pH values: 2.40, 2.60, 3.00, 3.40, 3.80, 4.20, 4.60, 4.80, and 5.20.

2-65. Dissolve exactly 10.0 mg bromophenol blue in 10 ml 95% (v/v) ethanol. The molecular weight of bromophenol blue is 670.

2-66. Number 8 16 × 150 mm test tubes and place them in a test tube rack. Into each tube *carefully* pipette 0.10 ml of the bromophenol blue solution.

2-67. Into each of the tubes carefully pipette 12.0 ml of one of the buffer solutions prepared in steps 2-62 and 2-64.

2-68. Mix thoroughly the contents of each tube by vortexing. At this point the tubes should range in color from bright yellow (pH 2.40) to deep purple (pH 5.20).

2-69. Determine the absorbance of a water blank and each of the nine samples prepared in step 2-68 first at a wavelength of 430 nm and then at a wavelength of 590 nm. If the absorbances are above the range of your spectrophotometer, add exactly 5.0 ml additional buffer of the appropriate pH values.

2-70. Prepare absorbance spectra of the pH 2.40, 3.80, and 5.20 samples. These spectra should be made in a dual beam spectrophotometer using a cuvette filled with glass-distilled water as a reference. The spectral measurements should be performed between 660 nm and 310 nm. From these spectra (Figure 2-20) it should be clear why wavelengths of 430 and 590 nm were selected in step 2-69.

2-71. The data collected in step 2-69 are used to calculate the pK_a of the indicator, bromophenol blue.

The following is a derivation of the method that will be used to make this calculation.

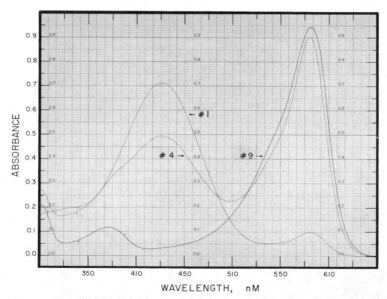

Figure 2-20. Absorbance spectra of bromophenol blue at three pH values. Curves numbered 1, 4, and 9 indicate the spectra performed on solutions at pH values of 2.40, 3.60, and 5.20, respectively.

$$\text{HInd} \rightleftharpoons \text{H}^+ + \text{Ind}^-$$

$$K_{\text{Ind}} = \frac{[\text{H}^+][\text{Ind}^-]}{[\text{HInd}]}$$

Let C = total concentration of the indicator (all species); that is,

$$C = \frac{\text{total grams bromophenol blue per liter of sample}}{\text{MW of bromophenol blue}}$$

ϵ = extinction coefficient of the mixture at a given wavelength, either 590 or 430 nm; that is,

$$\epsilon = \frac{\text{absorbance at 590 or 430 nm for a sample of any given pH}}{C}$$

For all the subsequent calculations, use absorbances at only one wavelength as just indicated.

C_1 = concentration of HInd

C_2 = concentration of Ind$^-$

ϵ_1 = extinction coefficient of HInd at λ (590 or 430 nm)

ϵ_2 = extinction coefficient of Ind$^-$ at λ (590 or 430 nm)

These extinction coefficients are determined using the absorbance data from the samples at pH 2.40 and 5.20, which are justly assumed to represent the pure compound, that is, bromophenol blue in the associated and dissociated forms, respectively.

$$\epsilon C = \epsilon_1 C_1 + \epsilon_2 C_2$$

and

$$C = C_1 + C_2$$

Therefore,

$$\epsilon (C_1 + C_2) = \epsilon_1 C_1 + \epsilon_2 C_2$$

$$\epsilon C_1 + \epsilon C_2 = \epsilon_1 C_1 + \epsilon_2 C_2$$

Dividing by C_1 gives

$$\epsilon \left(\frac{C_1}{C_1}\right) + \epsilon \left(\frac{C_2}{C_1}\right) = \epsilon_1 \left(\frac{C_1}{C_1}\right) + \epsilon_2 \left(\frac{C_2}{C_1}\right)$$

Substituting from the equilibrium equation above yields

$$\epsilon + \epsilon \frac{K_{\text{Ind}}}{[\text{H}^+]} = \epsilon_1 + \epsilon_2 \frac{K_{\text{Ind}}}{[\text{H}^+]}$$

$$\epsilon - \epsilon_1 = \frac{K_{\text{Ind}}}{[\text{H}^+]}(\epsilon_2 - \epsilon)$$

Taking the log of both sides of the equation and rearranging gives

$$\log(\epsilon - \epsilon_1) - \log(\epsilon_2 - \epsilon) = \log K_{Ind} + pH$$

Plot the pH of the samples used in Step 2-69 on the ordinate, and plot $\log[(\epsilon - \epsilon_1)/(\epsilon_2 - \epsilon)]$ on the abscissa. The various values of ϵ are calculated by dividing the total concentration of bromophenol blue (C) into the absorbance observed at each pH. The end points (pH = 2.40 and 5.20), however, cannot be used as part of this series. The Y intercept is the point where pH = pK. Convince yourself that

$$\log\left[\frac{(\epsilon - \epsilon_1)}{(\epsilon_2 - \epsilon)}\right] = \log\left(\frac{[Ind^-]}{[HInd]}\right)$$

A plot of these data is shown in Figure 2-21.

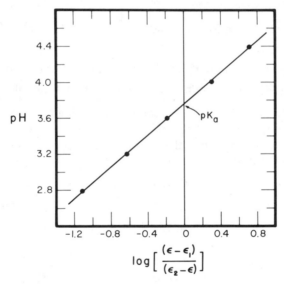

Figure 2-21. Determination of the pK_a of bromophenol blue through the use of spectrophotometric methods.

Spectral Characteristics of Biologically Significant Molecules

2-72. Turn on a dual beam recording spectrophotometer and allow it to warm up as directed by the manufacturer. Both the tungsten and deuterium bulbs should be turned on.

2-73. Prepare a solution of catalase by dissolving 1.25 to 1.50 mg protein in 1.0 ml 0.01M phosphate buffer adjusted to pH 7.0.

2-74. Prepare an absorbance spectrum of this solution in the above spectrophotometer operating between the wavelengths of 240 and 500 nm. Caution should be taken to change the light source at 340 nm from the tungsten to the deuterium source. Failure to do this will result in a very erratic spectrum below this wavelength.

2-75. A spectrum of catalase prepared in this matter is shown in Figure 2-22.

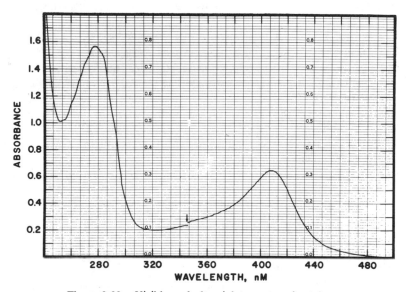

Figure 2-22. Visible and ultraviolet spectra of catalase.

2-76. Similar experiments may be performed using RNA, DNA, cytochrome C, hemoglobin, and the reduced and oxidized forms of NAD, FMN, and FAD.

REFERENCES

1. Allan G. Gornall, Charles J. Bardawill, and Maxima M. David, *J. Biol. Chem.*, **177**:751–766 (1949). Determination of Serum Proteins by Means of the Biuret Reaction.
2. Oliver H. Lowry, Nira J. Rosebrough, A. Lewis Farr, and Rose J. Randall, *J. Biol. Chem.*, **193**:265–275 (1951). Protein Measurement with the Folin Phenol Reagent.
3. Ennis Layne, *Methods Enzymol.*, **3**:447–454 (1957). Spectrophotometric and Turbidimetric Methods for Measuring Proteins.
4. P. S. Chen, Jr., T. Y. Toribara, and Huber Warner, *Anal. Chem.*, **28**:1756–1758 (1956). Microdetermination of Phosphorus.

5. Cyrus H. Fiske and Yella Pragada Subbarow, *J. Biol. Chem.*, **66**:375–400 (1925). The Colorimetric Determination of Phosphorus.

6. D. J. Merchant, R. H. Kahn, and W. H. Murphy, *Handbook of Cell and Organ Culture*, 2nd ed., Burgess, Minneapolis, 1969.

7. J. T. Park and M. J. Johnson, *J. Biol. Chem.*, **181**:149–151 (1949). A Submicro Determination of Glucose.

8. G. R. Penzer, *J. Chem. Educ.*, **45**:693–701 (1968). Applications of Absorption Spectroscopy in Biochemistry.

9. R. A. Friedl and M. Orchin, *Ultraviolet Spectra of Aromatic Compounds*, Wiley, New York, 1951.

10. K. P. Bauman, *Absorption Spectroscopy*, Wiley, New York, 1962.

11. M. G. Mellon, *Analytical Absorption Spectroscopy*, Wiley, New York, 1950.

12. G. H. Beaven et al., *Molecular Spectroscopy*, Heywood, 1961.

13. H. H. Jaffe and M. Orchin, *The Theory and Applications of Ultraviolet Spectroscopy*, Wiley, New York, 1962.

14. A. E. Gillam and E. S. Stern, *An Introduction to Electronic Absorption Spectroscopy*, 2nd ed., Arnold, 1957.

15. F. D. Snell and C. T. Snell, *Colorimetric Methods of Analysis*, Van Nostrand, Princeton, N.J., 1961.

Chapter 3

Radiochemistry

Tracer techniques have revolutionized biochemistry and molecular biology. For example, the availability of isotopically labeled compounds made it possible to demonstrate that macromolecules such as proteins, nucleic acids, and complex lipids are synthesized from simple cellular metabolites and provided many insights into the mechanisms and control of the synthetic events. The utility of radiochemical techniques is afforded by (1) their great sensitivity compared to other analytical methods (Table 3-1) and (2) the fact that they "label" the atoms of molecules without significantly altering their chemical properties, thus allowing them to be "traced" or followed from one molecule to another.

A discussion of radiochemical techniques must begin with an understanding of the atomic nucleus and the events that result from altering its composition. An atomic nucleus (below a mass of about 40) contains protons, having an atomic mass of one and a charge of plus one, and an equal number of neutrons, having an atomic mass of one and a charge of zero. The positive charge of each proton is balanced by an electron, having an atomic mass of zero and a charge of minus one, moving around the nucleus. An isotope is produced by bringing about an imbalance in the proton to neutron ratio in the nucleus. This may be done, for example, by bombarding the nucleus with neutrons or α particles (see Table 3-2). The isotopes thus produced may be stable or they may disintegrate, that is, spontaneously emit subatomic particles and/or radiation. The latter type of isotopes is denoted as radioactive. These definitions may be amplified by considering the synthesis and disintegration of ^{14}C. This isotope is produced by bombarding nitrogen atoms with neutrons (the neutrons are obtained as one of the fission products of uranium or plutonium in a nuclear reactor). The reaction is

$$^{14}N + {}^1_0n \longrightarrow {}^{14}C + {}^1_1p \tag{1}$$

65

Table 3-1. Sensitivity of Various Analytical Techniques

Analytical Technique	Limit of Sensitivity
Spectrophotometric	10^{15} molecules
Radiochemical	
^{14}C	10^{11} atoms
^{3}H	10^{9} atoms
^{32}P	10^{7} atoms

Table 3-2. Characteristics of Biologically Significant Isotopes

Isotope	$t_{1/2}$	Particle Emitted	Energy of Particle[a] (MeV)		Method of Production
^{45}Ca	165 days	β^-	0.25	(100%)	^{44}Ca$(n, \gamma)^{45}$Ca
^{14}C	5760 years	β^-	0.155	(100%)	^{14}N$(n, p)^{14}$C
^{60}Co	5.27 years	β^-	0.31	(100%)	^{59}Co$(n, \gamma)^{60}$Co
		β^-	1.48	(0.01%)	
		γ	1.17	(100%)	
		γ	1.33	(100%)	
^{3}H	12.26 years	β^-	0.018	(100%)	^{6}Li$(n, \alpha)^{3}$H
^{125}I	60 days	EC[b]		(100%)	^{123}Sb$(\alpha, 2n)^{125}$I
		γ	0.035	(7%)	
		γ	0.027		Te X-rays
^{131}I	8.04 days	β^-	0.25	(3%)	^{130}Te$(n, \gamma)^{131}$Te
		β^-	0.33	(9%)	$\overset{25\,m}{\searrow}$
		β^-	0.61	(87%)	^{131}I
		β^-	0.81	(1%)	
		γ	0.08	(2%)	U$(n, f)^{131}$Te
		γ	0.28	(5%)	
		γ	0.36	(80%)	
		γ	0.64	(9%)	
		γ	0.72	(3%)	
^{32}P	14.2 days	β^-	1.71	(100%)	^{31}P$(n, \gamma)^{32}$P
					^{32}S$(n, p)^{32}$P
^{33}P	25 days	β^-	0.25	(100%)	^{33}S$(n, p)^{33}$P
^{40}K	1.3×10^9 years	β^-	1.32	(89%)	Occurs
		EC		(11%)	naturally
		γ	1.46	(11%)	
^{35}S	87.2 days	β^-	0.167	(100%)	^{35}Cl$(n, p)^{35}$S
					^{34}S$(n, \gamma)^{35}$S

[a] E_{max} values given, percentage values in parentheses indicate the percentage of disintegrations giving rise to the particle or radiation designated.
[b] Electron capture.

This may also be written

$$^{14}N(n, p)^{14}C \qquad (2)$$

In this reaction a neutron is captured by the nitrogen nucleus and a proton is emitted. The ^{14}C nucleus contains six protons and eight neutrons. The excess of neutrons results in instability of the ^{14}C nucleus (this may be compared with ^{13}C whose nucleus contains six protons and seven neutrons and is stable). The unstable ^{14}C nucleus can become stable by disintegrating one neutron. This disintegration yields one proton, which remains within the nucleus converting the carbon atom (six protons) into a nitrogen atom (seven protons), and one electron or β particle, which is emitted.

$$^{14}_{6}C \longrightarrow {}^{14}_{7}N + \beta + \bar{\nu} \qquad (3)$$

Also emitted from the disintegrating nucleus is a neutrino ($\bar{\nu}$), an entity possessing little mass and no charge.

During the disintegration energy is released from the nucleus. The amount of energy is specific to the type of nucleus undergoing disintegration and is divided between the β particle and the neutrino.

$$E_n = E_\beta + E_{\bar{\nu}} \qquad (4)$$

where E_n is the total emitted energy and E_β and $E_{\bar{\nu}}$ are the energies of the β particle and neutrino. E_n is a constant for a given isotope, but E_β and $E_{\bar{\nu}}$ may each vary from zero to E_n. Therefore, if a large number of nuclei are disintegrating, the β particles emitted will not be of only one energy but rather will have a distribution of energies ranging from zero to E_n. This is graphically shown in Figure 3-1. E_{max} is the highest energy that a β particle, from the disintegration of a specific isotope, may have. In that case $E_{\bar{\nu}}$ is zero. The E_{max} of a variety of biologically significant isotopes is shown in Table 3-2.

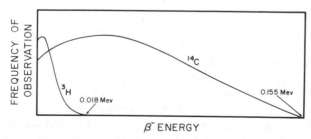

Figure 3-1. The observed energy distribution of a large number of β particles derived from ^{14}C and ^{3}H nuclei.

Radioactive disintegration or decay is a random process and may be expressed as a simple first order reaction. The mathematical expression describing the rate of such a reaction is

$$-\frac{dN}{dt} = \lambda N \tag{5}$$

where N is the number of radioactive atoms at time t and λ is the rate constant of disintegration or decay constant. If equation 5 is rearranged and integrated between the limits N_0 and N, the result is

$$\ln \frac{N_0}{N} = \lambda t \tag{6}$$

and

$$2.303 \log \frac{N_0}{N} = \lambda t \tag{7}$$

The exponential form of equation 7 is

$$N = N_0 e^{-\lambda t} \tag{8}$$

If λ and N_0 are known it is easy to calculate N for any time t. The decay constant of an isotope, however, is not usually available. A much more available characteristic is its half-life, denoted $t_{1/2}$. This is the time required for one-half of a given amount of an isotope to undergo disintegration. Substituting these values into equation 7, we have

$$2.303 \log \frac{1}{0.5} = \lambda t_{1/2} \tag{9}$$

or

$$0.693 = \lambda t_{1/2} \tag{10}$$

By using equation 10 it is possible to calculate λ if $t_{1/2}$ is known or vice versa. The half-lives of a number of isotopes are listed in Table 3-2. These values are usually determined by plotting $\log(N)$ as a function of time. This yields a linear plot whose slope may easily be shown to be $\lambda/2.303$. Such a plot is constructed in the experimental section of this chapter (Figure 3-40).

MEASUREMENT OF β RADIATION

Amounts of radioactivity are designated in terms of rads, roentgens, curies, disintegrations per unit time (minute), or counts per unit time (minute). The relationships between these various unit designations are summarized in Table 3-3. The first two units are rarely used except for measuring human exposure to ionizing radiation. It is worthy of emphasis

Table 3-3. Units of Radioactivity Measurement

Unit	Definition	Relationship to Other Units
Radiation absorbed dose (rad)	100 ergs/g (energy imparted to a unit mass of matter by ionizing radiation)	0.87 rad = r
Roentgen (r)	An exposure dose of X- or γ-radiation such that 1.61×10^{12} ion pairs are produced per gram of air	$\dfrac{r}{0.87} = \text{rad}$
Curie (c)	A quantity of radionuclide disintegrating at a rate of 3.7×10^{10} atoms per second	$c = 10^3\ mc$ $c = 10^6\ \mu c$
Millicurie (mc)	One-thousandth of a curie	$mc = 10^{-3}\ c$ $mc = 10^3\ \mu c$
Microcurie (μc)	One-millionth of a curie; 3.7×10^4 disintegrations per second (2.2×10^6 disintegrations per minute)	$\mu c = 10^{-6}\ c$ $\mu c = 10^{-3}\ mc$
Disintegrations per minute (dpm)	Number of atoms disintegrating per minute	$2.2 \times 10^6\ dpm = 1\ \mu c$
Counts per minute (cpm)	Number of β particles detected per minute by a detecting device	cpm = dpm × efficiency of counting device

that disintegrations per minute, or dpm, represent the actual number of β particles emitted per minute from a given sample and that cpm is the number of disintegrations that are counted or detected by a radioactivity detecting device. It is obvious that these two units are related by the efficiency of the detection device. If this is 100% then cpm = dpm; otherwise, cpm = dpm × efficiency of detection. It is sometimes useful to establish a relationship between the amount of a compound and the amount of radioactivity it contains. Therefore, the specific activity of a compound is defined as the amount of radioactivity per amount of compound. The most often used units of specific activity (abbreviated Sp. Act.) are curies/mole, mcuries/mmole, and μcuries/μmole.

There are three methods commonly used to ascertain the quantity of radioactive atoms in a given sample: film exposure, Geiger–Müller counting, and scintillation spectrometry. None of these techniques measures the number of β particles directly, but rather monitors the results of collisions between the β particles emitted from the radioactive atoms and some component of the assay system.

SCINTILLATION SPECTROMETRY

The most sophisticated technique is scintillation spectrometry. The basis of this method, in simplest terms, is to arrange for emitted β particles to collide with molecules that will give off light as a result of the collision and then measure the amount of light that is produced. The series of events involved in the production of light are depicted in Figure 3-2. The

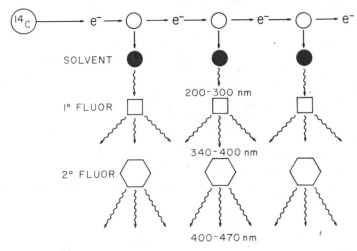

Figure 3-2. Interaction of β particles with aromatic solvents and subsequent fluor excitation. e^- represents the emitted β particles, $\bigcirc$ indicates a solvent molecule in its ground state, and $\bullet$ denotes solvent molecules in the triplet state. (From E. Rapkin, *Preparation of Samples for Liquid Scintillation Counting*, Picker Nuclear Corp., White Plains, New York.)

radioactive sample is dissolved or suspended in a scintillation system composed of solvent and primary and secondary scintillators. As shown in the figure, the emitted particle travels only a short distance before colliding with a solvent molecule. This collision results in transfer of a discrete amount of energy from the β particle to the solvent molecule. The transfer occurs most efficiently if an aromatic solvent is used, as demonstrated by the data shown in Table 3-4. Collision with solvent molecules and concomitant loss of energy are repeated until the β particle has lost sufficient energy to be captured. It should be noted that only about 5% of the total β particle energy is finally observed as light; the remainder is lost as lower energy (heat) quanta. The greater the β particle energy, the greater the number of solvent molecules with which it can collide; hence a greater amount of light is produced.

Table 3-4. Relative Counting Efficiency of Various Solvents[a]

Compound	Relative Efficiency (%)
Toluene	100
Methoxybenzene (anisole)	100
Xylene	97
1,3-dimethoxybenzene	81
1,4-dioxane	70
Ethylene glycol dimethyl ether	60
Acetone	12
Tetrahydrofuran	2
Ethanol	0
Ethylene glycol monomethyl ether	0
Ethylene glycol	0

[a] Data taken from J. D. Davidson and P. Feigelson, *Int. J. Appl. Radiat. Isot.*, **2**:1 (1957). Practical Aspects of Internal-Sample Liquid Scintillation Counting.

The energy embodied in the excited solvent molecule may be transferred to another solvent molecule or emitted as light; the latter process is termed phosphorescence. For phosphorescence to occur the excited toluene molecule, initially raised to a singlet state, must drop into its triplet state, as shown in Figure 3-3. The molecule remains in this state for a random period of 10^{-5} to 10^{-3} seconds. When it drops to the ground state its remaining excess energy is released as light. All the initial collisions and excitations caused by a β particle are completed in a relatively short time compared to that required for phosphorescence. However, the fact that excited toluene molecules spend varying amounts of time in the triplet state results in production of a sustained pulse of light rather than an instantaneous flash. The time course of the pulse is depicted in Figure 3-9A. The light emitted by the solvent is usually of a very short wavelength (260–340 nm; see Figure 3-4). Since it is not possible for existing instruments to detect light of this wavelength a second component must be added to the solvent. This compound, called a primary fluor, is capable of fluorescing, that is, absorbing light at one wavelength and reemitting it at a second, longer wavelength. The absorption and emission spectra of PPO (2,5-diphenyloxazole), the most widely used primary fluor, are shown in Figure 3-4. Note that the absorption spectrum of PPO completely overlaps the phosphorescent emission spectrum of toluene and that PPO emits the absorbed light at a longer wavelength. The transfer of energy from the β particle to the solvent to the primary fluor takes approximately 10^{-9}–10^{-3} seconds. Most scintillation spectrometers,

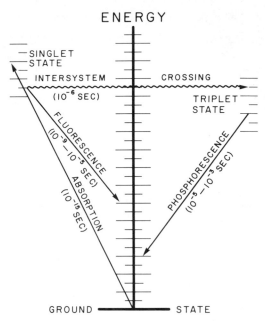

Figure 3-3. Modified Jablonski diagram showing the various processes that can occur upon excitation of an organic molecule. (From E. D. Bransome, *The Current Status of Liquid Scintillation Counting*, Grune and Stratton Inc., New York, 1970.)

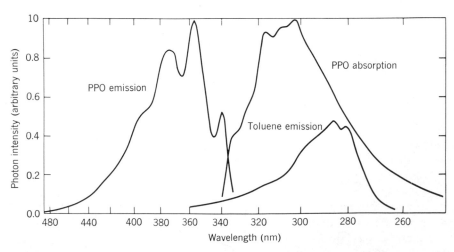

Figure 3-4. Emission spectrum of toluene and the emission and absorption spectra of PPO. (Courtesy of Beckman Instruments, Inc.)

the instrument used to measure the amount of emitted light, are sensitive to the fluorescent emission of the primary fluor. However, some of the older instruments require light of a still longer wavelength. Such a shift in the wavelength is accomplished by using a second compound (added to the system at a fraction of the concentration of the primary fluor), called a secondary fluor. The secondary fluor absorbs light at wavelengths equivalent to the emission spectrum of the primary fluor and emits fluorescent light with a maximum in the area of 420 to 440 nm. The fluorescent nature of the primary and secondary fluors is a result of their complex aromatic nature. The structures of the primary fluor, PPO, and secondary fluors, bis-MSB(p-bis-(o-methylstyryl)-benzene and dimethyl-POPOP(1,4-bis-2-(4-methyl-5-phenyloxazoyl)-benzene), are presented in Figure 3-5. The fluorescent maxima of most common primary and secondary fluors or scintillators are shown in Table 3-5.

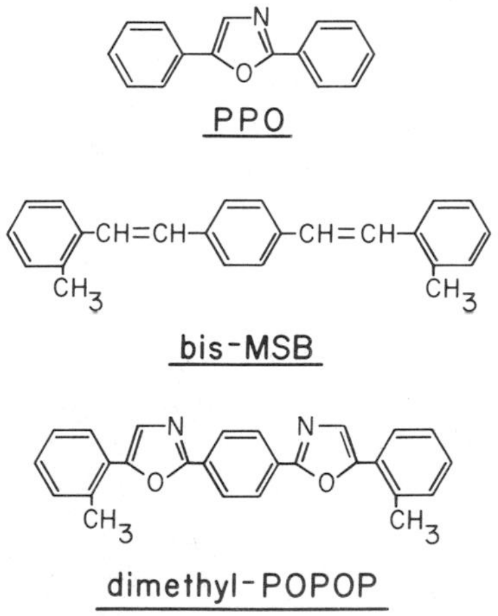

PPO

bis-MSB

dimethyl-POPOP

Figure 3-5. Structures of widely used primary and secondary fluors.

The amount of light yielded by fluors from the collisions of a β particle is diminishingly small and requires a highly sensitive means of detection. This is afforded by a scintillation spectrometer, shown in Figure 3-6. Its major components are diagramed in Figure 3-7. Although the scintillation spectrometer appears quite complex, it is in fact very simple; it is an assembly of components functioning together.

Most instruments currently available employ at least two photomultiplier tubes; usually one tube is situated on either side of the sample chamber. The photomultiplier tube is the sensing device of the instrument and can be discussed with the aid of the diagrams in Figure 3-8. The front

Table 3-5. Fluorescent Maxima of Various Scintillators in Toluene

Name	MW	Fluorescent Maxima (nm)		
		λ_1	λ_2	λ_3
Primary scintillator				
Naphthalene	128	325	336	351
Anthracene	178	405	428	454
Terphenyl	230	355		
PPO (2,5-diphenyloxazole)	221	365	380	
PBD [2-Phenyl-5-(4-biphenylyl)-				
1,3,4-oxadiazole]	298	364	377	
Butyl-PBD [2-(4-t-butylphenyl)-5-				
(4-biphenylyl)-1,3,4-oxadiazole]	354	367	382	
BBOT [2,5-bis(5-t-butylbenzoxazol-				
2-yl)thiophene]	430	438		
Secondary scintillator				
POPOP [1,4-bis(5-phenyloxazol-2-yl)-				
benzene]	271	420	441	
DMPOPOP [1,4-bis(4-methyl-5-				
phenyloxazol-2-yl)benzene]	392	430		
bis-MSB [p-bis(o-methylstyryl)benzene]	310	423		

of the tube is covered by a light sensitive material called a photocathode. When a pulse of fluorescent light from the secondary fluor strikes this surface a number of electrons are released. These electrons accelerate toward the positively charged first dynode (they accelerate through an electric field of approximately 100–200 V). To ensure that all the emitted electrons from the photocathode reach the dynode, a series of negatively charged focusing rings are employed. As each electron strikes the first dynode three to five electrons are released. Each of these electrons in turn strikes the second dynode and results in the release of three to five more electrons. This cascading process is continued 10 to 14 times. Therefore, for every electron striking the first dynode, approximately 10^6 electrons reach the fourteenth dynode. The time required for an electron to traverse the entire dynode cascade is of the order of 10^{-9} seconds. During this time the tube will not accept a second electron from the photocathode. This is called the dead time, or the time within which a second pulse of light is not detected if it occurs. The possibility, however, of missing a pulse during this time is quite small; this is discussed below.

The photomultiplier tube, which is capable of extremely high time resolution, detects a pulse of light rather than an instantaneous flash. Figure 3-9A depicts the varying intensity of light generated by a single β

Figure 3-6. A refrigerated scintillation spectrometer. The large bottom cabinet is a deepfreeze unit that contains the sample holders and photomultiplier detection units. The typewriter is used for data print out. (Courtesy of Packard Instrument Company, Inc.)

particle. The photomultiplier tube, responding to this pulse of light, produces an output signal (voltage) which duplicates the shape of the light pulse it observed (see Figure 3-9B). In other words the photomultiplier converts the light energy signal into an identical electrical signal that can be easily manipulated and measured.

The great sensitivity of a photomultiplier tube to light is obtained by

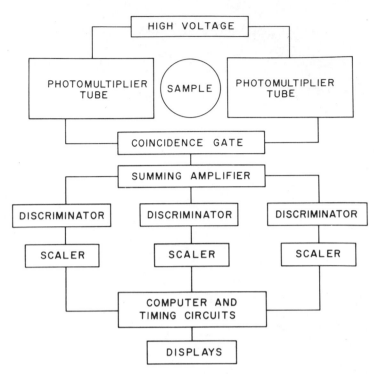

Figure 3-7. Block diagram of the basic electronic components of a scintillation spectrometer.

constructing the photocathode such that a very small amount of energy is required to eject a photoelectron. In this case the energy is provided by photons from the sample. However, energy in the form of heat (radiation of a longer wavelength) may also result in ejection of an electron. This emission of electrons from the photocathode as a result of thermal energy is called Joule–Thompson noise, or dark current, and may reach a level of 10^4 cpm. These emissions may be decreased by decreasing the thermal energy, that is, cooling the system. For every 13 degree drop in temperature, Joule–Thompson noise decreases by half. This is why many instruments have the counting chamber situated in a freezer (Figure 3-6) and operate at temperatures near 0°C. It is clear, however, that merely lowering the temperature does not solve the problem because it is not possible to obtain a background counting level at the desired 10–25 cpm by this method. A better solution to this problem is based on the fact that dark current arises spontaneously in a photomultiplier tube. If two photomultiplier tubes are simultaneously measuring the pulses of light

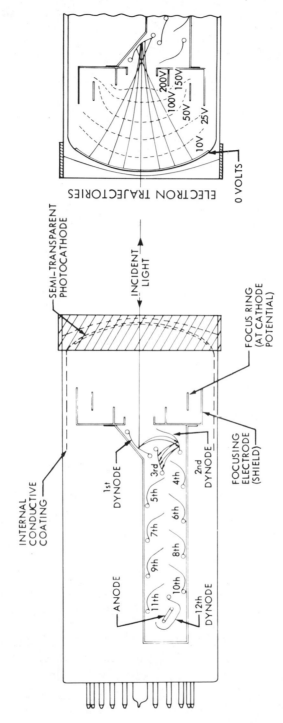

Figure 3-8. Beckman-RCA Bialkali 12-stage Photomultiplier Tube. (Courtesy of Beckman Instruments, Inc., Instruction Manual 1553-D.)

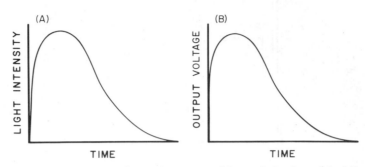

Figure 3-9. Time course of a light pulse generated by a single β particle (A) and the corresponding output signal from a photomultiplier tube (B).

generated by a sample, the thermally generated emissions from the two photocathodes would not be expected to occur at the same instant. Photocathode emissions resulting from photons generated by the sample, on the other hand, would be expected to occur simultaneously. Therefore, the output signal from two independently operating photomultiplier tubes is channeled through an electronic device called a coincidence gate. If a signal from each photomultiplier tube approaches the gate at precisely the same time (± 10 nseconds), they are summed together, passed through the gate, and processed further. These signals have a very high probability of being the result of a β particle emission. However, if signals from each of the photomultiplier tubes approach the gate separately, one at a time (as they do if they are the result of random thermionic noise), each signal is discarded without further processing.

Signals that pass through the coincidence gate proceed to the summing amplifier or pulse height analyzer (see Figure 3-7). This circuit is designed to determine the area under the pulse curve depicted in Figure 3-9B and generate a symmetrical pulse whose height is a function of that area. This is represented by a single line in subsequent figures. This device permits an area to be manipulated as a single number (the height of the symmetrical pulse) instead of a two-dimensional figure. If high energy β particles are being counted, the area and hence pulse height of the summing amplifier output signal are large. On the other hand, weak β particles generate only a very small pulse at this stage.

The output of the summing amplifier is fed simultaneously to two or three identical circuits designated as discriminators (Figure 3-7). A discriminator may be considered as a selection device; that is, it permits passage of only those signals that possess a pulse height within the range of that discriminator. The upper and lower limits of the range of operation may be designated via 10-turn potentiometers located on the control panel

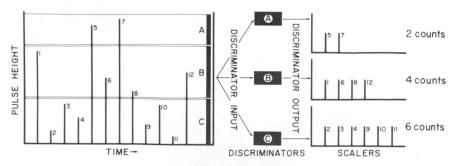

Figure 3-10. Operation of scintillation spectrometer discriminator circuits. The left graph represents the signals approaching the three discriminator circuits and the right panel depicts their output signals.

of the instrument. Figure 3-10 depicts the output of a summing amplifier as a function of time. Note that an entire spectrum of pulse heights is observed. The ranges of operation of three discriminators have been superimposed on the figure. Each signal is directed into each of the three discriminators. However, only a signal that lies within the range of a given discriminator appears in the output. Signals that are above or below the permissible range are discarded (see Figure 3-10). Since all three discriminators function totally independently they may possess ranges that do not overlap, as in Figure 3-10, or ranges that do overlap. What will be the consequences of discriminators that possess overlapping ranges of operation?

Connected to each discriminator is a scaler or electronic counter. This device merely keeps a tally of the number of pulses that appear in the output of the discriminator circuit. The scalers are controlled by timing and computer circuits. The desired counting time for each sample may be selected in two ways: (1) a preset number of minutes or seconds (designated as "preset time") or (2) the time required for a given number of counts to appear in each of the scalers (designated as "preset count"). When a sample first reaches the counting chamber, the timing circuits electronically set the timer and scaler to zero. When either the appropriate amount of time or number of counts has accumulated, the timing circuit terminates the operation. In more expensive instruments, the data in the timer and scalers are directed to a small computer which converts the counts taken to counts per minute. This computer also calculates a series of ratios whose significance is discussed later. Under command of the computer the data are printed out on paper or computer tape for further processing.

Many new instruments are equipped with a modification of the stan-

dard electronic clock. Dead time was defined above as that time (about 1 μsecond) during which a count is being registered by the instrument. It was noted that during this time the instrument would not register an event even if it originated from a β emission. Since the photomultiplier tube possesses reaction times on the order of nanoseconds, the majority of the 1 μsecond dead time of the instrument is produced by the electronic circuitry, for example, the summing amplifiers, discriminators, and scalers. It can be shown that a dead time of 1 μsecond would produce an undercount of 1% or 6000 cpm for a sample containing 600,000 cpm. To circumvent this problem, instrument designers have devised a clock that electronically switches off while a count is being processed and then switches back on again at the conclusion of the counting process. Therefore, time consumed in processing the instrument's response to a β particle (i.e., the time when the instrument cannot respond to a second β particle) is not recorded as part of the counting time. In this way the dead time error may be largely eliminated.

It is clear that the pulse height of a signal entering the discriminator is directly proportional to the energy of the β particle from which it originated. Also, as shown in Figure 3-1, β particles from a single isotope would be expected to possess a variety of energies ranging from zero to E_n. Therefore, it is possible to determine the energy distribution function of an isotope by appropriately adjusting the discriminators (the precise procedures are outlined in steps 3-13 to 3-19 in the experimental section). Briefly, this is done by dividing up the entire range of the discriminator circuit into very small intervals and determining the number of β particles (counts) whose energy falls within the interval selected per unit time. If, for example, the maximum range of the discriminator is 10 V, then the upper and lower discriminator levels are first set at 0.5 and 0 V, respectively, and the counts (β particles appearing in the interval, or window, as it is called) of 0 to 0.5 V per unit time are determined. On completion of this determination the upper discriminator level is set at 1.0 V and the lower at 0.5 V; once again the counts or β particles per unit time are determined. These procedures are then repeated throughout the entire 10.0 V range. If the number of counts, that is, β particles thus obtained, are plotted as a function of the average of the discriminator settings (this is actually the center of the interval selected), 0.25 and 0.75 V, respectively, for the present example, a plot similar to that in Figure 3-11 is observed. This plot is called a β spectrum. Note that its shape very closely approximates the idealized plot shown in Figure 3-1.

There is one set of adjustments that have thus far been neglected; they are the gain controls. In simplest terms the gain is an amplifier circuit that magnifies the size of an incoming signal. For example, a signal may

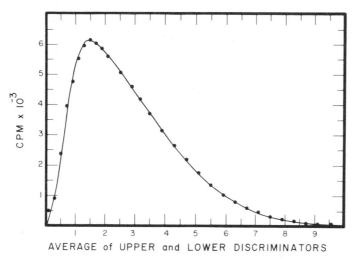

Figure 3-11. β spectrum of ^{3}H. The data for this figure were obtained as described in steps 3-13 to 3-19 of the experimental section.

approach the amplifier with a value of 1 mV and leave it with a value of 1000 mV. A gain control determines the number of times the incoming signal is magnified or amplified. Most contemporary scintillation counters provide two points at which the gain of the system may be regulated, (1) the high voltage gain regulating the output of the photomultiplier tube and (2) the discriminator gain regulating the input to each of the discriminator circuits. The gain of the photomultiplier tubes is usually controlled by regulating the high voltage supply of these tubes. It is easily seen that if the gain is increased at this point then all the signals from these tubes exhibit a greater value regardless of their origin, that is, ^{3}H, ^{14}C, or ^{32}P β particles. The effects of an increase in high voltage supply have been depicted in Figure 3-12B and C. Four of the signals that previously fell within the range set for low energy signals were increased to a point where they no longer fell within their initial range and were accepted in a higher energy range. If all the pulses reaching the surface of the phototube were magnified a fixed number of times, pulses that previously were below the sensitivity limit of the instrument would be counted. The effects of this regulation point are observed in all three channels of the counter.

The second gain control, which is exerted at each individual discriminator, produces a similar effect to that at the level of the photomultiplier tubes. In this case, however, only individual discriminator circuits are affected (Figure 3-12A and B). If the gain in channel B is decreased

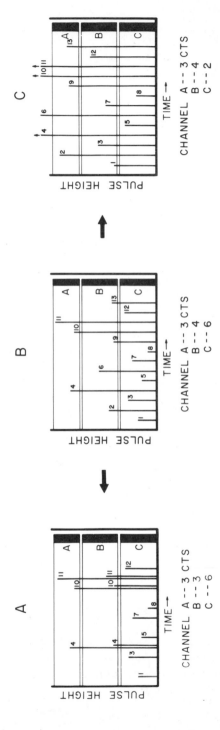

Figure 3-12. Effect of changing the gain on the distribution of signals within the discriminator circuits. Figure 3-12B depicts the starting configuration. Figure 3-12C demonstrates the effects of increasing either the photomultiplier (preamp) gain or all three discriminator circuit gains twofold. Figure 3-12A demonstrates the effects of decreasing the gain of only the B discriminator twofold. Two lines with the same number indicate that the same event appeared in two different channels. The output count distribution is shown below each diagram.

twofold, signals 4, 10, and 11 are registered in both scalers *A* and *B* because at the lower amplification of the *B* discriminator, they now fall within its range. Signals 2, 6, 9, and 13, however, are not registered in any scaler since these signals now fall below the range of the *B* discriminator and above that of the *C* discriminator. In practice the high voltage gain is usually set for an optimum functioning of the photomultiplier tubes. The discriminator gains, on the other hand, are adjusted such that the β spectrum of the particular isotope being counted just fills the full range of that discriminator. Since all three discriminators possess the same voltage range, signals originating from ^{3}H β particles require considerably more amplification to fill the window than those of ^{14}C. Figure 3-13 demonstrates the effects upon the β spectrum of decreasing the gain of a discriminator circuit. Note that changing the gain results in alterations of

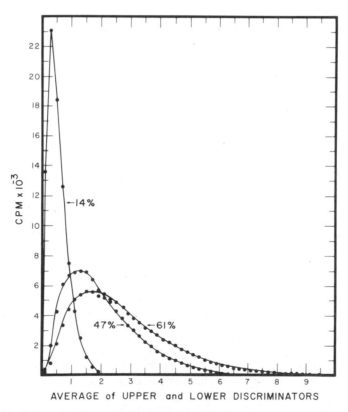

Figure 3-13. Effect of gain on the β spectrum of ^{3}H. These data were obtained using the methods described in steps 3-20 to 3-26 of the experimental section. The percentage values shown in this figure are the percentages of the maximum possible gain.

both the area under the curve and maximum energy points of the β spectrum.

Use of a Scintillation Spectrometer

The preceding discussion of a scintillation spectrometer's components provides the information necessary to properly adjust the instrument for efficient counting. Although the applicability of the following comments may vary slightly depending on the particular instrument and accessories being used, they illustrate the major operations involved. Adjustment of the high voltage supply or photomultiplier tube gain is usually performed by a service engineer and is not discussed here. With a sample of the isotope to be counted in the counting chamber, the channel to be used (i.e., set of discriminators and gain control) is selected. The upper and lower discriminators of that channel are set at their maximum and minimum values, respectively, (this would correspond to an interval of 10.0 and 0 V in the preceding discussion of discriminators). The interval of the discriminators is designated as a window, and when the interval is the maximum possible it is denoted as a full window. Use of a full window results in the inclusion of the entire β spectrum if the discriminator gain is properly adjusted. In other words the count rate (number of pulses entering the scalers per unit time) observed with a full window is the closest possible approximation to the area of the curve shown in Figure 3-1.

The next step is to adjust the gain (i.e., the amplification factor) so that the β spectrum of the isotope being studied just fills the full window. Different amounts of gain are required for different isotopes. This fact may be emphasized by considering the following set of direct proportions: the energy of a β particle determines how many solvent molecules it will interact with; this determines the intensity and duration of a phosphorescent flash of light, which determines the amount of light absorbed by the primary fluor, which determines the intensity and duration of the fluorescent flash of light, which determines the number of photoelectrons emitted from the photocathode of a photomultiplier tube, which determines the height and area of the pulse appearing at the input of the pulse height analyzer, which determines the height of the symmetrical pulse appearing at the output of the pulse height analyzer, which determines the voltage entering the discriminator circuits. If the discriminator interval is 10.0 V, the gain control must be adjusted so that a β particle having an $E_\beta = E_{max}$ results in a signal whose height is 10.0 V. If, for example, a gain setting of 1 is required for the ^{14}C to just fill the 10.0 V window then a gain setting of approximately 10 would be required for ^3H.

The precise gain setting that results in the β spectrum just filling the full window is called the balance point for that isotope. It is determined by increasing the coarse gain control in a stepwise fashion and determining the count rate per unit time at each setting (the discriminators are set to full windows during this procedure). When this is completed, the coarse gain control is set to the value that gave the highest rate. With the coarse gain set at the desired value the same procedure is performed with the fine gain control. A second method of determining the balance point is discussed in the experimental section (steps 3-68 to 3-80).

A plot of count rate versus gain exhibits a maximum rather than a plateau. In the above comments it was indicated that the discriminator had a maximum window of 0–10 V. In reality, however, the discriminator has a threshold sensitivity value, for example, 0.1 V. If a signal that is less than the threshold value (0.1 V) approaches the discriminator it is discarded regardless of the discriminator setting. As shown in Figure 3-1, the energies of the β particles increase from zero. Therefore, even after amplification some signals are below the threshold value and hence are lost. The number of signals falling below the threshold value, however, decreases as the gain is increased. Unfortunately, increasing the gain quickly reaches a point of diminishing return. This situation is illustrated in Figure 3-14. At a very low gain setting the height of a pulse arising from $^{14}C\beta$ particles possessing an energy equal to E_{max} falls far below 10.0 V

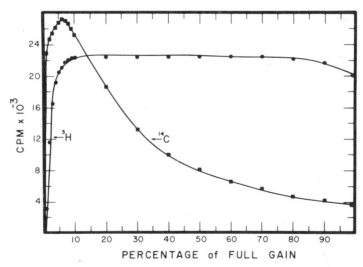

Figure 3-14. Effects of increasing the gain on the counting rate. These data were obtained using the procedures described in steps 3-3 to 3-12 in the experimental section.

(the upper limit of the discriminator). As the gain is increased the pulse height arising from this β particle moves closer to 10.0 V. At the balance point, that is, the point where the entire β spectrum just fits within the energy interval of 0 to 10.0 V, a β particle with an energy of E_{max} gives rise to a signal with a pulse height of 10.0 V. If, however, the gain continues to rise even higher, these signals possess values in excess of 10.0 V and thus are discarded. Therefore, although increasing the gain above the balance point rescues low energy events that would otherwise be lost, a concomitant loss in high energy events occurs and thus cancels the advantage won by rescuing the low energy events. Why does the ^{3}H curve plateau rather than decrease at high gain values?

With the gain set at the balance point the discriminator levels may be set. The positioning of the upper and lower levels is determined largely by two factors: (1) the desired level of background counts and (2) the number of isotopes that a sample contains. Background counts are the counts observed if a scintillation vial containing nonradioactive scintillation fluid (solvent, primary fluors, and, if needed, secondary fluors) is counted. These counts arise from cosmic radiation, naturally occurring radioactive ^{40}K in the glass of the vial, and thermionic noise that succeeded in penetrating the coincidence gate (see Table 3-6 for the quantitative contribution of each of these factors). Very low backgrounds are desired when samples containing very little radioactivity are being counted. Background counts may be decreased by using plastic vials, which do not contain ^{40}K, and by increasing the level of the lower discriminator

Table 3-6.   Factors Influencing Background Counting Rates

Sample	cpm Observed[a]		Source of Observed Increase in Counting Rate
	C^{14} Channel	H^3 Channel	
Empty chamber	3	5	Joule–Thompson dark current[b]
Empty glass vial	13	26	^{40}K in the glass of the vial
Empty plastic vial	7	8	
Toluene in glass vial	18	30	Cerenkov radiation
Toluene in plastic vial	12	16	
Scintillation fluid in glass vial	29	28	Cosmic and other environmental radiation
Scintillation fluid in plastic vial	24	13	

[a] Samples were each counted 20 minutes. Channel parameters were optimized for isotope designated using the full discriminator range.
[b] Photomultiplier cross talk due to ion recombination flashes was removed electronically in the instrument used in this experiment.

(thermionic noise usually gives rise to signals of low potential). It should be clear that decreasing the background counts by the latter procedure is done at the expense of discarding some portion of the low energy β events. The value of the upper discriminator has little effect on the background counting rate. Therefore, when a single isotope is being counted, this discriminator is set at its maximum value. In this way none of the events near the E_{max} of the β spectrum are lost. A discussion of the second factor influencing selection of appropriate discriminator level, that is, the simultaneous counting of multiple isotopes, must be deferred until after a discussion of counting efficiency. In new instruments these parameters are often preset internally by a service representative.

Counting Efficiency

Like most measuring instruments, the scintillation counter is not 100% efficient. The most sophisticated instruments presently available possess an efficiency near 60 and 90% for ^3H and ^{14}C, respectively. It will be recalled from earlier remarks that

$$cpm_{observed} = (dpm_{absolute\ activity})(efficiency\ of\ counting) \qquad (11)$$

Therefore, if the absolute radioactivity contained by a sample is to be accurately known, both the observed counting rate and the efficiency must be precisely determined.

Counting efficiencies less than 100% result from losses incurred within the measuring instrument and within the light generating system. Losses due to operation of the scintillation counter per se are not great. The majority of this loss occurs in the photocathode tube where the light produced by a β event may be too weak to result in ejection of a photoelectron, or more often, the photoelectron ejected gives rise to a pulse that is still below the threshold value of the discriminator circuits and hence lost as a component of the discarded "electronic noise." The overwhelming loss in counting efficiency occurs during the scintillation process itself. This loss, termed quenching, may arise in two ways, (1) chemical quenching and (2) color quenching. Table 3-7 lists a variety of common functional groups arranged according to the degree of quenching each brings about. Reagents that cause primary or chemical quenching, for example, strong acids or bases, prevent various organic molecules from phosphorescing and fluorescing. This may be done, for example, by chemical interaction with the scintillator, as is the case with hydroxyl ions converting dioxane to its nonphosphorescing epoxide derivative. Secondary or color quenching materials, on the other hand, do not affect the production of photons but rather absorb light produced in the scintillation

Table 3-7. Severity of Quenching Exhibited by Various Organic Compounds[a]

Strong Quenching Agents	Weak Quenching Agents	Agents that Neither Quench nor Transfer Energy
RSH^a	RCOOH	RH
$\underset{\substack{\parallel\ \ \ \ \parallel}}{RCOCR}$ (O O)	RNH_2	RF
	RCH=CHR	ROR
	RBr	RCN
$\underset{\parallel}{RCR}$ (O)	RSR	ROH
		RCl
$\underset{\parallel}{RCOR}$ (O)		
RNHR		
RCHO		
R_3N		
RI		
RNO_2		

[a] R is an aliphatic group.

process. Note, in Figures 3-2 and 3-3, that any material absorbing light of 200–440 nm would be expected to quench. This should emphasize that color quenching may occur at the level of solvent phosphorescence as well as primary and secondary fluor fluorescence.

Three methods have evolved to ascertain the degree of efficiency loss both within the instrument and as a result of quenching. These techniques are termed (1) internal standardization, (2) channels ratio quench correction, and (3) external standard channels ratio quench correction. Determination of counting efficiency by internal standardization may be performed in two steps. The sample is first accurately counted followed by the addition of a precisely known quantity of radioactivity to the vial (50,000–80,000 dpm ^{14}C or 100,000–150,000 cpm ^3H). It is important for the amount of added radioactivity to be considerably larger than that originally present in the vial. The sample is then counted a second time. The first count is the sample cpm and the second count is the sample cpm + (efficiency)(standard dpm). That is,

$$cpm_{second\ count} = cpm_{sample} + (dpm_{std})(efficiency) \tag{12}$$

or

$$efficiency = \frac{cpm_{second\ count} - cpm_{sample}}{dpm_{std}} = \frac{cpm_{std}}{dpm_{std}} \tag{13}$$

and

$$dpm_{sample} = \frac{cpm_{sample}}{efficiency} \qquad (14)$$

Example. A sample was counted in a scintillation counter and found to contain 58,413 cpm. 0.10 ml standard toluene (1.85×10^6 dpm/ml) was added and the sample was recounted and found to contain 173,113 cpm. What was the efficiency of counting for this sample and how much radioactivity (dpm) did the original sample contain?

$$efficiency = \frac{173,113 - 58,413}{185,000}$$

$$= \frac{114,700}{185,000}$$

$$= 0.62$$

$$dpm_{sample} = \frac{58,413}{0.62}$$

$$= 94,214$$

This method of standardization was one of the first developed; it was based on the assumption that the degree of quenching of the added standard isotope would be identical to that of the original sample. Although this method can be quite accurate, it suffers the following disadvantages: (a) for the technique to function successfully, it is necessary to pipette small volumes of an organic solvent in a highly accurate manner; (b) each sample must be manipulated and counted twice resulting in a large expenditure of time; and (c) the cost of the standard radioactive solutions make routine use of the technique prohibitively expensive.

The channels ratio method of quench correction is predicated on the fact that the quenching process results in a change in the β spectrum of an isotope. Figure 3-15 depicts the β spectra of an unquenched (curve C), a mildly quenched (curve B), and a strongly quenched sample (curve A). Quenching has two effects on the spectra: both the E_{max} and the area under the curve are successively decreased. Since the number of counts observed with full windows is really a reflection of the area under the β spectrum, decreases in this area from quenching are expected. The decrease in the E_{max}, however, requires further comment. If the path of energy transfer is reconsidered (Figure 3-2) it will be recalled that the number of solvent molecules (and subsequently fluor molecules) excited is a function of the energy of the β particle emitted. In other words, a β

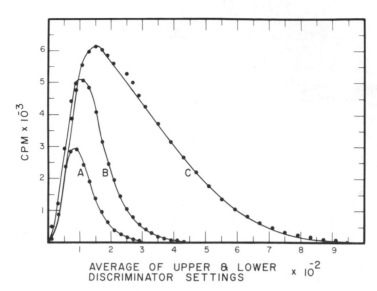

Figure 3-15. Effects of quenching on the β spectrum of ^{3}H. Curve C is an unquenched sample and curves B and A depict the results as quenching is heightened. These data were obtained using the procedures described in steps 3-34 to 3-43 in the experimental section.

particle with an energy of E_{max} would be expected to excite a large number of solvent molecules. If it is assumed that quenching of an excited molecule occurs randomly then it is reasonable that at least some of the many excited molecules arising from an E_{max} β particle are lost, resulting in the decrease of the E_{max} point. If the degree of spectral shift that results from quenching can be related to the efficiency of counting then it is possible to relate the amount of quenching to the efficiency. This is accomplished by counting the sample simultaneously in two channels. The windows of the two channels are positioned as shown in Figure 3-16. Channel B monitors the upper end of the spectrum while Channel A monitors the entire range of energies. As the degree of quenching is heightened, a greater and greater proportion of the counts are shifted out of Channel B (Figure 3-16). If the ratio of the counts in Channel B to those in Channel A is plotted as a function of the efficiency of counting in Channel A, the relationship depicted in Figure 3-17 is obtained. In practice a set of quenched standards (a series of vials all containing exactly the same and precisely known amount of radioactivity, but increasing amounts of a quenching agent) are counted as described above and plotted as described in Figure 3-17. An unknown sample is then counted in the two channels, and the resulting ratio is used in conjunction

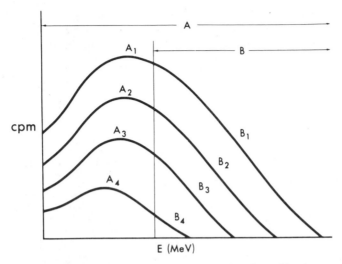

Figure 3-16. Channels ratio counting of quenched samples. Quenching increases sequentially from sample 1 to sample 4. Arrows above the figure indicate the range of the *A* and *B* channel discriminators. (Courtesy of Beckman Instruments, Inc.)

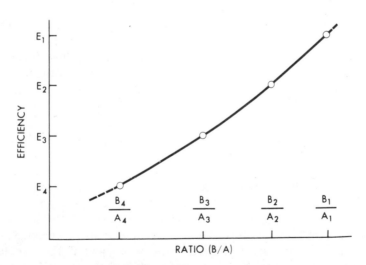

Figure 3-17. Channels ratio quench correction curve. This curve was constructed using the sample and discriminator settings described in Figure 3-16. (Courtesy of Beckman Instruments, Inc.)

with the correction curve to obtain the efficiency of counting. This information, with the observed counting rate in Channel A, can be used to determine the dpm of the sample according to equation 11.

Example.   A sample of ^{14}C was counted using the channels ratio technique and the following data were obtained:

$$\text{cpm in channel } A = 42{,}927$$

$$\text{cpm in channel } B = 19{,}983$$

What was the efficiency of counting and the absolute radioactivity (dpm) in the sample? These data give a channels ratio of

$$\frac{\text{cpm}_B}{\text{cpm}_A} = \frac{19{,}983}{42{,}927}$$

$$= 0.47$$

According to the plot depicted in the left panel of Figure 3-39, a ratio of 0.47 corresponds to an efficiency of 0.84 or 84%. Therefore,

$$\text{dpm}_{\text{sample}} = \frac{42{,}927}{0.84}$$

$$= 51{,}103$$

Although the channels ratio method of quench correction enjoys widespread acceptance, it has a severe limitation. The upper counting channel (B) monitors only a small portion of the β spectrum, resulting in registration of a correspondingly small fraction of the total counts at high levels of quenching. This precludes use of this method for samples containing very small amounts of radioactivity, because of the unacceptably long counting times that would be required to secure a statistically significant counting rate in the upper channel.

The external standard channels ratio method of quench correction was developed as a result of the deficiencies in the above procedures. This technique is merely a modification of the channels ratio method in which the degree of quenching is determined for an external radioactive source rather than for the radioactivity contained in the counting vial. The method is based on the fact that a charged particle or a γ-ray traveling through a medium will collide with the molecules of the medium, as shown in Figure 3-18. As a result of the collision an orbital electron of the target molecule (termed a Compton electron) is ejected, the direction of γ-ray movement is altered, and the wavelength of the γ-ray increases (its energy decreases by the amount imparted to the released electron). This process may be repeated many times as the γ-ray travels through the

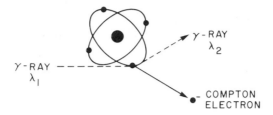

Figure 3-18. Production of Compton electrons. λ_2 is greater than λ_1 by the amount of energy lost in the collision.

medium. Since γ-rays easily penetrate glass vials, a source of γ-rays brought into proximity with a sample vial causes release of Compton electrons within the scintillation fluid. These electrons behave just as β particles; that is, they react with the solvent and fluors (Figure 3-2) to produce a flash of light. In other words, an effect identical to that obtained by addition of a large amount of radioactivity to the vial (internal standardization) is achieved even though no additions have been made.

In practice a sample is counted normally as described for the channels ratio method and then counted a second time with a source of γ-rays (usually ^{137}Cs or ^{133}Ba) positioned in the center of the counting chamber just below the sample vial (see Figure 3-19). Any quenching that occurs has the same effect on the efficiency and spectrum of the Compton electrons as it does on those of the sample β particles. Since this technique has incorporated the procedures of both internal standardization and channels ratio correction methods, it is no surprise that the data obtained must be treated as described above for both of these techniques. The first or normal count rate obtained in each channel arises only from the sample and may be represented as follows:

$$\text{cpm}_{\text{first count} (A)} = (\text{dpm}_{\text{sample}})(\text{efficiency in Channel } A) \qquad (15)$$

and

$$\text{cpm}_{\text{first count} (B)} = (\text{dpm}_{\text{sample}})(\text{efficiency in Channel } B) \qquad (16)$$

The second count rate obtained in the presence of the external standard is a composite of counts arising from the sample and from the external standard (via Compton electrons). It may be represented as follows:

$$\text{cpm}_{\text{second count} (A)} = (\text{dpm}_{\text{sample}})(\text{efficiency in } A) + (\text{dpm}_{\text{external std}})$$
$$\times (\text{efficiency in } A) \qquad (17)$$

$$\text{cpm}_{\text{second count} (B)} = (\text{dpm}_{\text{sample}})(\text{efficiency in } B) + (\text{dpm}_{\text{external std}})$$
$$\times (\text{efficiency in } B) \qquad (18)$$

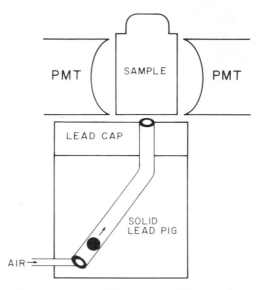

Figure 3-19. External standard assembly of a scintillation counter. The black circle represents the γ-ray producing source. When needed, the source is positioned below the sample by means of an air pump operating like a pea shooter. When the pump is turned off the source falls back to the bottom of the shielded container. Why does the delivery tube have a bend in it? PMT indicates the photomultiplier tubes.

Substituting equations 3-15 and 3-16 into 3-17 and 3-18, we obtain:

$$\text{cpm}_{\text{second count }(A)} = (\text{cpm}_{\text{first count }A}) + (\text{dpm}_{\text{external std}})(\text{efficiency in } A) \quad (19)$$

$$\text{cpm}_{\text{second count }(B)} = (\text{cpm}_{\text{first count }B}) + (\text{dpm}_{\text{external std}})(\text{efficiency in } B) \quad (20)$$

Therefore,

$$\text{cpm}_{\text{second count }(A)} - \text{cpm}_{\text{first count }(A)} = (\text{dpm}_{\text{external std}})(\text{efficiency in } A)$$
$$= \text{cpm}_{\text{external std }(A)} \quad (21)$$

$$\text{cpm}_{\text{second count }(B)} - \text{cpm}_{\text{first count }(B)} = (\text{dpm}_{\text{external std}})(\text{efficiency in } B)$$
$$= \text{cpm}_{\text{external std }(B)} \quad (22)$$

Dividing equation 3-22 by 3-21 gives the channels ratio for the external standard:

$$\frac{\text{cpm}_{\text{second count }(B)} - \text{cpm}_{\text{first count }(B)}}{\text{cpm}_{\text{second count }(A)} - \text{cpm}_{\text{first count }(A)}} = \frac{\text{cpm}_{\text{external std }(B)}}{\text{cpm}_{\text{external std }(A)}} \quad (23)$$

With a statistically significant channels ratio in hand, it is only necessary to relate this ratio to a specific efficiency. In the channels ratio technique this was done by determining the ratios and efficiency of counting for a set of quenched standards. In the present technique a set of quenched standards are also employed but in a slightly different manner. The quenched standards are counted twice, once in the absence of the external standard and once in its presence. As indicated in equations 3-15 and 3-16, the first count rate arises only from the sample. Therefore, this count rate along with the known amount of radioactivity contained in the vial provide the information necessary to calculate the efficiency of counting using equation 3-11. The arguments outlined above demonstrate how the external standard channels ratio value is obtained from the two counting rates (in the presence and absence of external standard). In the case of the channels ratio method both the efficiency and channels ratio values are derived from the same scintillation events (those arising from the sample β emissions). However, in the case of external standard channels ratio procedures, the efficiency value comes from one set of scintillation events (those arising from the sample β emissions) and the channels ratio value from another set (those arising from Compton electrons interacting with the quenched sample scintillation fluid plus those arising from the sample). The quench correction curve is constructed merely by performing this type of analysis on a series of quenched standards (Figure 3-20).

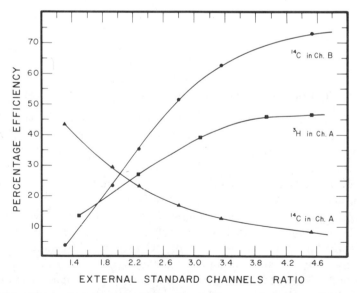

Figure 3-20. External standard channels ratio quench correction curves. These data were obtained using the procedures outlined in steps 3-50 to 3-67 of the experimental section.

Example. A sample of ^{14}C was counted using the external standard channels ratio technique and the following data were obtained.

Data from the first count (external standard absent)

$$cpm_A = 53,661$$

$$cpm_B = 66,065$$

Data from the second count (external standard present)

$$cpm_A = 163,661$$

$$cpm_B = 278,365$$

What is the efficiency of ^{14}C counting in channel B and what is the absolute radioactivity (dpm) in the sample? Use the quench correction curve in Figure 3-20.

$$\text{external standard channels ratio} = \frac{278,365 - 66,065}{163,661 - 53,661}$$

$$= \frac{212,300}{110,000}$$

$$= 1.93$$

From Figure 3-20, a ratio of 1.93 corresponds to an efficiency of 0.232 for ^{14}C in channel B. Therefore,

$$\text{dpm}_{\text{sample}} = \frac{66,065}{0.232}$$

$$= 284,762$$

Simultaneous Counting of Multiple Isotopes

Advances in the quality and sophistication of available scintillation counters permit simultaneous assay of two isotopes with a high degree of certainty and have greatly expanded the potential of isotopic methods. The methods used to count multiple isotopes are predicated on external standard channels ratio procedures. If the β spectra of ^3H and ^{14}C are superimposed on one another (see Figure 3-1), two facts are apparent: (1) a large fraction of the ^{14}C spectrum lies above the E_{max} for ^3H, and (2) the entire ^3H spectrum is contained within the ^{14}C spectrum. Therefore, it is possible to determine the amount of ^{14}C in a multiply labeled sample merely by adjusting discriminators on one of the counter's channels to values that will select β events with an energy greater than 0.018 and less than 0.155 MeV (there are no ^3H β particles in this energy range). For this discussion the channel set up in this way is called channel B. The complete overlap of the tritium β spectrum by that of ^{14}C precludes a direct determination of ^{3}H alone. The channel used for determining ^{3}H is called channel A. Its lower discriminator should be set sufficiently above zero to exclude undesirable "background noise." Its upper discriminator is decreased from full scale to a degree that a ^{14}C standard is counted with 10% or less efficiency in the channel operating at a ^{3}H balance point. Since channel A contains counts arising from both ^{3}H and ^{14}C, it is necessary to know the exact contribution of ^{14}C counts made to the total counts observed in the channel if those arising from ^{3}H β emissions are to be known. This information may be obtained by the following procedures. Through use of the external standard channels ratio technique described above, the external standard channels ratio is related to (1) the efficiency of ^{14}C counting in channel B, (2) the efficiency of ^{3}H counting in channel A, and (3) the efficiency of ^{14}C counting in channel A. Why can the efficiency of counting of ^{3}H in channel B be ignored? It is important to note that (1) the discriminator settings are never changed once they have been initially adjusted and (2) the efficiencies are determined using one isotope at a time; that is, the efficiencies of ^{14}C counting in channels A and B are determined using a set of ^{14}C quenched standards and that of ^{3}H in channel A is determined using a set of ^{3}H quenched standards. An example of these three quench correction curves appears in Figure 3-20. It is significant that the amount of ^{14}C found in channel A increases as quenching is heightened.

Once these standard curves have been prepared the multiply labeled sample may be counted in the presence and absence of the external standard. The counting rate in channel B (cpm ^{14}C) and the channels ratio value may be used in conjunction with the quench correction curve (Figure 3-20) to calculate the absolute dpm of ^{14}C using equation 3-24.

$$\text{cpm}_B = (\text{dpm } ^{14}C)(\text{efficiency of } ^{14}C \text{ counting in channel } B) \quad (24)$$

The counting rate in channel A on the other hand is a composite of ^{14}C and ^{3}H and may be expressed:

$$\text{cpm}_A = \text{cpm } ^{14}C + \text{cpm } ^{3}H \quad (25)$$

or

$$\text{cpm}_A = (\text{dpm } ^{14}C)(\text{efficiency of } ^{14}C \text{ counting in channel } A)$$
$$+ (\text{dpm } ^{3}H)(\text{efficiency of } ^{3}H \text{ counting in channel } A) \quad (26)$$

Since we have already calculated dpm ^{14}C above, equation 3-26 has only one unknown (the efficiencies are obtained from Figure 3-20 at the appropriate channels ratio values), the dpm of ^{3}H in the sample. If the amount of radioactivity in the sample is sufficiently low that the counter background must be considered, then this value should be determined for both channels A and B. The values obtained should then be subtracted from the total cpm observed in channel A or B before any of the above computations are performed.

Example.   A sample containing both ^{14}C and ^{3}H was counted using the external standard channels ratio technique and the following data were obtained.

Data from the first count (external standard absent)

$$A = 77,832$$
$$B = 33,665$$

Data from the second count (external standard present)

$$A = 197,832$$
$$B = 93,665$$

Use the quench correction curve in Figure 3-20 and calculate the efficiency for ^{3}H and ^{14}C in channel A, the efficiency for ^{14}C in channel B, and the absolute amounts of ^{14}C and ^{3}H (dpm) in the sample.

$$\text{external standard channels ratio} = \frac{197,832 - 77,832}{93,665 - 33,665}$$

$$= \frac{120,000}{60,000} = 2.00$$

From the quench correction curve in Figure 3-20 the following efficiencies correspond to a ratio of 2.00.

$$\text{efficiency for } ^{14}\text{C in channel } B = 0.263$$
$$\text{efficiency for } ^{14}\text{C in channel } A = 0.267$$
$$\text{efficiency for } ^{3}\text{H in channel } A = 0.210$$

$$\text{dpm } ^{14}\text{C} = \frac{33,665}{0.263}$$

$$= 128,003$$

$$\text{cpm}_A = \text{cpm } ^{14}\text{C} + \text{cpm } ^{3}\text{H}$$

$$\text{cpm } ^{3}\text{H} = \text{cpm}_A - (\text{dpm } ^{14}\text{C})(\text{efficiency for } ^{14}\text{C in channel } A)$$

$$= 77,832 - (128,003)(0.267)$$

$$= 43,655$$

$$\text{dpm } ^{3}\text{H} = \frac{43,655}{0.210} = 207,882$$

Why is the channels ratio not affected by the nature of the isotope contained in the vial?

Sample Preparation for Scintillation Counting

A large number of scintillation fluid recipes have been developed and a few of these are listed in Table 3-8 along with their salient characteristics. In our experience the most widely applicable of these are solutions 1 and 10. In most of these formulations, compounds other than the primary and secondary fluors have been added to the solvent. These additional materials are included because they either permit greater additions of aqueous solution to the scintillation fluid (i.e., ethylene glycol, methanol, triton X-100, and Carb-O-Sil) or improve the transfer of energy from the β particle to the solvent and fluors (i.e., naphthalene). The reason for such a wide variety of recipes is the desire to count materials that either are not miscible with organic solvent systems normally used or highly quench the system. Kinard (26) developed a means of quantitating the ability of various preparations to meet these two requirements. He defined a "figure of merit,"

$$\text{figure of merit} = (E)(H) \tag{27}$$

where E is the efficiency of counting for ^3H and H is the percentage of the total sample that is aqueous. For example, the figure of merit of a sample containing 23% aqueous solution and counting with an efficiency of 18% is

$$\text{figure of merit} = (23)(18)$$
$$= 414$$

The higher the figure of merit possessed by a given fluid, the more adequately it is supposedly fulfilling the requirements cited above.

The high cost of preparing scintillation fluid raises the question—how much fluid is necessary to efficiently count a sample? The data shown in Table 3-9 indicate that 5.0 ml in a standard scintillation vial is sufficient for counting in most instruments. This amount may be further decreased, using "mini vials" and plexiglass adapters that are now commercially available (see Figure 3-21). It should be emphasized, however, that these volumes apply only to unquenched samples. If a constant amount of quenching agent is added to successively smaller and smaller amounts of scintillation fluid, its effective concentration in the sample increases, yielding more severely quenched samples. Therefore, if the samples are quenched or aqueous, (1) more scintillation fluid is required to minimize the effect of the quenching agent or (2) more drastic mathematical correction of the data is required.

Table 3-8.   Useful Scintillating Solutions

Solutions[a]	Comments
Basic Scintillator Solutions	
1. PPO, 4 g POPOP, 0.1 g Toluene or xylene, 1000 ml	—
2. POPOP, 0.1 g/l *p*-Terphenyl Toluene or xylene (saturated solution at counting temperature)	—
3. PPO, 10 g POPOP, 0.25 g Naphthalene, 100 g Dioxane, 100 ml	—
Systems for Counting Water and Aqueous Solutions	
4. PPO, 5 g POPOP, 0.1 g Toluene or xylene, 1000 ml	Holds up to 0.3 ml H_2O in 10 ml solvent, with up to 4 ml absolute ethanol counting efficiency less than 10%
5. POPOP, 0.1 g/l *p*-Terphenyl Toluene or xylene (saturated solution at counting temperature)	As above; without POPOP it is useful for those acids which react with PPO, POPOP
6. PPO, 10 g POPOP, 0.25 g Naphthalene, 100 g Dioxane, 1000 ml	Holds 3 ml H_2O at a counting efficiency of 8% for 15 ml total volume
7. PPO, 10 g POPOP, 0.25 g Naphthalene, 200 g Dioxane, 1000 ml	Holds 1 ml H_2O at a counting efficiency of 15% for 15 ml total volume
8. PPO, 6.5 g POPOP, 0.13 g Naphthalene, 104 g Methanol, 300 ml Dioxane, 500 ml Toluene, 500 ml	Holds 10% H_2O or conc. HCl in 10 ml solvent plus 3 ml $0.5M$ ethanolic KOH; maximum counting efficiency to 15%
9. PPO, 4 g POPOP, 0.2 g Naphthalene, 60 g Ethylene glycol, 20 ml Methanol, 100 ml Dioxane, 880 ml	Holds 10% H_2O at counting efficiency of 10% (Bray's solution)

Table 3-8. (*continued*)

Solutions[a]	Comments
10. PPO, 5.5 g POPOP, 0.1 g Toluene, 667 ml Triton X-100, 333 ml	Aqueous samples up to 12% water and 23–50% salts and particulates are tolerated. Discontinuous between 12 and 23% water
11. PPO, 5 g POPOP, 0.1 g Carb-O-Sil, 40 g Toluene, 1000 ml	Particular samples, also samples that tend to adsorb on vial walls

Systems for Counting in the Presence of Hyamine

12. PPO, 9 g POPOP, 0.2 g Toluene, 1000 ml	Holds up to 3 ml hyamine (wet tissue digest), may be blended into 10 ml solvent with 2–5 ml absolute ethanol. Counting efficiency less than 5%
13. PPO, 10 g POPOP, 0.25 g Naphthalene, 100 g Dioxane, 500 ml Toluene, 500 ml	Holds up to 5 ml hyamine (wet tissue digest); may be blended into 10 ml solvent with 1 ml dioxane. Counting efficiency about 5% with tissue digest, 2.5% with blood digest

[a] The formulations are taken from the literature and some predate the secondary scintillators dimethyl-POPOP and bis-MSB. These secondary scintillators may be used in any of the formulations given and are especially recommended for quenched solutions, in which an increased amount of secondary scintillator is desired.

Table 3-9. Effect of Scintillation Fluid Volume on Efficiency of Counting

Volume of Scintillation Fluid Used (ml)	cpm Observed
1.0	15,241
2.5	15,981
5.0	16,162
10.0	16,736
15.0	15,968
20.0	14,995

Figure 3-21. Plastic adapters for use of mini vials in a standard scintillation counter. (Courtesy of Packard Instrument Co., Inc.)

DETERMINATION OF RADIOACTIVE CARBON DIOXIDE

Gaseous carbon dioxide is a natural product of many biologically significant reactions. As such its radiochemical determination deserves special consideration. There are essentially two methods of trapping the gas in a form that can be counted. The first method is based on the following reactions:

$$CO_2 + OH \rightleftharpoons HCO_3^- \qquad (28)$$

$$HCO_3^- + OH^- \rightleftharpoons CO_3^{2-} + H_2O \qquad (29)$$

$$Ba^{2+} + CO_3^{2-} \rightleftharpoons BaCO_3 \qquad (30)$$

In practice the gas is allowed to react with base and the carbonate formed is precipitated as barium carbonate. This precipitate may be collected by filtration and its radioactivity determined by means of a gas flow counter.

The second method of recovery is also predicated on the reaction of CO_2 with base. However, in this method the bases most often used are ethanolamine, ethylenediamine, phenethylamine, or hyamine hydroxide [*p*-(diisobutylcreoxyethoxyethyl)dimethylbenzyl ammonium chloride]. All these compounds are quaternary ammonium bases and have the advantage of being soluble in methanol, and the methanolic solutions are in turn miscible in standard scintillation fluids. A convenient vessel in which to perform CO_2 generating reactions is shown in Figure 3-22.

Figure 3-22. Reaction vessel used to perform CO_2 generating reactions. The vessel is composed of a 25 or 50 ml Erylenmeyer flask, a serum vial rubber cap, and a plastic center well. (Courtesy of Kontes Glass Co.)

0.20 ml methanolic hyamine hydroxide is placed in the plastic center well. The vessel is assembled and the reaction mixture injected into the flask with a Hamilton syringe. At the conclusion of the incubation period 0.3 ml of a nonvolatile acid, such as perchloric, is added followed by an additional incubation period of 45 minutes at 30°C, during which the CO_2 is quantitatively absorbed by the hyamine. The vessel is carefully disassembled, and the cup of hyamine is cut from its stem (using wire cutters) and allowed to drop into a vial containing scintillation fluid. This vial is vigorously shaken to remove the hyamine from the cup. The sample must then be allowed to stand several hours or preferably overnight in order to permit low energy chemiluminescence to decay. Such chemiluminescence is most easily observed in the lower energy ³H channel of the spectrometer. As shown by the data in Table 3-10 this requires more than 2.5 hours.

GAS FLOW OR GEIGER COUNTING

The second major technique of radioactivity quantitation is gas flow or Geiger counting. This method is based on the production of ion pairs when β particles are permitted to pass through a noble gas such as argon

Table 3-10. Chemiluminescence of Aqueous Scintillation Fluids by Hyamine Hydroxide

Bray's		Acquasol	
Time[a] (minutes)	cpm Observed ^{3}H Channel	Time[a] (minutes)	cpm Observed ^{3}H Channel
2	836,840	2	1,115,810
12	11,400	6	75,030
19	3,070	14	11,940
32	710	26	3,350
45	230	40	1,240
61	150	56	1,060
93	110	87	450
119	45	114	371
151	37	146	301

[a] Time elapsed after addition of 0.2 ml hyamine hydroxide to 15 ml scintillation fluid.

or helium. Ion pairs result from collisions between β particles and monatomic gas molecules. A gas flow counter is simply a device composed of a lead shielded chamber (see Figure 3-23), within which ion pairs are produced and detected, and the electronic circuits needed to record their production.

Ion pairs produced in a closed chamber and in the absence of any electric field drift aimlessly about until they randomly approach one another close enough to reunite yielding once again a neutral molecule. A small electric field may be set up within the chamber by making the walls negatively charged with respect to a positive element or anode suspended into the center of the chamber, as shown in Figure 3-24. In the presence of this small electric field ions migrate slowly toward either the walls or central anode. This configuration is termed an ionization chamber. If the voltage difference between the chamber walls and anode is increased, a Townsend avalanche occurs. As depicted in Figure 3-25, electrons produced from β particle collisions are accelerated by the electric field and may strike second helium or argon molecules, thus producing secondary electrons which may in turn produce tertiary electrons in cascade fashion. All these electrons move very rapidly toward the central anode and produce a signal upon arrival. The size of the signal is a function of the number of electrons in the cascade, which is a function of the electric field. Operation under these conditions is within what is termed a proportional region; the size of the output signal is directly

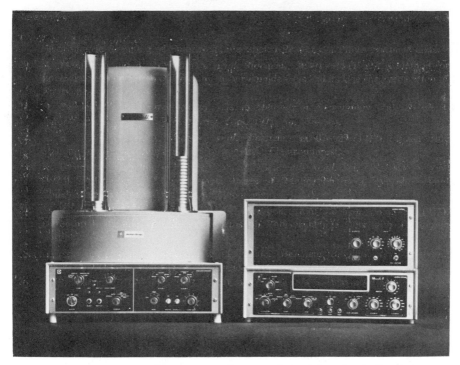

Figure 3-23. A gas flow counter. (Courtesy of Amersham Searle Corp.)

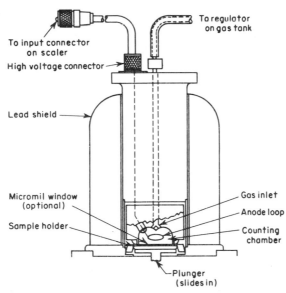

Figure 3-24. Counting chamber of a gas flow counter. (Courtesy of Amersham Searle Corp.)

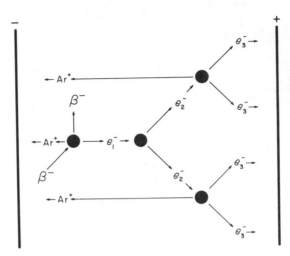

Figure 3-25. Schematic representation of the events leading to a Townsend avalanche. Filled circles are argon molecules.

proportional to the voltage difference applied to the chamber. This is shown in Figure 3-26. Further increase in the chamber potential produces a maximum avalanche and results in an output signal independent of the potential of the β particle initiating the event. This is termed the Geiger region of the response curve and is the range where most present instruments operate. Notice that use of the Geiger region precludes the simultaneous counting of two isotopes.

Thus far we have discussed the onset of a β particle induced output signal but have not considered what happens thereafter. Electrons move much more rapidly toward the central anode than do the positively charged ions toward the walls. This results in a sheath of positive ions being produced around the anode. This sheath effectively nullifies the electric field and results in failure to accelerate a subsequently produced primary electron. As the sheath of positive ions moves toward the cathodic wall, the chamber can again respond to β particle collisions, a process requiring 200 to 300 μseconds. This dead time is substantially greater than that observed for the scintillation spectrometer and requires correction of the data at counting rates in excess of 10,000 cpm. It constitutes one of the major factors contributing to the instrument's decreased counting efficiency. As a positive ion comes in contact with the chamber wall, it picks up an electron. However, the resulting molecule is not stable. It decays to a stable molecule with the release of soft X-radiation. This radiation produces Compton electrons which behave

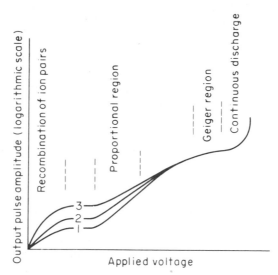

Figure 3-26. Output pulse amplitude as a function of the voltage applied to the chamber. The three curves designate the results observed with sequentially higher energy β particles. (From H. H. Willard, L. L. Merritt, and J. A. Dean, *Instrumental Methods of Analysis*, 4th ed., Van Nostrand, Princeton, N.J., 1965.)

just as legitimate, primary electrons and as such maintain the chamber in an unacceptable state of continuous discharge. To prevent this, a small percentage (1–4%) of an organic gas, such as isobutane, is added to the noble gas. The unstable noble gas molecules interact with the organic molecules to yield stable argon or helium and ionized isobutane. However, isobutane, upon receiving electrons from the chamber wall, is not unstable and hence quenches the X-radiation, thereby preventing continuous chamber discharge.

The electronic circuitry of a gas flow counter is composed of a high voltage power supply, an amplifier for the chamber signal, a timing circuit, and a small computer used to convert the registered counts to cpm. All these have been described with regard to scintillation counters. The only setup procedure necessary for such a counter is determination of the voltage required for Geiger region operation. This is accomplished by placing a known amount of radioactivity in a planchet (see Figure 3-27), the disposable vessel which holds the sample, and counting it at a variety of voltage settings. The desired setting, as shown in Figure 3-41, is approximately one-third of the way through the Geiger region of the voltage response curve.

Figure 3-27. Planchet—the device used to hold samples for gas flow counting. Planchets may be made of copper, aluminum, or stainless steel and are usually 3 cm in diameter and 0.3 cm deep. (Courtesy of Sigma Chemical Co.)

STATISTICS OF COUNTING

If the amount of radioactivity in a given sample is to be known with certainty, account must be taken of the fact that nuclear disintegration is a random, infrequently occurring event. The infrequency of ^{14}C disintegration is emphasized by its 5760 year half-life. The count of a radioactive sample is an estimate of the number of disintegrations occurring per unit time. If, however, a sample is counted a large number of times the successive estimates are observed to differ significantly. This variation is amply demonstrated by the data shown in Figure 3-28A. A sample was counted 1894 times and the number of times a given counting rate, χ_i, was observed was plotted as a function of χ_i. Two significant questions arise from consideration of these data: (1) which counting rate is correct, and (2) how far from the correct counting rate was the first estimate? The second question is important because most samples routinely assayed for radioactivity are counted only once. The answers to these questions may be found in a discussion of the statistics involved in the counting process.

The collection of data in Table 3-11 can be used to provide an illustration of the basic terms used in statistics. In the first column of figures are the observed estimates, χ_i, for a sample of radioactivity. The average or mean value of these estimates, $\bar{\chi}$, is 26,644 and is a better estimate of the true amount of radioactivity in the sample than any of the observed data. The mean value approaches the true amount of radioactivity in the sample as the number of independent estimates approaches infinity. A measure of the scatter observed in the 10 estimates is shown in the second column. These values, termed the deviations, are obtained by subtracting the mean from each of the estimates.

$$\text{deviation} = \text{estimate } (\chi_i) - \text{mean } (\bar{\chi}) \tag{31}$$

Notice that the number of times the estimate is greater than the mean is

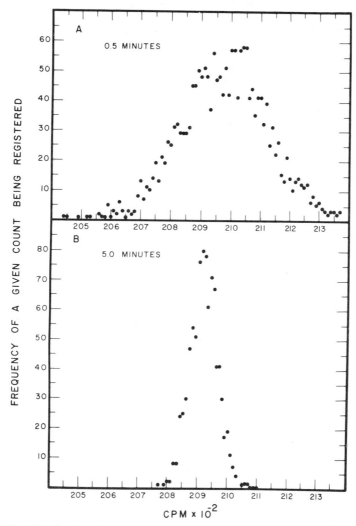

Figure 3-28. Distribution of observed counting rates (registered as cpm) from a sample counted repeatedly for 0.5 minutes (*A*) or 5.0 minutes (*B*).

about equal to the number of times it is less. This is also true of the data shown in Figure 3-28. A more useful measure of the scatter is a value termed the standard deviation, designated *s* if a small number of values are considered and σ if the number is very large.

$$s = \left[\frac{1}{n-1} \sum_{i=1}^{i=n} (\chi_i - \bar{\chi})^2 \right]^{1/2} \qquad (32)$$

Table 3-11. Calculation of the Standard Deviation for a Series of Determinations

χ_i Estimates	$\chi_i - \bar{\chi}$ Deviations	$(\chi_i - \bar{\chi})^2$
26,930	+286	81,796
26,230	−414	171,396
26,440	−204	41,616
26,485	−159	25,281
26,515	−129	16,641
26,895	+251	63,001
27,085	+441	194,481
26,465	−179	23,041
26,510	−134	17,956
26,885	+241	58,081

Mean, $\bar{\chi} = 26,644$

$\Sigma(\chi_i - \bar{\chi})^2 = 702,290$

$s = 279$

$$\sigma = \left[\frac{1}{n} \sum_{i=1}^{i=n} (\chi_i - \bar{\chi})^2 \right]^{1/2} \tag{33}$$

In the example of Table 3-11 the standard deviation is

$$s = [\tfrac{1}{9}(702,290)]^{1/2}$$
$$= 279$$

In Figure 3-28A it can be seen that all the observed counts cluster around a central value, the mean. This distribution of data closely approximates the graphic solution of the Poisson distribution equation,

$$P_{\chi_i} = \frac{(\bar{\chi})^{\chi_i}(e)^{-\bar{x}}}{\chi_i!} \tag{34}$$

for a given value of $\bar{\chi}$ and an infinite number of χ_i values, as shown in Figure 3-29. The Poisson distribution equation was derived for an infrequently occurring, random event, and permits calculation of the probability, P_{χ_i}, of finding a given value, χ_i, when the mean value is $\bar{\chi}$. The breadth of the curve shown in Figure 3-29 is described by the standard deviation value cited above. For the Poisson equation

$$\sigma = \left[\sum_{i=1}^{i=n} (\chi_i - \bar{\chi})^2 \cdot \frac{(\bar{\chi})^{\chi_i}(e)^{-\bar{x}}}{\chi_i!} \right]^{1/2} \tag{35}$$

$$\sigma = (\bar{\chi})^{1/2} \tag{36}$$

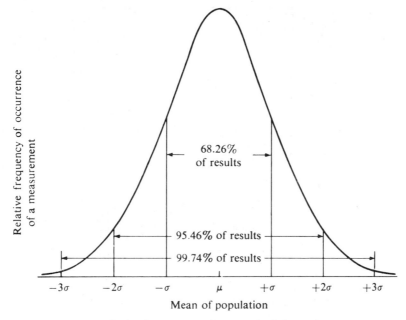

Figure 3-29. Normal distribution curve. (From J. S. Fritz and G. H. Schenk, Jr., *Quantitative Analytical Chemistry*, 2nd ed., Allyn and Bacon, Boston, 1972.)

Figure 3-30 relates the percentage of observed values of χ_i that are found outside the interval $\bar\chi \pm a\sigma$. At 1.0σ, 32% of the observed values of χ_i will be found outside the interval, $\bar\chi \pm 1.0\sigma$. In other words, a single observation has a 68% chance of falling in the interval $\bar\chi \pm 1.0\sigma$ or a 95% chance of falling in the interval $\bar\chi \pm 1.96\sigma$.

$\bar\chi$ is not usually available because most experimental samples are counted only once. As a result of this limitation it has become customary to indicate a proportional error of a given count. At a confidence level of 95% (i.e., there is a 95% chance the observed value will fall within the interval $\bar\chi \pm 1.96\sigma$) the proportional error may be calculated from the relationship

$$\text{proportional error} = \frac{a(\chi_i)^{1/2}}{\chi_i} \tag{37}$$

where a is the number of σ's desired and χ_i is the observed value. If a is 1.96 and an χ_i is 5000 cpm, the proportional error would be

$$\text{proportional error} = \frac{(1.96)(5000)^{1/2}}{5000}$$

$$= 0.0277$$

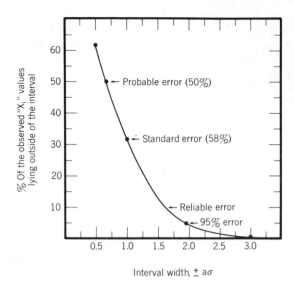

Interval width, ± aσ

Figure 3-30. Relationship between the size of σ and the percentage of observed values lying outside the interval ±aσ, where a is the valve indicated on the abscissa.

or 2.77%. A graph of the counts registered in a single determination as a function of the proportional, or percent, error is shown in Figure 3-31. As shown here the error decreases from 19.6 to 1.24% as the number of counts accumulated per determination increases from 100 to 25,000 cpm. It should be emphasized that the fewer counts accumulated per determination, the greater the probable error in the observed rate. Notice the decrease in the breadth of the curve shown in Figure 3-28A and B when

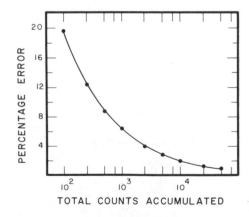

TOTAL COUNTS ACCUMULATED

Figure 3-31. Calculated proportional error as a function of the counts accumulated.

the counting time was increased tenfold; that is, more counts were accumulated per determination. As a general rule it is always advisable to count each sample long enough to accumulate a minimum of 10,000 counts, giving a probable error of about 2%.

LABELING PROCEDURES

Judicious use of labeling procedures can widely expand an investigator's potential. However, the exact procedures to be used must be given careful consideration and their experimental implications understood if full potential of this technique is to be realized. Our discussion is limited to incorporation of radioactive material into the cellular constituents of microorganisms because these systems exemplify most of the problems commonly encountered. Three major components must be monitored and regulated during a radiochemical experiment: (1) the cells being labeled, (2) the chemical nature of radioactive material, and (3) the labeling format.

Primary consideration must be given to the system being labeled. It is imperative that it be in a steady state condition unless the experiment specifically dictates otherwise. If, for example, the object is to follow incorporation of ^{3}H-uridine into *E. coli* RNA, it is necessary to establish clearly that the culture is in balanced growth throughout the course of the experiment. It is additionally advisable to begin the experiment early enough in log phase so that the culture remains in steady state for a period of time 150–200% greater than that required to complete the experiment, thus assuring a comfortable safety margin. Failure to heed this precaution produces results that are compromised by not knowing whether the observed results are due to experimental perturbation of the system or to the culture undergoing drastic physiological changes as it makes the transition between log and stationary phase.

The second consideration is the chemical nature of the radioactive material. The material presented to the experimental organism must be taken up or quickly metabolized to something else that can. For example, some strains of *Saccharomyces cerevisiae* are unable to accumulate exogenously provided uridine but can take up uracil quite normally. If, on the other hand, uridine is added to a culture of *E. coli*, 23% of it is extracellularly converted to uracil in the first 3 minutes of incubation. Figure 3-32 depicts the incorporation of uridine and uracil into TCA precipitable material in *E. coli* and *S. cerevisiae*, respectively. In both instances a lag is observed prior to linear incorporation. This lag may vary from 17 seconds for *E. coli* uridine incorporation to several hours for

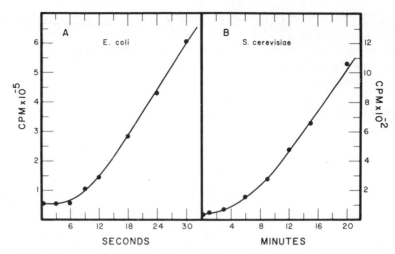

Figure 3-32. Incorporation of ³H-uridine and ³H-uracil into TCA precipitable material in *E. coli* and *S. cerevisiae*, respectively. The specific activities of the two radioactive molecules were quite different so comparison of the absolute number of counts accumulated is meaningless.

some metabolites of *S. cerevisiae*, and must be carefully determined so that it can be accounted for in the experimental design. Figure 3-33 depicts the time course of uridine incorporation into 10 ml cultures of *E. coli* presented with the same amount of ³H-uridine along with increasing amounts of nonradioactive carrier. As the amount of carrier is increased the period of linear incorporation is also increased. It is usually advisable to provide sufficient carrier so that linear incorporation occurs for a period at least twice as long as necessary to complete the experiment.

There are three different labeling formats, that is, ways of labeling cellular constituents, each providing quite distinct information. These methods are designated pulse labeling, equilibrium labeling, and pulse-chase labeling. Each technique involves a number of assumptions and requirements. The first and second techniques are compared in detail here because data obtained from them may be related mathematically. This comparison uses the following model of procaryotic transcription. It is assumed that addition of inducer to a steady state culture of *E. coli* results in the immediate onset of transcription of an operon by one RNA polymerase molecule. After 2 seconds (this time is inaccurate but serves our illustrative purpose) of transcription by the first RNA polymerase, sufficient space is available at the initiation site for a second RNA polymerase molecule to bind and also begin transcription. In like manner a total of 10 polymerase molecules may be accommodated on this operon.

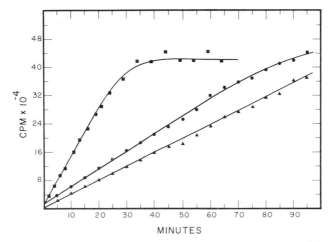

Figure 3-33. The effect of concentration on uridine uptake into TCA precipitable material in *E. coli*. Each of three cultures received 100 μc ³H-uridine and sufficient nonradioactive uridine to yield final concentrations of 1 (■), 5 (●), or 10 (▲) μg/ml.

When the first RNA polymerase reaches the end of the operon it drops off and releases a free RNA chain, which for the moment we inaccurately assume has not undergone any degradation.

Pulse labeling can be illustrated using the above model (Figure 3-34). The experiment (see Figure 3-35) is initiated by addition of inducer, and at various times thereafter samples of the induced culture are transferred to vessels containing a small amount of very high specific activity substrate (³H-uridine in this example). Material of high specific activity must be used if sufficient amounts of radioactivity are to be incorporated in the product over so short a labeling period. The period of incubation in the presence of radioactive substrate is dictated by the half-life of the molecule being labeled. Duration of the labeling period should optimally be no more than 10% of the half-life values observed for the molecule being labeled; for example, the incubation period used for pulse labeling *lac* messenger RNA at 30°C in *E. coli* should be no more than 0.25 minutes because the messenger has a half-life of about 2.5 minutes. The incubation period is usually terminated by killing the cells (addition of ice and CN⁻ works well in this example). The pertinent molecules are then isolated and the amount of radioactivity they contain is determined. Two things must be remembered with regard to a pulse labeling procedure: (1) the process being studied is occurring both before and during the labeling period, and (2) only the molecules or parts of molecules synthesized during the labeling period are radioactive. This technique may be com-

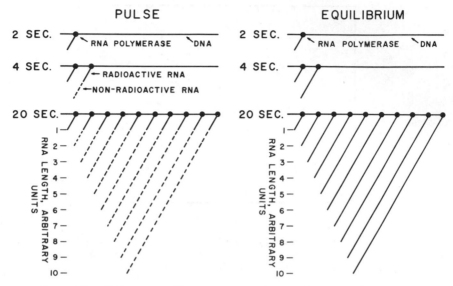

Figure 3-34. Pulse and equilibrium labeling of a specific mRNA from *E. coli.*

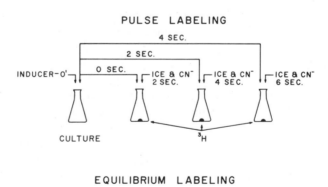

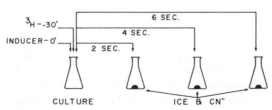

Figure 3-35. Experimental manipulations involved in pulse and equilibrium labeling protocols.

pared to a man passing a home under construction on his way to work. He sees what is happening for only a few minutes each day. If, however, the number of his trips past the construction site is increased, the man will have a more complete idea of how the house was constructed. Return now to the sample experiment shown in the left panel of Figure 3-34. If inducer is added to the culture at zero time and a sample is immediately transferred to a vessel containing ³H-uridine for an incubation period extending from 0 (time of inducer addition) to 2 seconds, only one RNA polymerase molecule is transcribing the DNA and only one unit of RNA is made (see the "2 second line" of Figure 3-34). This result is plotted as the first point in Figure 3-36 (assuming 10 cpm per unit of RNA). If a second sample is removed at 2 seconds and incubated in the presence of radioactivity between 2 and 4 seconds, the situation described in the "4 second line" of Figure 3-34 is observed. Between 0 and 2 seconds the first RNA polymerase has synthesized one unit of RNA, but since no radioactive material was present the RNA is not radioactive. Between 2 and 4 seconds the first RNA polymerase has synthesized one unit of radioactive RNA, and the second RNA polymerase has produced one unit of radioactive RNA. Notice that the rate of RNA synthesis between 2 and 4 seconds is twice that occurring between 0 and 2 seconds because two RNA polymerase molecules are making RNA. This argument may be repeated 10 times. During the period between 18 and 20 seconds, 10 RNA

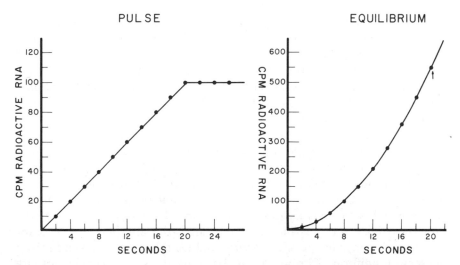

Figure 3-36. Comparison of data obtained by labeling the mRNA described in Figure 3-34 by pulse and equilibrium techniques. Subsequent to the time indicated by an arrow, the equilibrium plot is linear.

polymerase molecules are simultaneously transcribing the DNA and the rate of RNA production is tenfold greater than in the period of 0 to 2 seconds. This linear increase in the rate of synthesis is shown in Figure 3-36. However, at 20 seconds the DNA is saturated and can no longer accept additional RNA polymerase molecules. Therefore, between 20 and 22 seconds the same amount of RNA is synthesized as between 18 and 20 seconds. In other words, as shown in Figure 3-36, the rate of synthesis becomes constant. It should be clear from this example that a pulse experiment monitors the rate of synthesis of a given molecule. As the duration of the labeling period decreases to zero the observed rate of synthesis approaches the true instantaneous rate of synthesis (ds/dt). This is why it is desirable to use as short a labeling time as possible. Figure 3-34 emphasizes a second characteristic of pulse labeling experiments. If a macromolecule such as RNA is labeled, the radioactivity is not uniformly distributed throughout the molecule. On the contrary, it is localized at the end being synthesized during the short labeling period. This forms the basis for use of pulse labeling techniques to visualize rapidly degraded molecules. How may this comment be rationalized and on what assumption does it depend?

An equilibrium experiment is performed (see Figure 3-35) by first incubating a culture in the presence of a large amount of low to moderate specific activity radioactive material. Low specific activity material is used here owing to the extended labeling period. After permitting all the appropriate precursor pools to reach constant specific activity, the experiment is initiated (inducer addition in this example). Samples of this culture are thereafter removed and labeling is terminated by addition of cyanide and ice. The pertinent molecule is isolated and the amount of radioactivity it contains is determined. The period of preincubation in the presence of radioactivity is critical since failure of the metabolite precursor pools to reach constant specific activity results in the labeled molecules possessing a specific activity which depends on the time the sample was removed from the culture and terminated. Such a condition of multiple variables (the physiological phenomenon being studied and a time dependent change in precursor specific activity) is totally unacceptable. The preincubation time required ranges from a few minutes to many hours depending on the organism and system studied. Two things must be remembered with regard to an equilibrium labeling procedure: (1) the process being studied is initiated only after the appropriate precursor pools reach constant specific activity, and (2) the molecules synthesized possess a uniform distribution of label of constant specific activity throughout. Let us return to the sample experiment shown in the right panel of Figure 3-34. This *E. coli* system was preincubated for a minimum of 30 minutes with ^{3}H-uridine prior to initiation of the experiment. At zero time inducer is added and 2 seconds later the first sample is removed. During this time the first polymerase molecule has synthesized one unit or 10 cpm (see Figures 3-34 and 3-36) of RNA. A second sample is removed at 4 seconds after inducer addition. In 4 seconds the first and second polymerase molecule has synthesized two and one units of RNA, respectively, or a total of three units (30 cpm). This may be repeated up to 20 seconds and, as shown in Figure 3-36, the amount of RNA observed in each successive sample increases exponentially. However, between 20 and 22 seconds the amount of RNA synthesized is the same as that synthesized between 18 and 20 seconds because the DNA has become saturated with polymerase molecules. Thereafter, radioactivity continues to accumulate linearly in this particular species of RNA. From this example it should be clear that an equilibrium experiment monitors the accumulation of a given molecule. Figure 3-34 emphasizes that the label is uniformly distributed throughout the synthesized molecule, in contrast to the pulse format of labeling.

An important relationship exists between the data derived from pulse and equilibrium labeling experiment. If one assumes that no degradation of the pertinent molecule occurs, integration of the curve derived from the pulse labeling experiment should yield a curve similar to that obtained from the equilibrium labeling experiment. Conversely, differentiation of the curve obtained from the equilibrium experiment yields the curve obtained from the pulse experiment. Precise interconversion of the data, however, requires that differences in specific activity be accounted for. Precursor specific activity is constantly changing throughout the labeling period of most pulse experiments because the incubation periods are much shorter than the time required for precursor pools to reach constant specific activity (see Figure 3-32). However, as long as the labeling time in a pulse format is maintained precisely constant from sample to sample, this changing specific activity is in most cases of little consequence to the primary interpretation of the experiment. Time dependent specific activity changes would have to be evaluated, however, if it became desirable to precisely integrate the curve derived from pulse data.

Thus far it has been assumed that the degradation rate of the molecule being studied is zero. Degradation during a pulse experiment is usually insignificant. However, in the equilibrium format it is always a dominant consideration. The data obtained experimentally from an equilibrium experiment are a measure of accumulation minus the amount degraded. Therefore, this component must be taken into account when rectifying data obtained using the two different labeling formats.

A pulse-chase experiment is merely a variation of the pulse labeling technique. It involves initiating a physiological event (for example, by addition of inducer), incubating the culture for a short time in the presence of high specific activity radioactive material, removing or diluting the radioactive material 500- to 1000-fold with nonradioactive material, and sampling the labeled culture at various times after the radioactive material was initially added. The usefulness of this technique resides in its ability to establish precursor–product relationships. Consider the conversion of *A* to *D* with *B* and *C* as intermediates. If a very small amount of radioactive *A* (such that it completely reacts in a few seconds) is provided to a steady state system, the amounts of radioactivity found in each of the four intermediates as a function of time are as depicted in Figure 3-37. What would be the expected time dependent

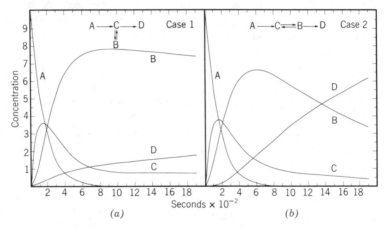

Figure 3-37.   Time dependent appearance and disappearance of intermediates in branched and sequential reactions initiated by addition of a small amount of *A*. The curves reflect concentration changes in the constituent (*A*, *B*, *C*, or *D*) indicated on each curve.

distribution of radioactivity if a large excess of radioactive *A* had been provided instead? A note of caution is appropriate regarding the interpretation of data derived from this type of experiment. Although the results obtained from pulse labeling a particular system may support or contest precursor–product relationships, they may rarely be used as unequivocal proof for such arguments. It is necessary to use these tools in concert with other approaches if a sound argument is to be constructed.

EXPERIMENTAL

Preparation of Aqueous and Organic Scintillation Fluid

3-1.   Dissolve 4.0 g PPO (2,5-diphenyloxazole) and either 0.050 g POPOP [p-bis(5-phenyloxazol-2-yl)benzene] or 0.080 g bis-MSB [p-bis(o-methylstyryl)benzene] in 1 liter of a solution composed of 667 ml toluene and 333 ml Triton X-100. Make certain that the contents are completely dissolved before the solution is used. This solution should be stored in a brown bottle in a cool dark space.

3-2.   Dissolve 4.0 g PPO and either 0.050 g POPOP or 0.080 g bis-MSB in 1 liter toluene. Make certain that the contents are completely dissolved before the solution is used. This solution should be stored in a brown bottle in a cool dark place.

Determination of the Scintillation Counter Balance Point

3-3.   Identify all the controls on a liquid scintillation counter using the instrument manual as a guide.

3-4.   Place 15 ml standard scintillation fluid into a scintillation vial.

3-5.   Add 0.50 ml radioactive (^{14}C) toluene (containing approximately $1-2 \times 10^5$ dpm/ml) to the vial. The amount of radioactivity per unit volume should be accurately known. Precisely standardized toluene solutions may be obtained commercially and may be used in place of those described here.

3-6.   Set the upper discriminator level of the channel being used at its maximum setting and the lower discriminator at its minimum setting.

3-7.   Place the sample in the counter and, using the manual counting mode, count the sample for 0.5 minutes at each of the coarse gain control settings.

3-8.   Plot the cpm observed at each setting as a function of the gain. A plot similar to this is depicted in Figure 3-14.

3-9.   Set the coarse gain control to the value that gave the highest counting rate and recount the sample at each of the fine gain control settings. The entire range of the fine gain control is equivalent to a one-step increase in the coarse gain control.

3-10.   Repeat step 3-8 using the data obtained in step 3-9.

3-11.   The values yielding maximum counting rates in these figures are the coarse and fine gain settings of the balance point.

3-12.   Repeat steps 3-4 to 3-11 inclusively, using ^{3}H-toluene in place of the ^{14}C-toluene. These data also appear in Figure 3-14 for comparison.

Determination of a β Spectrum

3-13. Repeat steps 3-4 and 3-5 using a solution of 4×10^5 to 5×10^5 cpm/ml ^{3}H.

3-14. Set the gain control to the balance point for the isotope being used, ^{3}H in this case (step 3-12).

3-15. Place the sample in the counter and, using the manual counting mode, count the samples for 0.5 minutes at each pair of discriminator settings listed in Table 3-12.

3-16. Repeat step 3-13 using ^{14}C-toluene in place of ^{3}H-toluene.

3-17. Set the gain control to the balance point for the isotope being used, ^{14}C in this case (step 3-11).

3-18. Repeat step 3-15.

3-19. Plot the counts per minute observed in steps 3-15 and 3-18 as a function of the average of the upper and lower discriminator values. Why are the shapes of the plots so similar even though the maximum β energy of these isotopes differs by a factor of more than 10? A plot from this type of experiment is shown in Figure 3-11.

Effect of Gain on a β Spectrum

3-20. Set the gain control to the ^{3}H balance point (steps 3-4 to 3-11).

3-21. Set the upper and lower discriminators at their maximum and minimum levels, respectively.

3-22. Place a ^{3}H sample in the counter and, using the manual counting mode, count the sample for 0.5 minutes at each pair of discriminator settings listed in Table 3-12.

3-23. Decrease the gain by 25% of the balance point value and repeat step 3-22.

3-24. Decrease the gain by 50% of the balance point value and repeat step 3-22.

3-25. Decrease the gain by 75% of the balance point value and repeat step 3-22.

3-26. Plot the counts per minute observed in steps 3-22 to 3-25 as a function of the average of the upper and lower discriminator values. Data from this type of procedure appear in Figure 3-13.

Alternative Method of Determining the Balance Point of an Isotope

3-27. Repeat steps 3-4 and 3-5 using ^{14}C-toluene.

3-28. Adjust the upper discriminator of the channel being used to its maximum value.

Table 3-12. Discriminator Settings Used to Determine a β Spectrum

Lower Discriminator Setting (% of Maximum Range)	Upper Discriminator Setting (% of Maximum Range)	Average of Discriminator Settings
0	2.0	1.0
4.0	6.0	5.0
8.0	10.0	9.0
12.0	14.0	13.0
16.0	18.0	17.0
20.0	22.0	21.0
24.0	26.0	25.0
28.0	30.0	29.0
32.0	34.0	33.0
36.0	38.0	37.0
40.0	42.0	41.0
44.0	46.0	45.0
48.0	50.0	49.0
52.0	54.0	53.0
56.0	58.0	57.0
60.0	62.0	61.0
64.0	66.0	65.0
68.0	70.0	69.0
72.0	74.0	73.0
76.0	78.0	77.0
80.0	82.0	81.0
84.0	86.0	85.0
88.0	90.0	89.0
92.0	94.0	93.0
96.0	98.0	97.0

3-29. Adjust the lower discriminator of that channel to a value that is 96 to 98% of its maximum possible value.

3-30. Adjust both the coarse and fine gain settings to zero.

3-31. Count the sample repeatedly for 0.5 minutes, increasing the coarse gain control one step each time until more than 20 counts are accumulated in 0.5 minutes.

3-32. Stop counting and decrease the coarse gain control one step to the last setting at which less than 20 counts were accumulated in 0.5 minutes.

3-33. Repeat steps 3-31 and 3-32 using the fine gain control.

These are the values of the balance point. Notice that a balance point is defined as the gain giving the highest efficiency of counting with a full window, or as the gain that permits the full β spectrum to just fit within the voltage range of the full window.

Effect of a Quenching Agent on a β Spectrum

3-34.   Number consecutively the caps of 10 scintillation vials.

3-35.   Place 15 ml toluene scintillation fluid in each vial.

3-36.   Add very carefully 0.200 ml of ^{3}H-toluene containing approximately 1.0×10^6 cpm/ml to each vial. The dpm/ml of this solution should be accurately known.

3-37.   Add one of the following volumes of acetone of each of the vials: 0.0, 0.1, 0.2, 0.4, 0.6, 0.8, 1.0, 1.2, 1.4, 1.6, and 1.8 ml. Also a set of commercially prepared quenched standards may be used here.

3-38.   Cap the vials and equilibrate them in the counter for approximately half an hour. Why can the increase in volume be neglected without causing a large error?

3-39.   Set the upper and lower discriminators at their maximum and minimum levels, respectively.

3-40.   Set the gain controls to the balance point of the isotope being counted.

3-41.   Count each of the samples from step 3-38 for 1 or 2 minutes.

3-42.   Plot the counts per minute observed as a function of the acetone volume used. Data from this type of experiment appear in Figure 3-38.

3-43.   Repeat steps 3-14, 3-15, and 3-19 for the vials containing 0.0, 0.2, and 1.8 ml of acetone. Data from this type of experiment are depicted in Figure 3-15.

Counting Quenched Samples Using the Channels Ratio Technique

3-44.   Adjust the gain of two channels (designated A and B) to the balance point for ^{14}C using the procedures outlined in steps 3-27 to 3-33. The two values are not necessarily the same. Why?

3-45.   Adjust the upper discriminator of one channel (Channel A) to its maximum value and its lower discriminator to a value sufficiently above zero to yield the desired level of background counts.

3-46.   Adjust the upper discriminator of the second channel (Channel B) to its maximum value and its lower discriminator to a value sufficiently above zero so that only about 25 to 50% of the count rate observed in Channel A for an unquenched standard appears in Channel B.

3-47.   Count a set of quenched ^{14}C standards (see steps 3-35 to 3-38 for their preparation) in both channels.

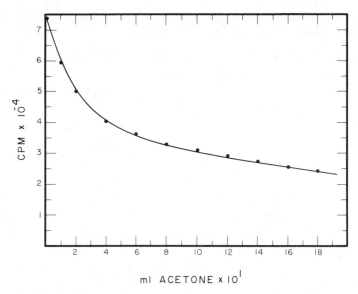

Figure 3-38. Effects of increasing amounts of acetone, a good quenching agent, on the efficiency of counting of ³H.

3-48. Plot the efficiency of counting observed in Channel A as a function of the ratio cpm$_B$/cpm$_A$. A plot of this type is shown in Figure 3-39.

3-49. Count a set of unknown variably quenched samples using the procedures described in steps 3-44 to 3-47 and then, using the quench correction curve obtained in step 3-48, calculate dpm ^{14}C in each sample.

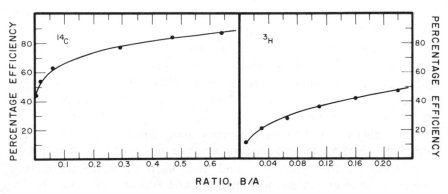

Figure 3-39. Channels ratio quench correction curves for ^{14}C and ^{3}H.

Radioactivity Determinations of Multiply Labeled Samples Using the External Standard Channels Ratio Technique

3-50.   Adjust the gain of one channel to the balance point of ^{3}H. This is designated Channel A. See steps 3-27 to 3-33.

3-51.   Adjust the gain of a second channel, designated Channel B, to the balance point of ^{14}C. See steps 3-27 to 3-33.

3-52.   Set the upper discriminator of Channel B to its maximum value.

3-53.   Set the lower discriminator of Channel B to a value that is sufficiently high so that if a ^{3}H standard is counted in Channel B an insignificant fraction (less than 0.1%) of ^{3}H counts appear in this channel.

3-54.   Set the lower discriminator of Channel A to a value that is sufficiently above zero so that the desired level of background counts is obtained (the lower the level of background desired, the higher the discriminator setting required).

3-55.   Set the upper discriminator of Channel A to a value that is sufficiently below the maximum so that less than 10 to 12% of the counts from a ^{14}C standard are observed in this channel.

3-56.   Count a set of ^{3}H and ^{14}C quenched standards (see steps 3-35 to 3-38 for preparation of the two sets of standards) in each channel in the presence and absence of the external standard.

3-57.   Plot the efficiency (in channel A) of each quenched ^{3}H standard as a function of the external standard channels ratio value it yielded. This type of plot is shown in Figure 3-20.

3-58.   Plot the efficiency (in both channels A and B) of each quenched ^{14}C standard as a function of the external standard channels ratio value it yielded. These types of plots are also shown in Figure 3-20.

3-59.   Label the caps of 16 scintillation vials consecutively from 1 to 16.

3-60.   Very carefully pipette the volumes of water shown in Table 3-13 into each of the appropriate vials.

3-61.   Very carefully pipette each of the volumes of aqueous ^{14}C and ^{3}H solutions shown in Table 3-13 into the appropriate vials.

3-62.   Add 15 ml aqueous scintillation fluid, cap the vials tightly, and thoroughly mix the contents by shaking. This step may also be performed prior to the above additions.

3-63.   Count each of the vials using the counter parameters established in steps 3-50 to 3-55, in the presence and absence of the external standard.

3-64.   Calculate the cpm of ^{14}C and ^{3}H in each vial using the quench correction curve obtained from steps 3-57 and 3-58 and equations 3-24 to 3-26.

Table 3-13. Radioactivity Additions for Double Label Counting Experiment

Vial Number	Volume ^{3}H (ml) $(4.0 \times 10^5$ dpm/ml)	Volume ^{14}C (ml) $(4.0 \times 10^5$ dpm/ml)	Volume Water (ml)
1	0.10	0.10	1.40
2	0.20	0.20	1.20
3	0.30	0.30	1.00
4	0.40	0.40	0.80
5	0.50	0.50	0.60
6	0.60	0.60	0.40
7	0.70	0.70	0.20
8	0.80	0.80	0.00
9	0.80	0.10	0.70
10	0.70	0.20	0.70
11	0.60	0.30	0.70
12	0.50	0.40	0.70
13	0.40	0.50	0.70
14	0.30	0.60	0.70
15	0.20	0.70	0.70
16	0.10	0.80	0.70

3-65.   Uncap each of the 16 vials counted in step 3-63 and add 0.5 to 2.5 ml of acetone to each vial. The amount of acetone in each vial should be different but the precise amount need not be known.

3-66.   Repeat steps 3-63 to 3-64.

3-67.   Compare the two sets of data and note any differences. The data from this type of experiment appear in Table 3-14.

Alternative Method of Determining the cpm of ^{14}C and ^{3}H in Multiply Labeled Samples

3-68.   The following method is a convenient way of determining the cpm of ^{3}H and ^{14}C in a multiply labeled sample. It may be used, however, only if (1) the ^{14}C standard and samples are prepared identically and (2) the degree of quenching is precisely the same in all of the samples. The most useful application of this technique is for counting samples derived from the fractionation of sucrose gradients or polyacrylamide gels. The gain and discriminator controls are adjusted in the same manner described for multiply labeled samples in steps 3-50 to 3-55.

3-69.   A sample of standard ^{14}C is prepared *exactly* the same as the experimental samples. If, for example, polyacrylamide gels are going to be fractionated and counted, the standard ^{14}C solution is polymerized into acrylamide gel and one gel slice is used as a ^{14}C standard.

Table 3-14. Determination of ^{3}H and ^{14}C in Multiply Labeled Quenched and Unquenched Samples

Vial Number	Unquenched Samples		Quenched Samples		Percentage Difference[a]
	dpm ^{14}C	dpm ^{3}H	dpm ^{14}C	dpm ^{3}H	
1	18,420	27,696	19,571	28,739	4
2	34,368	54,850	31,435	66,550	18
3	52,096	83,708	54,028	92,612	10
4	70,002	111,000	69,958	122,109	9
5	86,742	137,949	88,847	148,831	7
6	105,102	164,266	105,812	182,616	10
7	122,602	191,042	128,003	207,881	8
8	140,222	212,506	142,450	234,816	10
9	18,939	221,642	18,116	245,541	10
10	35,017	189,598	35,000	208,655	9
11	53,260	160,947	54,158	178,052	10
12	69,634	137,677	69,992	152,748	10
13	87,041	107,344	90,711	117,625	9
14	104,890	80,450	110,500	89,435	10
15	122,075	54,300	124,370	62,095	13
16	140,421	26,445	142,033	32,333	18

[a] Difference between the dpm of ^{3}H calculated in the presence and absence of acetone.

3-70.   Count the ^{14}C standard sample in both Channels A and B.

3-71.   Calculate the ratio $cpm_A/cpm_B = x$.

3-72.   Count all the multiply labeled experimental samples in both channels.

3-73.   Multiply the cpm observed in channel B by x and subtract this value from the observed cpm in Channel A, that is, cpm ^{3}H $= cpm_A - x\, cpm_B$.

The following experiment demonstrates the validity of this method.

3-74.   Place 10,000 to 15,000 cpm ^{3}H-toluene into each of six scintillation vials containing 10 ml organic scintillation fluid.

3-75.   Repeat step 3-70.

3-76.   Repeat steps 3-60 and 3-70.

3-77.   To each of the vials prepared in step 3-74 add one of the following concentrations of ^{14}C-toluene (4.15×10^5 cpm/ml): 0.1, 0.2, 0.4, 0.6, 0.8, and 1.0.

3-78. Add nonradioactive toluene to each vial to yield a final total addition of 1.0 ml.

3-79. Repeat steps 3-70 to 3-73.

3-80. Table 3-15 shows data from this type of experiment.

Table 3-15. Shortened Method of Determining cpm of ^{3}H and ^{14}C in Multiply Labeled Samples

Sample Number	First Count ^{3}H alone		Second Count ^{3}H plus ^{14}C		Amount of ^{3}H in Channel A calculated from second count
	cpm A	cpm B	cpm A	cpm B	
^{14}C standard	13,406	116,506	—	—	—
1	12,786	117	16,163	30,415	12,663
2	12,686	119	19,498	60,523	12,534
3	12,592	127	26,304	118,944	12,617
4	12,622	118	33,597	177,367	13,188
5	12,757	136	40,431	239,931	12,823
6	12,573	114	47,245	297,158	13,051

Determining the Half-Life of ^{32}P

3-81. Place 15 ml scintillation fluid in a scintillation vial.

3-82. Add 0.100 ml ethanolic solution of $H_3{}^{32}PO_4$ containing approximately 100,000 cpm.

3-83. Set up a scintillation counter to count ^{32}P. This procedure was described for ^{14}C and ^{3}H above.

3-84. Equilibrate the vial in the counter for 30 minutes and then count the vial for 1–5 minutes.

3-85. Note the time of day and repeat step 3-84 at the same time every third day for approximately 4 to 6 weeks.

3-86. Plot the data obtained (log cpm) from step 3-85 as a function of the time. Such a plot appears in Figure 3-40.

Determination of the Plateau Value on a Gas Flow Counter

3-87. Identify all the controls on the instrument using the instruction manual as a guide.

3-88. Turn on the instrument and allow it to warm up for 20–30 minutes. Permit the sample chamber to purge with the counting gas mixture during this time.

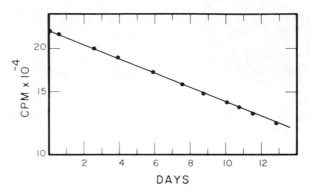

Figure 3-40. Time dependent decay of ^{32}P.

3-89.   Adjust the high voltage to zero.

3-90.   Count a standard sample for 1 minute and then increase the voltage by 50 V.

3-91.   Repeat step 3-90 until the observed cpm plateaus.

3-92.   Continue the procedure until the counts begin to again increase and stop. Further voltage increases will result in continuous discharge and damage to the chamber.

3-93.   Plot the cpm observed as a function of voltage as shown in Figure 3-41. The appropriate voltage setting for normal operation of the instrument is indicated in this figure.

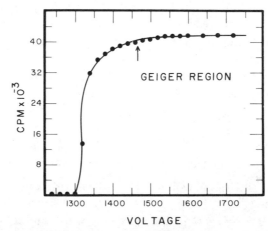

Figure 3-41. Determination of the optimal operating voltage (arrow) within the Geiger counting region for gas flow counting.

Determination of the Instrument Dead Time

3-94. From a piece of cardboard or $\frac{1}{16}$ in. plastic cut a circle 3.0 cm in diameter (this will fit most planchets). If necessary decrease the size of plastic disk so that it can be easily removed from the planchet.

3-95. Suspend a small amount of uranyl nitrate in airplane glue, spread a film of this mixture on the disk, and allow it to dry.

3-96. Cut the disk in half.

3-97. Count each half of the disk separately 10 times for 5 minutes each.

3-98. Count both halves together 10 times for 5 minutes each.

3-99. Calculate the mean of each set of measurements.

3-100. The dead time, τ, may be calculated from these data using the formula

$$\tau = \frac{m_1 + m_2 - m_{1,2} - b}{m_{1,2}^2 - m_1^2 - m_2^2}$$

where m_1, m_2, $m_{1,2}$, and b represent the cpm observed for the first half of the disk alone, the second half of the disk alone, and both halves of the disk together, respectively; b is the number of counts observed when no sample was in the instrument, that is, the background count.

3-101. In this experiment the following data were obtained:

	Mean of 10 determinations
Half 1	35,390
Half 2	34,641
Halves 1 and 2	64,077

$$\tau = \frac{35,390 + 34,641 - 64,077 - 4}{(64,077)^2 - (35,390)^2 - (34,641)^2}$$

$$= \frac{5954}{1.6534 \times 10^9}$$

$$= 3.60 \times 10^{-6}$$

3-102. With this value of τ, calculate the true count rate, n, for 5 to 10 hypothetically observed values, m ranging from 2,000 to 30,000 cpm. This may be done using the formula

$$n = \frac{m}{1 - m\tau}$$

The derivation of this formula may be found in reference 5.

3-103. Construct a graph similar to that shown in Figure 3-42. This graph may be used to correct all the experimental data obtained for coincidence losses as long as the same Geiger–Müller tube is always used.

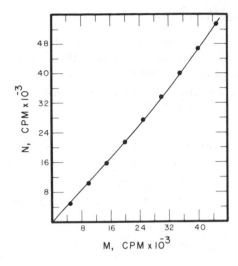

Figure 3-42. Correction of experimental data obtained by gas flow counting for the dead time of the instrument.

Incorporation of ³H-Leucine into *E. Coli* Proteins

3-104. Grow a 25 ml culture of wild type *E. coli* to a cell density of 45 Klett units (this is early log phase) on Ozeki's minimal medium (27) in a shaking water bath adjusted to 30°C.

3-105. Prepare 0.5–1.0 ml of a ³H-leucine solution containing 25 μg leucine per 0.1 ml at a specific activity of 1 μc/μmole.

3-106. Dissolve 10 mg rifampicin (rifamycin SV) in 1.0 ml dimethyl sulfoxide. Prepare also a 5% (w/v) solution of trichloroacetic acid (TCA).

3-107. Place exactly 0.1 ml ³H-leucine solution (step 3-105) in a 125 ml Erlenmeyer flask and place the flask in the bath.

3-108. When the culture reaches a cell density of 45 Klett units, transfer exactly 6.0 ml to the 125 ml flask and immediately begin timing the experiment using a stopwatch or digital laboratory timer.

3-109. At 3, 6, 9, and 12 minutes after cells were added to the radioactive leucine solution, remove 0.2 ml samples and place them in ice cold 13 × 100 mm test tubes. Immediately after transfer, add 5 ml cold 5% TCA to the test tubes.

3-110. At 13 minutes after initiation of the experiment, transfer 2.5 ml of the culture to a second 125 ml flask containing 0.05 ml of the rifampicin solution prepared in step 3-106. Mix the solution thoroughly.

3-111. Continue removing samples from the untreated half of the culture at the following times: 15, 17, 19, 21, 23, 25, and 27 minutes. Treat these samples as described in step 3-109.

3-112. Remove samples from the rifampicin treated culture at the following times: 16, 18, 20, 22, 24, 26, and 28 minutes. Treat these samples as described in step 3-109.

3-113. When all the samples have been collected, place them in a vigorously boiling water bath for 20 to 30 minutes. This treatment is required to release amino acids from tRNA.

3-114. Cool the tubes and collect the precipitate on 25 mm glass fiber filters. The Millipore filter apparatus works well for this operation.

3-115. Wash the test tube four times with 5 ml cold 5% TCA. Pass these wash solutions through the filter also.

3-116. Dry the filters in a 70°C oven for 1 hour. All the TCA must be removed by drying or erratic results will be observed.

3-117. Transfer the filters to scintillation vials and add 5 to 10 ml scintillation fluid. Determine the amount of radioactivity contained in each vial only after a 16 to 24 hour incubation period.

3-118. Data from this type of experiment are shown in Figure 3-43.

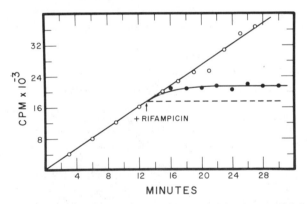

Figure 3-43. Incorporation of ³H-leucine into TCA precipitable material from untreated (○) and rifampicin treated (●) cultures of *E. coli*. Dashed line indicates the position of data expected on treating the culture with chloramphenicol instead of rifampicin.

3-119. This experiment may be repeated and half the culture treated with 0.05 ml chloramphenicol (375 μg/0.05 ml) in place of rifampicin. The dashed line in Figure 3-43 demonstrates the expected position of the data. Why do the rifampicin and chloramphenicol curves differ?

REFERENCES

1. Carlos G. Bell and F. Newton Hayes, Eds., *Liquid Scintillation Counting*, Pergamon, New York, 1958.

2. J. B. Birks, *The Theory and Practice of Scintillation Counting*, Pergamon, New York, 1964.

3. E. Rapkin, *Int. J. Appl. Radiat. Isot.*, **15**:69 (1964). Liquid Scintillation Counting 1957–1963: A Review.

4. V. P. Guinn and C. D. Wagner, *Atomlight* (a publication of New England Nuclear Corp.), No. 12 (April 1960). A Comparison of Ionization Chamber and Liquid Scintillation Methods for Measurement of Beta Emitters.

5. B. L. Funt and A. Hetherington, *Int. J. Appl. Radiat. Isot.*, **13**:215 (1962). The Kinetics of Quenching in Liquid Scintillators.

6. E. D. Bransome, Ed., *The Current Status of Liquid Scintillation Counting*, Grune & Stratton, New York, 1970.

7. A. Feinendegen, *Tritium Labeled Molecules in Biology and Medicine*, Academic Press, New York, 1967.

8. R. K. Swank, *Nucleonics*, **12**(3), 15 (1954). Recent Advances in Theory of Scintillation Phosphors.

9. M. Furst and H. Kallmann, *Phys. Rev.*, **94**:503 (1954). Energy Transfer by Means of Collision in Liquid Organic Solutions under High Energy and Ultraviolet Excitations.

10. E. A. Dawes, *Quantitative Problems in Biochemistry*, 5th ed., Williams and Wilkins, Baltimore, 1972.

11. R. J. Herberg, *Anal. Chem.*, **35**:786 (1963). Statistical Aspects of Liquid Scintillation Counting by Internal Standard Technique.

12. R. J. Herberg, *Anal. Chem.*, **33**:1308 (1961). Counting Statistics for Liquid Scintillation Counting.

13. J. Sharpe and V. A. Stanley, *Int. At. Energy Agency*, **1**:211 (1962). Photomultipliers for Tritium Counting, in Tritium in the Physical and Biological Sciences (Proceedings of a Symposium on the Detection and Use of Tritium in the Physical and Biological Sciences).

14. H. H. Seliger and C. A. Ziegler, *Nucleonics*, **14**(4):49 (1956). Liquid Scintillator Temperature Effects.

15. C. P. Petroff, P. P. Nair, and D. A. Tumer, *Int. J. Appl. Radiat. Isot.*, **15**:491 (1964). The Use of Siliconized Glass Vials in Preventing Wall Adsorption of Some Inorganic Radioactive Compounds in Liquid Scintillation Counting.

16. E. Rapkin and J. A. Gibbs, *Int. J. Appl. Radiat. Isot.*, **14**:71 (1963). Polyethylene Containers for Liquid Scintillation Spectrometry.

17. T. Higashimura, O. Yamada, N. Nohara, and T. Shidei, *Int. J. Appl. Radiat. Isot.*, **13**:308 (1962). External Standard Method for the Determination of the Efficiency in Liquid Scintillation Counting.

18. J. K. Weltman and D. W. Talmadge, *Int. J. Appl. Radiat. Isot.*, **14**:541 (1963). A Method for the Simultaneous Determination of H^3 and S^{35} in Samples with Variable Quenching.

19. R. S. Hendler, *Anal. Biochem.*, **7**:110 (1964). Procedure for Simultaneous Assay of Two β-Emitting Isotopes with the Liquid Scintillation Counting Technique.

20. E. T. Bush, *Anal. Chem.*, **35**:1024 (1963). General Applicability of the Channels Ratio Method of Measuring Liquid Counting Efficiencies.

21. G. A. Bruno and J. E. Christian, *Anal. Chem.*, **33**:650 (1961). Correction for Quenching Associated with Liquid Scintillation Counting.

22. E. T. Bush, *Anal. Chem.*, **36**:1082 (1964). Liquid Scintillation Counting of Doubly-Labeled Samples. Choice of Counting Conditions for Best Precision in Two-Channel Counting.

23. G. T. Okita, J. J. Kabara, F. Richardson, and G. V. LeRoy, *Nucleonics*, **15**(6):111 (1957). Assaying Compounds Containing H^3 and C^{14}.

24. B. Scales, *Anal. Biochem.*, **5**:489 (1963). Liquid Scintillation Counting: The Determination of Background Counts of Samples Containing Quenching Substances.

25. G. A. Bray, *Anal. Biochem.*, **1**:279 (1960). A Simple Efficient Liquid Scintillator for Counting Aqueous Solutions in a Liquid Scintillation Counter.

26. F. E. Kinard, *Rev. Sci. Instr.*, **28**:293 (1957). Liquid Scintillator for the Analysis of Tritium in Water.

27. T. G. Cooper, P. Whitney, and B. Magasanik, *J. Biol. Chem.*, **249**:6548 (1974). Reaction of *lac*-specific Ribonucleic Acid from *Escherichia coli* with *lac* Deoxyribonucleic Acid.

Chapter 4

Ion Exchange

Ion exchange may be defined as the reversible exchange of ions in solution with ions electrostatically bound to some sort of insoluble support medium. The ion exchanger is the inert support medium to which is covalently bound positive (in the case of an anion exchanger) or negative (in the case of a cation exchanger) functional groups. Any ion electrostatically bound to the exchanger is referred to as a counterion. The value of this technique in the isolation and separation of charged compounds is that conditions can be found under which some compounds are electrostatically bound to the ion exchanger whereas others are not.

The essential features of ion exchange are summarized in Figure 4-1. One begins with an exchanger prepared in such a way as to be fully charged (Figure 4-1A). The mixture of ionic species to be separated is incubated in contact with the exchanger for sufficient time to allow the following equilibria to be attained (Figure 4-1B):

$$Exch^- X^+ \rightleftharpoons Exch^- + X^+$$
$$Exch^- + YH^+ \rightleftharpoons Exch^- YH^+$$
$$Exch^- + Z^+ \rightleftharpoons Exch^- Z^+$$

where $Exch^-$ is the charged cation exchanger and X^+, YH^+, and Z^+ are cations. Neutral molecules and anions do not bind at all to this column. Following the electrostatic binding of species possessing a net charge opposite that of the exchanger, the like charged and uncharged species are washed from the medium. Bound ions YH^+ and Z^+ are then sequentially eluted (Figure 4-1C and D) either by washing the medium with increasing concentrations of X^+ and hence increasing the probability that X^+ will replace YH^+ or Z^+ in the above equilibria, or by increasing the pH and hence converting YH^+ and Z^+ to Y^0 and ZOH, respectively. When the concentration of X^+ is increased, the strength of binding depends on the

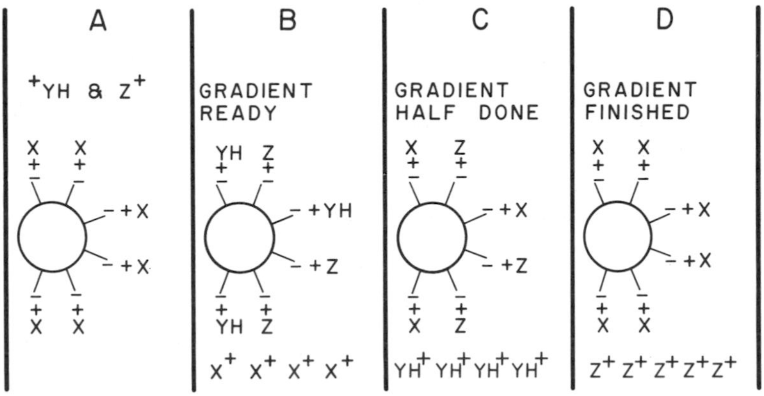

Figure 4-1. Operation of a cation exchange column. (*A*) The exchanger has been prepared with X as the starting counterion. YH^+ and Z^+ are the cations to be separated. (*B*) Cations YH^+ and Z^+ have been bound to the column displacing equivalent amounts of counterion X. This is the situation just prior to beginning the elution gradient. (*C*) Part of the gradient (increasing concentrations of X^+) has passed through the column, displacing one of the ions to be separated, YH^+. (*D*) The gradient has been completed with the second ion, Z^+, also being displaced.

quantity of charge possessed by YH^+ and Z^+; the greater the charge of the species (YH^+ or Z^+), the higher the concentration of X^+ required to elute them. When the pH is changed, the binding depends on the pK of the ionic species ($YH^+ + OH^- \rightleftharpoons Y^0 + HOH$ or $Z^+ + OH^- \rightleftharpoons ZOH$); the higher the p$K$ of the ion (Z^+ or YH^+), the higher the pH required to bring about elution.

Thus far only molecules capable of possessing a single type of charge have been considered, but the same principles would apply if the desired species were large molecules, such as proteins, which are capable of possessing both negative and positive charges. In this case the molecules could bind to either anion or cation exchangers since they possess both types of charge. However, the strength of binding would depend upon the pH, as shown in Figure 4-2. The molecule can, within limits, be made more negative (resulting in a more tenacious binding to anion exchangers) if the pH is raised, and more positive (resulting in a more tenacious binding to cation exchangers) if the pH is lowered. At one specific pH, the isoelectric point, the molecule contains an equal number of positive and negative charges. This amphoteric nature of proteins may be manipulated to great advantage when using ion exchange chromatography for purification purposes. For example, the pH of a protein mixture may be lowered to the point where the desired protein behaves as a cation. If the

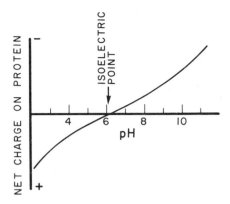

Figure 4-2. The effect of pH upon the net charge of a protein.

preparation is chromatographed on a cation exchange column, under these conditions of pH, many of the anionic protein species are lost. If this is followed by an increase in pH to convert the desired protein to its anionic form, the preparation may be chromatographed on an anionic exchange column with the concomitant loss of many of the cationic species. It should be pointed out that anion and cation exchange chromatography used sequentially often afford a large degree of purification even if the pH of the preparation cannot be varied.

THE EXCHANGER

The selection of an exchanger that is best suited to a particular application is largely an empirical process. However, the various media can be classified into broad groups and a number of generalizations are possible. The nature of the supporting matrix usually determines its flow properties, ion accessibility, and chemical and mechanical stability. Charged groups that are covalently bonded to the matrix determine the type and strength of binding. Basically three major groups of materials are used in the construction of ion exchangers: polystyrene or polyphenolic resins, various types of cellulose, and polymers of acrylamide and dextran. A variety of commercially available exchangers of these materials are listed in Tables 4-1 and 4-2, along with the charged functional group that each carries; exchangers are available ranging from strongly basic to strongly acidic (Figure 4-3).

Dowex, one of the most important exchangers used in the separation of small molecules, is made, by reacting styrene with varying proportions of divinyl benzene to yield a beadlike polymer, as shown in Figure 4-4. The

Table 4-1. Cation Exchange Resins

Type and Exchange Group	Bio-Rad Resin[a]	Dowex Resin[b]
Strongly acidic, phenolic type, $RCH_2SO_3^-H^+$	Bio-Rex 40	
Strongly acidic, polystyrene type, $C_6H_5SO_3^-H^+$	AG 50W-X1	50-X1
	AG 50W-X2	50-X2
	AG 50W-X4	50-X4
	AG 50W-X5	50-X5
	AG 50W-X8	50-X8
	AG 50W-X10	50-X10
	AG 50W-X12	50-X12
	AG 50W-X16	60-16
Intermediately acidic, polystyrene type, $C_6H_5PO_3^{2-}(Na^+)_2$	Bio-Rex 63	
Weakly acidic, acrylic type, $RCOO^-Na^+$	Bio-Rex 70	
Weakly acidic chelating resin, polystyrene type, $C_6H_5CH_2N(CH_2COO^-H^+)_2$	Chelex 100	A-1

[a] Bio-Rad Laboratories, 32nd and Griffin Ave., Richmond, Calif. 94804.
[b] Dow Chemical Co., 2030 Abbott Road Center, Midland, Mich. 48640.

reaction with divinyl benzene results in the cross-linkage of the polystyrene. The degree of cross-linkage is usually indicated by the designation X-1, 2, 4, 8, 12, etc., where the number signifies the percentage of the total polymer that is divinyl benzene. The consequences of increasing the amount of cross-linkage are summarized in Table 4-3. Decreasing the amount of cross-linkage would have effects opposite to those listed. Selection of the proper amount of cross-linkage is usually a compromise between the desired characteristics of high and low degrees of cross-linkage. As a starting point, 8% cross-linkage is widely used.

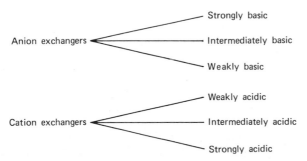

Figure 4-3. Classification of ion exchangers based on the strength of their charged functional groups.

The size (diameter) of the resin beads (referred to as mesh size) determines the flow rate, equilibration time, and capacity of the exchanger. The larger the mesh size (decrease in bead size), the larger the capacity and counterion equilibration time but the lower the flow rate. As a rule, 100–200 mesh is appropriate for most analytical applications. Smaller mesh sizes are useful for large scale, low resolution applications, whereas the larger mesh sizes are especially suited to high resolution analytical separations.

Introduction of charged groups into the matrix is accomplished by reacting the support medium with various reagents which covalently bond

Table 4-2. Anion and Mixed Ion Exchange Resins

Type and Exchange Group	Bio-Rad Resin	Dowex Resin
Anion Exchange Resins		
Strongly basic, polystyrene type		
$C_6H_5CH_2N^+(CH_3)_3Cl^-$	AG 1-X1	1-X1
	AG 1-X2	1-X2
	AG 1-X4	1-X4
	AG 1-X8	1-X8
	AG 1-X10	1-X10
	AG 21K	21K
$C_6H_5CH_2N^+(CH_3)_2(C_2H_4OH)Cl^-$	AG 2-X4	2-4
NH^+Cl^-	AG 2-X8	2-X8
	AG 2-X10	
	Bio-Rex 9	
Intermediately basic, epoxypolyamine,		
$RN^+(CH_3)_3Cl^-$ and $RN^+(CH_3)_2(C_2H_4OH)Cl^-$	Bio-Rex 5	
Weakly basic, polystyrene, or phenolic		
polyamine, $RN^+HR_2Cl^-$	AG 3-X4	3-X4
Ion Retardation (Zwitterion) Resin		
Strongly basic anion plus weakly acidic		
cation exchange resin, $C_6H_5N^+(CH_3)_3Cl^-$		
and $RCH_2COO^-H^+$	AG 11A 8	
Redox Resin		
$—NR_2Cu^0$		
Mixed Bed Resins		
$C_6H_5SO_3^-H^+$ and $C_6H_5CH_2N^+(CH_3)_3OH^-$	AG 501-X8	
$C_6H_5SO_3^-H^+$ and $C_6H_5CH_2N^+(CH_3)_3OH^-$		
plus indicator dye	AG 501-X8 (D)	
$C_6H_5SO_3^-H^+$ and		
$C_6H_5CH_2N^+(CH_3)_2(C_2H_4OH)OH^-$		

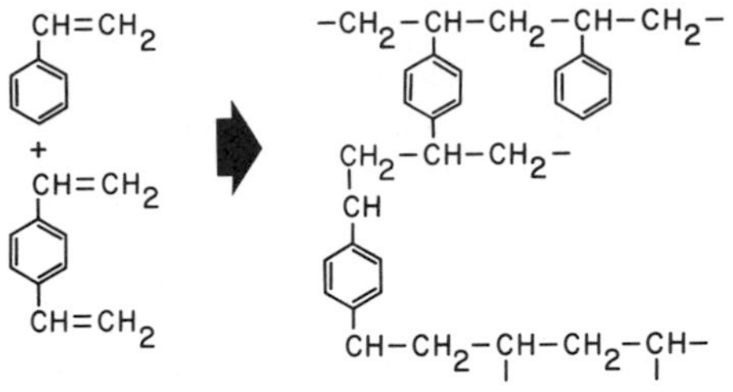

Figure 4-4. Synthesis of polystyrene resins from styrene and divinyl benzene.

functional groups to the polymer. For example, Dowex 50, a strongly acidic cation exchanger, is produced by reacting the polymer (Figure 4-4) with chlorosulfonic acid or fuming sulfuric acid (Figure 4-5). The other charged groups of Tables 4-1 and 4-2 are introduced by similar techniques. The strength of the acidity or basicity of these groups and the number per unit volume of resin determine the type and strength of binding that an exchanger possesses. For example, introduction of a strongly acidic group, such as sulfonic acid, results in an exchanger with a large negative charge which binds cations tenaciously. Introduction of weakly acidic or basic groups results in an exchanger whose charge depends on the pH of the environment. Selection of the appropriate type of functional group and the number of charges per unit volume of resin (charge density) is based on the strength of binding that is desired. If the ionic species to stable to drastic changes of pH and ionic strength, a strongly acidic or basic exchanger of high charge density can be used. If, however, the desired molecule is sensitive to these variations, a weaker exchanger of low charge density is advisable because the desired ionic species can be eluted by much more gentle conditions of pH and ionic strength.

If apparent irreversible binding between a given ion and exchanger is observed, a weaker exchanger of lower charge density should be used. This is often seen when the ionic species being isolated is a large molecule possessing a high charge density. For this reason, as well as their limited

Table 4-3. Effects of Increasing Cross-Linkage in Polystyrene or Phenolic Resins

1. Permeability decreases, that is, the interior regions of the matrix become accessible to smaller and smaller molecules.
2. Capacity (equivalents of ion bound per unit weight of exchanger) increases because highly cross-linked particles swell only slightly, resulting in more exchangeable sites per unit volume than with a resin having a lower amount of cross-linkage.
3. Selectivity increases owing to high density of ionic sites in the resin.
4. Equilibration times increase because diffusion through resin decreases.

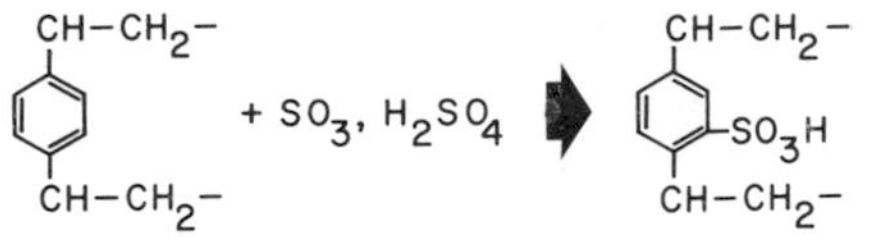

Figure 4-5. Addition of anionic groups to a polystyrene resin.

permeability toward large molecules, use of polystyrene and polyphenolic ion exchange resins with macromolecules is in most cases precluded.

The unsuitability of Dowex type exchangers for the isolation and separation of macromolecular polyelectrolytes has resulted in development of a second class of exchangers which use cellulose as the principal support medium. The advantages of using cellulosic exchangers for macromolecular applications result from (1) the support medium having a much greater permeability to macromolecular polyelectrolytes and (2) only a few percent of the available positions in the medium (hydroxyl groups of the sugar residues composing cellulose) being substituted, thus producing an exchanger with a much lower charge density than the Dowex type resins, where substitution is nearly complete. These exchangers employ cellulose isolated from cotton, softwood, and hardwood as the supporting matrix. The major exchangers of this type are DEAE cellulose (produced by reacting purified cellulose with chloro-triethylenediamine), CM cellulose (produced by treatment of the cellulose with chloroacetic acid), and phosphocellulose. Table 4-4 lists other types of substituted celluloses that are available. As for the Dowex resins, there is a wide range of exchangers available; general considerations in selecting the one most appropriate were discussed above.

Recently a new type of cellulosic exchanger, called microgranular cellulose, has become available. These exchangers differ from the older, fibrous types in that the cellulose has been treated to remove the amorphous portions of cellulose, leaving only the more uniformly sized microcrystalline regions. These regions have been chemically cross-linked to prevent excessive swelling and result in dense rod shaped particles which yield columns with high capacities and charge densities. The practical effects of processing the exchanger in this way correspond to the effects observed when the degree of cross-linkage is increased in Dowex resins. Therefore, selection between the two possibilities, fibrous and microgranular celluloses, is made on the basis of those items summarized in Table 4-3. For example, if high flow rates are required owing to the lability of a preparation, it may be necessary to sacrifice the increased resolution of the microgranular celluloses to gain the increased flow rates (decreased contact time between the preparation and exchanger) of the lower resolution fibrous types.

Table 4-4. Cellulosic Ion Exchangers

Ion Exchanger	Ionizable Group	Structure
	Anion Exchangers	
Intermediate base		
AE	Aminoethyl	$-OCH_2CH_2NH_2$
Strong base		
DEAE	Diethylaminoethyl	$-OCH_2CH_2N(C_2H_5)_2$
TEAE	Triethylaminoethyl	$-OCH_2CH_2N(C_2H_5)_3$
GE	Guanidoethyl	$-OCH_2CH_2NH\overset{\overset{\displaystyle NH}{\|}}{C}NH_2$
Weak base		
PAB	*p*-Aminobenzyl	$-OCH_2-\langle\bigcirc\rangle-NH_2$
Intermediate base		
ECTEOLA	Triethanolamine coupled to cellulose through glyceryl and polyglyceryl chains mixed groups (mixed amines)	
DBD	Benzylated DEAE cellulose	
BND	Benzylated naphthoylated DEAE cellulose	
PEI	Polyethyleneimine adsorbed to cellulose or weakly phosphorylated cellulose	
	Cation Exchangers	
Weak acid		
CM	Carboxymethyl	$-OCH_2COOH$
Intermediate acid		
P	Phosphate	$-O\overset{\overset{\displaystyle O}{\|}}{\underset{\underset{\displaystyle OH}{\|}}{P}}OH$
Strong acid		
SE	Sulfoethyl	$-OCH_2CH_2\overset{\overset{\displaystyle O}{\|}}{\underset{\underset{\displaystyle O}{\|}}{S}}OH$
SP-Sephadex	Sulfopropyl	$-C_3H_6\overset{\overset{\displaystyle O}{\|}}{\underset{\underset{\displaystyle O}{\|}}{S}}OH$
Strong base		
QAE-Sephadex	Diethyl(2-hydroxypropyl) quaternary amino	$-C_2H_4N^+(C_2H_5)_2$ $\underset{\underset{\underset{\displaystyle OH}{\|}}{\displaystyle CH_2CHCH_3}}{\|}$

A final class of exchangers is the ion exchange gels which are produced in a way similar to the cellulosic exchangers with the exception that the cellulose support medium is replaced by small polydextran or polyacrylamide beads. These exchangers have the advantage, by virtue of their molecular sieving properties (discussed in Chapter 5), of separating on the basis of size as well as charge. In practice these exchangers are used in a manner similar to the cellulosic exchangers. They need not, however, be precycled before use (see following section). It is noteworthy that proteins possessing molecular weights greater than 200,000 make very little use of the sieving characteristics of this type of exchanger.

PREPARATION OF THE EXCHANGE MEDIUM

Successful performance of ion exchange techniques is largely based on proper preparation of the exchanger. This preparation is essentially the conversion of an exchanger from the form provided by the manufacturer to a form that can be used for the separation of various ions. There are four basic procedures involved in preparing an exchanger: (1) removal of impurities that result from incomplete purification of the exchanger by the manufacturer or slight decomposition of the exchanger during storage; (2) swelling the medium (termed precycling) so that a greater percentage of the exchanger's charged groups are exposed to the suspending solution; (3) removal of the fines (very small particles of exchanger); and (4) conversion of the counterion of the exchanger (the ion electrostatically bound to the charged functional groups of the resin) to that desired for a given application. As indicated above, gel type exchangers need be subjected to only the last two steps, whereas the other ion exchangers must be subjected to all four procedures.

The primary function of washing the exchanger is to remove impurities that it may contain. This is a critical step, regardless of exchanger type, because the purity of commercially available exchangers varies greatly. Even those exchangers, which are provided in a high state of purity (i.e., Whatman DE 52), are subject to some oxidative degradation. Washing procedures for the polyphenolic and polystyrene resins should include washes with water, hydrochloric acid, ethanol, and a high concentration of the counterion to which the exchanger is being converted. In the case of the cellulosic exchangers, the alcoholic wash is omitted and further washing of the medium is accomplished during precycling.

The purpose of precycling an exchanger is to expose the charged groups that are bound to the matrix. The cellulosic exchangers are discussed here because precycling is most critical to these media. However, these comments apply also to the polyphenolic and polystyrene resins. At the molecular level, cellulosic exchangers are carbohydrate polymers that have substituted hydroxyl groups. During the drying step of the manufacturing process, water is removed from the exchanger and extensive intermolecular hydrogen bonding of the hydroxyl groups occurs. Such bonding results in very dense packing of the carbohydrate polymers, which in turn results in burial of many of the charged functional groups. Suspending an exchanger in water breaks a small number of these hydrogen bonds (exposing a few of the functional groups), but much stronger treatment is required to fully swell the matrix and hence break the remaining hydrogen bonds. This swelling is accomplished by suspending the exchanger in either strong acid or strong base (Table 4-5). In the case of DEAE, an anion exchanger, treatment with

Table 4-5. Order of Treatment of Ion Exchangers with Acid and Base

Exchanger	1st Treatment	2nd Treatment
Anion	$0.5N$ HCl	$0.5N$ NaOH
Cation	$0.5N$ NaOH	$0.5N$ HCl

hydrochloric acid results in the conversion of all the diethylaminoethyl groups to their charged species ($C_2H_4N^+H(C_2H_5)_2$). Placing like positive charges on the functional groups results in mutual repulsion and hence a maximum amount of swelling (exposure of the charged groups). For a cation exchangers such as CM cellulose, the resin is treated with strong base. This results in the conversion of all the carboxylic acid groups to charged carboxylate ions. Once the matrix is fully swelled the acid or base is washed away and the resin is treated with base in the case of an anion exchanger or acid in the case of a cation exchanger. This converts the functional groups of the DEAE and CM cellulose to their free base and free acid forms, respectively. These are the forms that can be most easily equilibrated with the desired counterions. Since strong acid and base tend to decompose the exchanger, treatment with these agents, as described in the experimental section, should be confined to concentrations equal to or smaller than $0.5N$ and periods less than 2 hours. Some exchangers such as DEAE readily bind heavy metals. If such metal ions have a deleterious effect on the ions being separated, as is the case if the ions are proteins containing SH groups, then the swelled exchanger should be treated with a $0.01M$ EDTA solution as a final step in the precycling procedure.

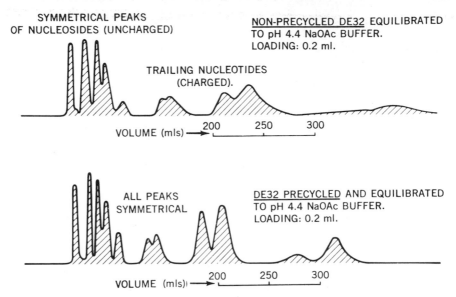

Figure 4-6. The effects of precycling an ion exchange resin on its resolution capabilities. (Copyright W. and R. Balston, Ltd., Maidstone, Kent, England.)

Failure to precycle an exchanger results in drastically reduced capacity and resolution (Figure 4-6). The loss in resolution becomes more acute as the size and charge densities of the molecules being separated increases. Therefore, for protein separations on cellulosic exchangers precycling is a critical step.

After the exchanger has been thoroughly swelled, the fines should be removed. Fines are minute particles of exchanger generated during manufacture of the exchanger or as a result of excessively vigorous stirring during washing and precycling procedures. This process should always be performed after the matrix is fully swelled to prevent the trapping of fines in the unswelled portion of the exchanger. Removal of these fine particles is accomplished by repeatedly suspending the exchanger in a large volume of water, allowing approximately 90 to 95% of the exchanger to settle, and decanting off slow sedimenting material. Failure to remove the fines usually results in decreased flow rates and poor resolution.

The final operation in the preparation of an ion exchanger is equilibration of the exchanger with the appropriate counterions. The actual conversion of the exchanger from one counterion to another is effected by passing a large volume of the desired counterion, in high concentration, over the resin. Table 4-6 summarizes a number of counterion conversions for polyphenolic and polystyrene resins along with methods for determining when the conversion is complete. Following conversion of the exchanger to the desired form, excess counterions are removed by washing the exchanger with large volumes of water or dilute buffer.

There are a large number of buffers that may be selected to maintain the pH of a chromatographic medium. Four general considerations are pertinent to this selection. (1) Cationic buffers should be used with anion exchangers and anionic buffers should be used with cation exchangers. If the buffer ions carry charges opposite to those of the exchanger's charged groups they will participate in the ion exchange process, resulting in local variations of pH and decreased resin capacity. (2) The pK of the chosen buffer should lie within ± 0.7 units of the pH at which the system will be buffered. Too drastic a departure from the pK results in a very low buffer capacity and hence the possibility of pH variation. (3) The pH of the buffer system should be chosen so that the ions to be separated possess the same charge as the counterions of the exchanger. This consideration is of particular significance in the case of proteins whose isoelectric points usually lie within the pH range used for ion exchange chromatography. (4) The buffer system selected must not interfere with analysis of the fractions yielded. For example, if separation of a number of proteins is being carried out and the Lowry procedure (Chapter 2) is to be used for an assay of the protein concentration, Good's buffers, which are cyclic peptides, should be avoided. Likewise, if the separated ions are to be concentrated and used further, as is the case if ion exchange chromatography is used to purify a substrate, then the buffer used should be volatile to facilitate its removal from the purified preparation (see Chapter 1 for a list of these buffers).

CHROMATOGRAPHY

Four items must be considered in the chromatography of a group of ions: (1) the volume and shape of the column, (2) the shape and size of the gradient to be employed, (3) the rate of elution, and (4) the size of the fractions to be collected.

The Column

The volume of exchanger used should be at least 2 to 5 fold greater than that needed to bind all of the sample. Huge excesses, however, should be

Table 4-6. Conversion of Ion Exchange Resins from One Counterion to Another[a]

Resin	Conversion	Reagent	Vol. Soln/Vol. Resin	Flow Rate[c] (ml/min/cm² of bed)	Type of Exchange[b]	Test For Completeness of Conversion	Vol. Rinse Water/Vol. Resin	Test for Completion of Rinsing
AG 50	$H^+ \longrightarrow Na^+$	1N NaOH	2	2	N	ph 9[h]	4	pH < 9
Bio-Rex 40	$H^+ \longrightarrow Na^+$	1N NaOH	4	2	N	pH > 4.8[h]	4	pH < 9
	$Fe^{3+} \longrightarrow H^+$	6N HCl	4	2	IX	Fe^{3+} [g]	4	Cl^-
AG 1	$Cl^- \longrightarrow OH^-$	1N NaOH[d]	20		IX	Cl^- [f]	4	pH < 9
	$OH^- \longrightarrow$ formate	1N formic acid	2	2	N	pH < 2	4	pH > 4.5
	$Cl^- \longrightarrow$ formate	Use $Cl^- \longrightarrow OH^-$, then $OH^- \longrightarrow$ formate	20 / 2		IXN			pH > 4.8 / pH > 4.8
AG 2	$Cl^- \longrightarrow OH^-$	1N NaOH[d]	2	2	IX	Cl^- [f]	4	pH < 9
	$Cl^- \longrightarrow NO_3^-$	0.5N NaNO₃	5		IX	Cl^- [f]	4	
AG 3 or Bio-Rex 5	$Cl^- \longrightarrow OH^-$	0.5N NaOH[d]	2	1	IX	Cl^- [f]	4	pH < 9
Bio-Rex 63	$H^+ \longrightarrow Na^+$	0.5N NaOH	3	2	N	pH > 9		pH > 4.8
Bio-Rex 70	$H^+ \longrightarrow Na^+$	0.5N NaOH	3	1	N	pH > 9		
Chelex 100[e]	$Cu^{2+} \longrightarrow H^+$	1N HCl	3	1	IX	Cu^{2+} [g]	4	pH < 8
AG 11A8[i]	$Na^+Cl^- \longrightarrow$ self-absorbed	H₂O	20 / 4	2	IR	Cl		
	$H^+Cl \longrightarrow$ self-absorbed	(a) NaOH (b) H₂O	20			pH		pH < 10

148

[a] Typical conversions are listed. The same reagents can be used to convert from other ionic forms. Two step regeneration (IXN) is included because of ease of conversion and saving of expensive reagents.

[b] N = neutralization; IX = ion exchange; IR = ion retardation; IXN = two step process: ion exchange to acid or base form followed by neutralization with appropriate base of acid or base of salt. Example: (Step 1) resin-Cl + NaOH $\longrightarrow$ resin-OH (IX); (Step 2) resin-OH + H-formate $\longrightarrow$ resin-formate + $H_2O(N)$.

[c] For 50–100 or finer resin. For 20–50 mesh about $\frac{1}{3}$ the flow rate is recommended.

[d] Use USP or CP grade (low chloride).

[e] Chelex 100 may lose iminodiacetic acid groups on long standing. To remove odor, heat to 80°C for 2 hours in $3N$ NH_4OH and rinse.

[f] Test for Cl$^-$ in effluent: acidify sample with a few drops of conc. HNO_3. Add 1% $AgNO_3$ solution. White ppt indicates Cl$^-$, yellow Br$^-$ or too basic (Ag_2O precipitate).

[g] Test for Fe^{3+} or Cu^{2+} in effluent: to 2 drops effluent on spot plate add 2 drop conc. HCl and 2 drops fresh 5% potassium ferrocyanide solution. Blue color indicates Fe^{3+}. Brown color is Cu^{2+}. Most spot tests can be performed directly on the resin; e.g., a particle of resin containing 2 ppm iron will turn blue when subjected to ferrocyanide tests.

[h] Test for pH 4.8: pH paper or methyl orange (red, pH 1; yellow, pH 4.8). Test for pH 9: pH paper or thymolphthalein (blue, pH 10; colorless, pH 9).

[i] Request Bio-Rad Tech. Bull. 113, "Desalting With AG 11A8 Ion Retardation Resin," for detailed regeneration procedures.

avoided because they may result in broadening of the eluted peaks and hence decrease resolution.

The shape of the column also bears on the degree of broadening observed. A long narrow column may yield considerably less resolution than a column with a higher diameter to height ratio. The reason for the decrease in resolution can be explained by considering the elution process in the presence of a steep gradient. When the counterion concentration in the eluting gradient becomes high enough to completely mobilize the adsorbed ions, these ions move down the column at the same rate as the solution. Indeed, these ions are free in solution. If they must cover a long distance from the point of desorption from the exchanger to the point of collection, each desorbed species has an opportunity to diffuse. This diffusion results in peak broadening and occurs especially during separation of species with widely different net charges. These effects are shown in Figure 4-7 for the separation of catalase from glucose oxidase. It is clear that in this case doubling the column length decreased the resolution to a point where the two activities could no longer be separated. Therefore, if a greater volume of exchanger is necessary, it is advisable to increase the volume of the column by increasing its diameter.

There are circumstances, however, when increasing the column length increases its resolution. This occurs when the column is washed continuously with a solution of one concentration rather than a gradient of rapidly increasing concentrations. In this case the salt concentration of the eluting solution is chosen such that it binds to the column with about the same tenacity as the species being separated. When the desired

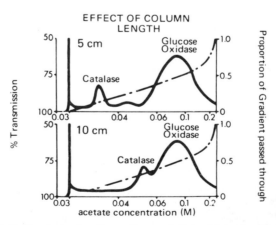

Figure 4-7. Effect of column length on the resolution obtained with DEAE cellulose exchange medium. (Copyright W. and R. Balston, Ltd., Maidstone, Kent, England.)

species are mobilized they have a very good chance of being rebound to the ion exchanger. Just how good a species' chances are of rebinding depends on its tenacity of binding. Increasing the number of times that slightly different species may rebind to the exchanger allows these slight differences to be made manifest. Hence increasing the column length, and as a result the number of opportunities of rebinding, increases resolution. This would also apply in the case of a very shallow gradient.

The Gradient

The gradient employed to elute various ions from an exchanger is the most important parameter to be manipulated during ion exchange chromatography. Its size (volume with respect to the column volume) and shape (variability in salt concentration) profoundly affect the resolution of the column.

A gradient is simply a mechanical means of constantly changing the salt concentration of a solution that is being passed through the column. In this discussion a linear gradient is used as an example. The usual device for construction of linear gradients is shown in Figure 4-8. It is composed of two vessels of equivalent volume connected to one another. Stopcocks are situated at the exit of each vessel, and the vessel connected to the column is provided with some means of being stirred. In the stirred vessel is placed a solution whose salt concentration is that desired at the beginning of the gradient. In the other vessel is placed a solution whose salt concentration is much higher and is that desired at the end of the gradient. The depth of solution in each vessel must be exactly the same because gradient formation is initiated by gently opening the stopcock between the two vessels. As solution from the stirred vessel slowly passes through the column the depth of solution in the stirred vessel gradually decreases. However, since the two vessels are connected, the more concentrated salt solution from the unstirred vessel enters the stirred vessel in order to maintain equivalent heights of solution in each vessel. As a result the salt concentration in the solution of the stirred vessel is constantly and linearly increased. For linear gradients the concentration, c of the eluent passing over the column at a given time is related to the volume of solution, v, which has already passed from the gradient mixer over the column by the expression

$$c = \left(\frac{C_a - C_b}{V}\right)(v) + C_b$$

where C_a and C_b are the concentrations of solute in the unstirred and stirred vessels of the gradient mixer, respectively, and V is the total

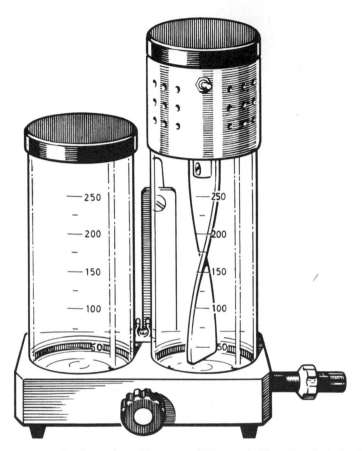

Figure 4-8.   Gradient mixer. (Courtesy of Pharmacia Fine Chemicals, Inc.)

volume of the gradient (usually 5 to 10 times the volume of the column).

Five gradient shapes that may be employed are (1) linear, (2) convex, (3) concave, (4) complex, and (5) stepped. Figure 4-9 depicts the type of apparatus needed to produce the first three gradient shapes and the behavior of the salt concentration as a function of the effluent volume that can be expected with each apparatus. Complex gradients are produced using a multichambered mixing device described in detail in reference 20. Stepped gradients, which find greatest application in large scale preparative procedures where behavior of the desired ions on a given exchanger have been well characterized, are produced by washing the ion exchange column sequentially with successively more concentrated solutions. The shape of gradient that is most desirable depends entirely on the particular

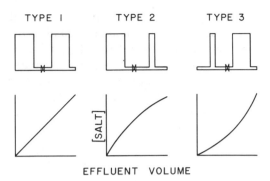

Figure 4-9. Three devices for production of gradients and the variety of gradient obtained with each one.

application, and therefore little generalization is possible. It is usually best to start with a linear gradient and proceed by trial and error. Decisions concerning appropriate concentrations to be used in a stepped gradient should be made conservatively and on the basis of data obtained by using linear gradients.

In addition to the shape of a gradient it is also possible to vary its volume with respect to the volume of the column. Such manipulation is in fact manipulation of the effective slope of the gradient. In general, decreasing the slope of a gradient (increasing its total volume) results in increased resolution. This is exemplified in Figure 4-10, which depicts the results obtained when a mixture of glucose oxidase and catalase is separated using DEAE cellulose and three linear gradients with volumes of 150, 300, and 600 ml. The increase in resolution is clearly evident. A

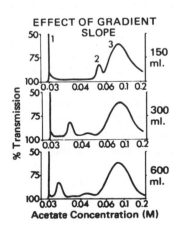

Figure 4-10. Effect of gradient slope on the resolution of catalase and glucose oxidase chromatographed on a DEAE cellulose column. (Copyright W. and R. Balston, Ltd., Maidstone, Kent, England.)

liability of decreasing the slope of a gradient (not apparent in Figure 4-10) is that the eluted protein peaks are usually broadened. This results in greater dilution of the sample. Therefore, if satisfactory resolution is obtained, but the observed peaks are too broad, increasing the slope of the gradient sharpens them considerably. An increase in slope may be accomplished by decreasing the total volume of the gradient as discussed above or by maintaining a constant volume and increasing the concentration of salt in the high salt reservoir (i.e., the reservoir that is not connected to the column).

Composition of the gradient is rather restricted, falling essentially into the following three categories. (1) Gradients of increasing ionic strength are usually composed of a simple salt such as sodium or potassium chloride, dissolved in a dilute solution of buffer. Whenever possible it is advisable not to use the salts of weak acids and bases as agents for increasing the ionic strength because they will participate in the buffering of the system and alter the pH as the gradient proceeds. There are times, however, when this cannot be avoided. The most significant example is the situation encountered when a gradient composed of volatile salts is desired; all the volatile salts are salts of weak acids and bases. (2) Gradients of changing pH are usually produced by mixing two buffers of different pH and/or capacity. Selection of the buffers to be used in this type of gradient is quite critical and special attention must be paid not only to the pH of the components, but also to their buffer capacity over the range of pH values being used. If allowances are not made for these factors the pH gradient observed in practice is very different from that expected on the basis of the pH values of the initial components. Unlike an ionic strength gradient, which is always increasing regardless of the type of exchanger used, a pH gradient should increase from a low value if a cation exchanger is used and decrease from a high value if an anion exchanger is used. The actual range of pH values that are permissible depends on the effects of pH upon the ions being isolated. (3) An additional component of gradient solutions must be considered in the chromatography of proteins. A variety of nonionic compounds, such as glycerol, sucrose, polyethylene glycol, and mercaptoethanol, may be present to stabilize the proteins during the chromatographic process. If such materials are included in the gradient solutions, the column should be fully equilibrated with them before use. This is also true when using the substrate of a given enzyme at low concentration for stabilization purposes (see Chapter 10).

Column Elution

The rate of elution is largely dictated by the exchanger used. There are two things to be kept in mind when selecting the flow rate of a column. (1) Ions are adsorbed to an exchanger more slowly than they are desorbed. Therefore, it is advisable to load the sample onto the exchanger at a considerably slower rate than will be used for its elution. (2) Too great a departure from an experimentally determined optimum rate of elution greatly decreases the resolution of the column. If the flow rate is faster than the rate of desorption, the peak being eluted smears and if it is too slow, diffusion occurs. The result in both cases is loss in resolution.

Sample Size

The final area of concern in the chromatographic process is the size of the fractions. Decreasing the fraction size does not increase the resolution of the column, but does allow realization of the full resolution capability of the column to be expressed.

ION EXCHANGE TECHNIQUES FOR THE ASSAY OF ENZYMES

Generally ion exchange techniques are considered as qualitative tools. If they are applied with appropriate care, however, they yield quantitative results as depicted in Figure 4-11. Each of the malate and succinate points in this figure was obtained by summing the amount of radioactivity

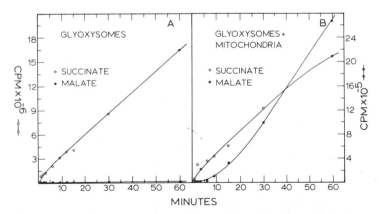

Figure 4-11. Data obtained through use of Dowex-1-formate columns for the separation of succinate, malate, and isocitrate. [From T. G. Cooper and H. Beevers, *J. Biol. Chem.*, **244**:3507–3513 (1969).]

located in appropriate areas (see Figure 4-15) of Dowex-1-formate chromatograms. The usefulness of ion exchange techniques for assay of an enzyme occurs most often in situations where the substrate of an enzymatic reaction is charged and the product uncharged or, conversely, the substrate is uncharged and the product is charged. Such assays necessitate using a large number of ion exchange columns simultaneously. This is easily accomplished employing disposable Pasteur pipettes for columns and the apparatus shown in Figure 4-12 as a column support.

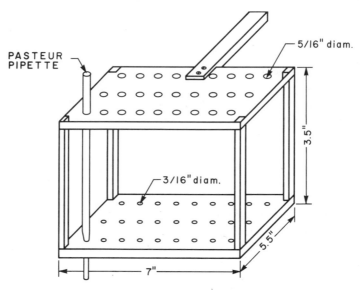

Figure 4-12.   Column support for Pasteur pipette mini-columns.

An alternative to this is use of ion exchange filter paper disks. Such filter paper is cellulose to which either positive or negative functional groups have been covalently bonded as described earlier. The technique involves placing a sample of the reaction mixture on the filter disk, which is then thoroughly washed to remove uncharged materials. One advantage to this procedure is the fact that all the numbered filtered disks may be washed simultaneously by swirling them in a beaker of solution. Following the wash procedure, filters are dried and the radioactivity they contain is determined by standard procedures.

EXPERIMENTAL

The Separation of Organic Acids on Dowex Resins

Preparation of the Anionic Exchange Resin

4-1.   Prepare 1 lb Dowex-1-chloride X-8 resin, 200–400 mesh. The same initial procedures can be used for any of the polystyrene type resins.

4-2.   Suspend the resin in 2 liters distilled water.

4-3.   When the resin has almost completely settled, decant off the liquid containing the fines (small particles that settle only very slowly).

4-4.   Repeat step 4-3 five to seven times.

4-5.   Rinse the resin twice with 1 liter 95% ethanol to remove ethanol-soluble impurities.

4-6.   Rinse twice with 1 liter $2N$ HCl.

4-7.   Suspend the resin in $2N$ HCl and heat to 100°C. Repeat this operation three to four times or until supernatant is clear and colorless, using fresh HCl each time. Allow the resin to cool for 1 hour each time before repeating the heating step.

4-8.   Divide the resin into two equal parts; treat one part according to steps 4-9 to 4-15 and the second part according to step 4-16.

4-9.   Suspend the resin in 500 ml $2N$ acetic acid and heat to 60–70°C for 5 minutes.

4.10.   Repeat step 4-9 three times or until the supernatant is clear.

4-11.   Pack the resin into a large column (on the order of 3.5×60 cm).

4-12.   Pass 3 liters $1M$ sodium acetate over the column.

4-13.   Pass 1 liter $0.1M$ acetic acid over the column.

4-14.   Wash the column with water until the pH of the effluent is the same as that of the water being added.

4-15.   Collect resin into a bottle and store at 4°C.

4-16.   Repeat steps 4-9 to 4-15, but use formic acid and sodium formate, at the concentrations indicated, in place of acetic acid and sodium acetate.

Pouring a Column

4-17.   In the following experiments use a 1.0×30.0 cm column containing a fritted glass base (Kimax 28570 or equivalent).

4-18.   Attach a 5 cm piece of $\frac{3}{16}$ in. i.d. rubber tubing to the bottom of the column.

4-19.   Into the other end of the rubber tubing insert a 6 in. disposable Pasteur pipette which has its top 8 cm cut off.

4-20. Position the column vertically on a ring stand.

4-21. Add sufficient glass beads to the column to make a layer 0.5 to 1.0 cm deep. Wash the glass beads beforehand using the following procedure. Soak fine glass beads (0.2–0.3 mm diameter) for 72 hours in a fourfold (v/v) excess of concentrated HCl. Wash the beads with water until pH of the wash is neutral. Dry beads at 120°C overnight in shallow dishes. These beads are most conveniently prepared in 1–10 lb batches.

4-22. Fill the column half full of water and add sufficient resin to make a bed height of 10 cm.

4-23. Wash the resin off the sides of the column with water. If the resin is not sufficiently fined, small particles will stick to the glass, resulting in poor resolution. Pack the resin by placing a 5–10 cm column of water on top of the resin and forcing the water through the resin under positive air pressure (the positive pressure can be provided by connecting the top of the column to a regulated air line). Caution should be used at this point, because if the connections are too tight or the pressure too high the column may burst. The safest way to connect the air line to the column is via a rubber stopper that fits loosely in the top of the column. If the pressure builds to an excessive level, the stopper will pop out, thus serving as a safety valve.

4-24. Allow the column to drain and, when the level of the liquid reaches the top of the resin, add the mixture of organic acids (5 mg each in 5 ml solution adjusted to pH 7.9) to be separated.

4-25. Allow the solution to drop once again to the top of the resin.

4-26. Just as the level of the liquid drops below the surface of the resin, gently add 5 ml water and repeat step 4-25.

4-27. When this wash is completed add water to a depth of 3 cm from the top of the column.

4-28. Clamp the rubber tube at the bottom of the column (arrow in Figure 4-13) and transfer the column to a fraction collector. This may be done before loading the sample if desired.

Development of the Column

4-29. Assembly of the column and gradient maker are shown in Figure 4-13.

4-20. Place a rubber stopper (size 00 or 000 depending on size of glass tube) in the top of the column. This stopper should contain a hole through which is inserted a small glass tube.

4-31. Connect the glass tube protruding from the stopper to a type 1 gradient maker (see Figure 4-9 for the various types) by means of a piece

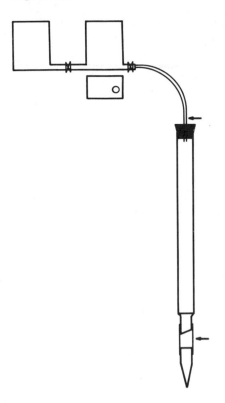

Figure 4-13. Assembly of an ion exchange column, gradient mixer, and magnetic stirring motor. Arrows indicate points where the column is closed with pinch clamps.

of rubber tubing not longer than 25 to 35 cm. The smaller the diameter of this tube, the better. The glass tube cited in step 4-30 may be replaced with an 18 gage needle. This will permit the use of small diameter plastic tubing and Teflon stopcocks in the assembly (Ace Glass, Incorporated, catalogue numbers 5851 and 5852 or equivalent).

4-32. Check to see that the stopper can be inserted tightly enough so that it has no air leaks. What would happen if it leaked during the chromatography?

4-33. Prior to addition of the components of the gradient into the gradient maker, fill one side of the maker with water. Open the valve between the vessels and allow the water to run from one side to the other for about 10 seconds. While the water is still flowing, close the valve between the two vessels of the gradient maker. This procedure removes any bubbles trapped in the valve.

4-34. Open the valve between the stirred side of the gradient maker and the column and allow the tube leading to the column to fill. The distance between the top of the liquid in the gradient maker and the outlet of the column should be adjusted to give a flow rate of 40–50 ml/hour. This may be done by raising or lowering the gradient maker with respect to the column.

4-35. Clamp the bottom of the rubber tubing just above the connection with the glass tube which protrudes from the rubber stopper (arrow in Figure 4-13).

4-36. Gently empty the water from both sides of the gradient maker.

4-37. The gradient to be used is a 500 ml, 0–5N linear formic acid gradient. Place 250 ml water in the stirred side (side connected to the column) of the gradient maker and 250 ml 5N formic acid in the unstirred side. Be sure that the gradient maker is perfectly level.

4-38. Position a magnetic stirrer below the gradient maker, leaving a $\frac{1}{2}$ in. air space between the two (this air space prevents transmission of heat from the stirring motor to the gradient solutions). Center the stirring flea in the vessel to be stirred by positioning the stirring motor appropriately. This may be disregarded if a top-stirring model is used (see Figure 4-8).

4-39. Turn on the magnetic stirrer and adjust the speed so that the contents of the vessel are well stirred but no vortex is forming. What would be the result of vortex formation? Up to 30 minutes may be required for variation in the speed of the stirring motor to cease. During the 30 minute period the speed of the motor will slowly but continuously increase. It is therefore advisable to continuously operate the motor for $\frac{1}{2}$ to 1 hour before it is needed.

4-40. Insert the stopper in the top of the column and check to be sure that the connection is both air- and watertight.

4-41. Open the bottom of the column and let the solution begin to drop into the tubes.

4-42. After a few seconds unclamp the rubber tube connecting the column and gradient maker.

4-43. Gently open the valve between the stirred side of the gradient maker and the tubing.

4-44. When it is certain that everything is flowing well, gently open the valve between the two sides of the gradient maker. Why is it advisable to do this last?

4-45. Check to see that the solution is flowing between the two sides of the gradient maker. Do this by looking for the schlieren patterns that exist at the entry of the formic acid into the water.

4-46. Once gradient formation has begun, leaving everything alone. The key to making smooth, linear gradients is not to touch the apparatus once the gradient has begun to form.

4-47. Recheck the speed of the stirring motor during the run. The speed must be gradually and gently decreased as the level of the liquid in the gradient maker decreases.

4-48. The fraction sizes should be approximately 5.0 ml or 200–250 drops each.

Analysis of the Fractions

4-49. Place the eluted fractions in test tube racks.

4-50. Place these racks in a 37°C water bath and insert the manifold tubes shown in Figure 4-14 so that the ends of the tubes are 2 to 3 cm above the liquid surface. With a large water bath up to 200 tubes (20 manifolds) can be evaporated at once. Evaporation may also be accomplished with a commercial concentrator such as the Brinkmann SC/48 Sample Concentrator.

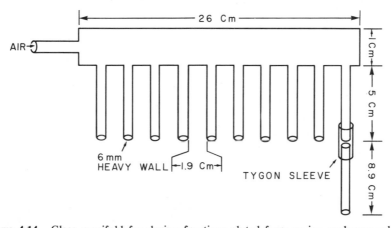

Figure 4-14. Glass manifold for drying fractions eluted from an ion exchange column.

4-51. Gently turn on the air until the surface of the liquid is disturbed but does not spatter.

4-52. Leave the tubes evaporating overnight (8–12 hours is sufficient). This cannot be done when acetate and lactate are being isolated because they will evaporate along with the formic acid. At this point samples may be stored in a dry form.

4-53. Add exactly 2 ml water to each tube.

4-54. Add 2 drops of 5% phenolphthalein (5 g phenolphthalein dissolved in 100 ml absolute ethanol) and titrate each tube to an end point with a standard solution of 5×10^{-3} to $10^{-2} M$ NaOH.

4-55. If radioactive organic acids were used, proceed through step 4-53 and then transfer an appropriately sized sample to a scintillation vial.

4-56. Add 5 to 15 ml of one of the aqueous scintillation fluids discussed in Chapter 3 and count the vials in a scintillation counter.

4-57. Data yielded from this type of experiment appear in Figure 4-15. A large variety of acids have been chromatographed in order to provide a basis for selecting those most useful to the reader.

Separation of Amino Acids from Organic Acids on Dowex Resins

Preparation of the Cationic Exchange Resin

4-58. Prepare 1 lb of Dowex-50 X-8 as described in steps 4-2 to 4-7 above.

4-59. Pack the resin in a large column (3.5×60 cm or equivalent).

4-60. Wash with water until the pH of the effluent is the same as that of the water added.

4-61. Collect the resin in a brown bottle and store at 4°C.

Pouring and Development of the Columns

4-62. Pour a 1×10 cm column of the cation exchanger using steps 4-17 to 4-23 above.

4-63. When the level of the liquid added in step 4-62 reaches the top of the resin, add the mixture of amino and organic acids (5 mg each in 5 ml solution adjusted to pH 7.0) to be separated.

4-64. Allow the solution to drop once again to the top of the resin. Collect the effluent.

4-65. Gently add 5 ml water and repeat step 4-64.

4-66. Repeat step 4-65 three times.

4-67. The combined effluents from steps 4-64 to 4-66 contain, quantitatively, all the organic acids added to the column. This can be verified by using a radioactively labeled organic acid as one of the acids to be separated.

4-68. The combined organic acid containing effluents may be passed through a Dowex formate column for separation (steps 4-1 through 4-57) without further treatment.

4-69. The amino acids may be eluted as a group by repeating steps 4-64 to 4-67 using 20 ml $4N$ NH₄OH in place of water.

4-70. Evaporate the combined effluents from step 4-69 to dryness using a flash evaporator (Buchler Instruments, NO. FE-2, PTFE-1GN or equivalent).

4-71. Dissolve the residue in 10 ml water.

4-72. Pass the solution from step 4-71 through a 1×10 cm column of

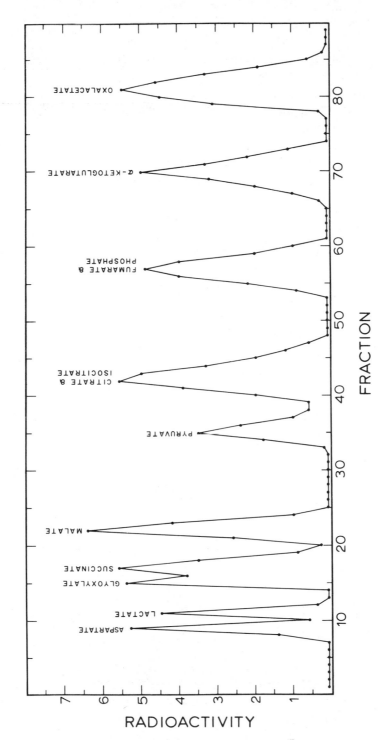

Figure 4-15. Separation of common organic acids using Dowex-1-formate as the ion exchange resin.

Dowex-1-acetate X-8 (from step 4-15). The column may be prepared by repeating steps 4-17 to 4-28.

4-73. Repeat steps 4-64 to 4-66.

4-74. All the amino acids except glutamate and aspartate appear quantitatively in the effluent.

4-75. Glutamate and aspartate may be sequentially eluted (see Figure 4-16) by subjecting the column to linear gradient elution using a 600 ml 0–2N acetic acid gradient. This is done by applying steps 4-29 through 4-48, but using 2N acetic acid instead of 5N formic acid.

Analysis of the Amino Acid Fractions

4-76. Repeat steps 4-49 through 4-53 on the fractions yielded from step 4-75.

4-77. The fractions may be analyzed spectrophotometrically using the ninhydrin reaction or spectrofluorometrically using fluorescamine (Chapter 6). If radioactive glutamate or aspartate is used, the radioactivity in an appropriate aliquot should be determined with a scintillation counter (Chapter 3).

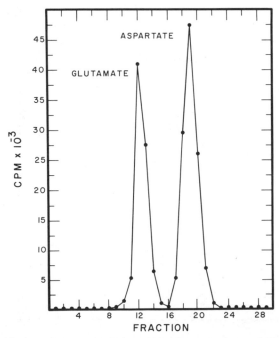

Figure 4-16. Elution of glutamate and aspartate from a column of Dowex-1-acetate.

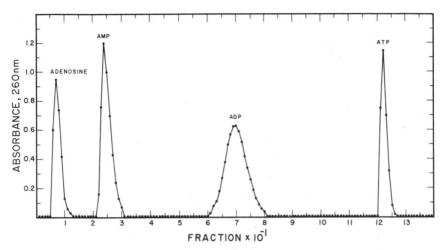

Figure 4-17. Elution of adenosine, AMP, ADP, and ATP from a column of Dowex-1-formate.

Separation of Nucleotides on Dowex Formate Columns

4-78. Prepare Dowex formate columns according to steps 4-1 to 4-8, 4-16, and 4-17 to 4-23.

4-79. Allow the column to drain and, when the level of the liquid reaches the top of the resin, add the mixture of the nucleotides (10 mg each of adenosine, AMP, ADP, and ATP in 1 ml solution adjusted to pH 7.9) to be separated.

4-80. Uncharged adenosine may be washed from the column with water (see Figure 4-17) using steps 4-65 to 4-66.

4-81. Repeat steps 4-29 to 4-48. This procedure ($0–5N$ linear formic acid gradient) removes the AMP and ADP.

4-82. Wash the column (using steps 4-65 to 4-66) using $5N$ formic acid containing $0.8N$ ammonium formate. This removes the ATP.

4-83. A sample from each fraction is transferred to a cuvette and its absorbance at 260 nm is determined. Data from this type of experiment appear in Figure 4-17. The orcinol reaction may be used to identify the nucleotide elution peaks (see Chapter 2 for the appropriate procedures).

Assay of Acid Phosphatase Using Mini-Ion Exchange Columns

4-84. Using a clean, wooden applicator stock, position a very small piece of nonabsorbent cotton in the narrow end of a 6 in. disposable Pasteur pipette, but do not pack the cotton. It is important that the cotton be nonabsorbent because absorbent cotton significantly decreases flow rates.

4-85. Place eight pipettes prepared in this manner into a holder such as that shown in Figure 4-12.

4-86. Place approximately 1 ml water into each pipette. This can be done with a wash bottle.

4-87. Add 1 ml of a heavy suspension of Dowex-1-formate (prepared as described in steps 4-1 to 4-8 and 4-16) to each of the columns.

4-88. Wash the resin from the sides of the column with water and allow the column to settle. Do not, however, allow the column to dry out. This can be prevented by small additions of water.

4-89. Dissolve 30.4 mg glucose-6-phosphate in 1 ml water (the final concentration is 100mM).

4-90. Dissolve 1.0 mg commercially prepared acid phosphatase (0.5 units/mg) in 1.0 ml 0.1M citrate buffer (pH 5.5).

4-91. Prepare a solution of ^{14}C-glucose-6-phosphate which contains 0.5 to 1.0 μc of carrier-free, radioactive material/0.025 ml.

4-92. Place the following components in a conical test tube: water, 0.75 ml; glucose-6-P from step 4-89, 0.1 ml; and ^{14}C-glucose-6-P from step 4-91, 0.025 ml

4-93. Add 0.05 ml of the enzyme solution prepared in step 4-90 to the assay mixture, mix thoroughly, and begin timing the incubation with a stopwatch or suitable timer.

4-94. At 0, 1, 3, 6, 9, 12, 15, 18, and 21 minutes remove a 0.1 ml sample from the reaction mixture and transfer it to a cold 13 × 100 mm test tube containing 1 ml 0.01M KOH Take care to use a clean pipette for each sample to avoid contaminating the reaction mixture with KOH.

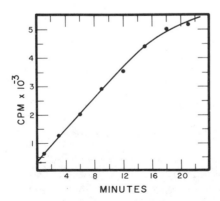

Figure 4-18. Time dependent production of ^{14}C-glucose from ^{14}C-glucose-6-phosphate. The arrow indicates the amount of radioactivity in the sample removed at zero time (just before addition of enzyme).

4-95. Transfer each sample to a mini-column using a Pasteur pipette. When this solution has washed onto the column, wash each tube with 1 ml water and transfer this to the column also. As these two solutions leave the column collect them in a scintillation vial. Add 10 to 15 ml aqueous scintillation fluid to the vial and set it aside.

4-96. Repeat this washing procedure twice using 2 ml water each time. This should give a total of three scintillation vials per sample.

4-97. Allow the scintillation vials to stand for 16 hours before determining the amount of radioactivity they contain.

4-98. Data yielded from this type of experiment are shown in Figure 4-18. The same type of assay may also be used for $5'$-nucleosidase catalyzing the reaction AMP $\longrightarrow$ adenosine $+ P_i$.

REFERENCES

1. J. Leggett Bailey, *Techniques in Protein Chemistry*, 2nd ed., American Elsevier, New York, 1967.
2. G. R. Bartlett, *J. Biol. Chem.*, **234**:449–458 (1959). Human Red Cell Intermediates.
3. G. R. Bartlett, *J. Biol. Chem.*, **234**:459–465 (1959). Methods for the Isolation of Glycolytic Intermediates by Column Chromatography with Ion Exchange Resins.
4. G. R. Bartlett, *J. Biol. Chem.*, **234**:466–469 (1959). Phosphorus Assay in Column Chromatography.
5. I. Calmon and T. R. E. Kressman, *Ion Exchangers in Organic and Biochemistry*, Interscience, New York, 1957.
6. T. G. Cooper, *J. Biol. Chem.*, **246**:3451 (1971). The Activation of Fatty Acids in Castor Bean Endosperm.
7. T. G. Cooper and H. Beevers, *J. Biol. Chem.*, **244**:3507 (1969). Mitochondria and Glyoxysomes from Castor Bean Endosperm. Enzyme Constituents and Catalytic Capacity.
8. H. A. Flaschka and J. R. Barnard, *Chelates in Analytical Chemistry*, Vol. 1, Marcel Dekker, New York, 1967.
9. R. K. Gerding and R. G. Wolfe, *J. Biol. Chem.*, **244**:1164 (1969). Malic Dehydrogenase. VIII. Large Scale Purification and Properties of Supernatant Pig Heart Enzyme.
10. F. Helfferich, *Ion Exchange*, McGraw-Hill, New York, 1962.
11. Erich Heftmann, Ed., *Chromatography*, 2nd ed., Reinhold, New York, 1967.
12. J. Inczedy, *Analytical Application of Ion Exchangers*, Pergamon, New York, 1966.
13. J. X. Khym and L. P. Zill, *J. Am. Chem. Soc.*, **74**:2090 (1952). The Separation of Sugars by Ion Exchange.
14. R. Kunin, *Ion Exchange Resins*, Wiley, New York, 1958.
15. E. Lederer and M. Lederer, *Chromatography*, Elsevier, Amsterdam, 1957.
16. J. A. Marinsky, *Ion Exchange*, Vol. 1, Marcel Dekker, New York, 1966.

17. C. J. O. R. Morris and P. Morris, *Separation Methods in Biochemistry*, Pittman, New York, 1964, pp. 228–364.

18. I. I. Ohms, J. Zec, J. V. Benson, and B. Patterson, *Anal. Biochem.*, **20**:51–57 (1967). Column Chromatography of Neutral Sugars: Operating Characterisristics and Performance of a Newly Available Anion-Exchange resin.

19. J. K. Palmer, *Conn. Exp. Stn. Bull.*, **589**:3–31 (1959). Determination of Organic Acids by Ion Exchange Chromatography.

20. E. A. Peterson, *Cellulosic Ion Exchangers*, American Elsevier, New York, 1970.

21. T. Shima, S. Hasegawa, S. Fujimura, H. Matsubara, and T. Sugimura, *J. Biol. Chem.*, **244**:6632–6635 (1969). Studies on Polyadenosine Diphosphate-ribose.

22. I. Zelitch, *J. Biol. Chem.*, **233**:1299–1303 (1958). The Role of Glycolic Acid Oxidase in the Respiration of Leaves.

23. I. Zelitch, *J. Biol. Chem.*, **240**:1869–1876 (1965). The Relation of Glycolic Acid Synthesis to the Primary Photosynthetic Carboxylation Reaction in Leaves.

Chapter 5

Gel Permeation
Chromatography

Complementing methods for separation of macromolecules on the basis of charge are separation procedures based on molecular size. The most often used of these procedures is molecular sieve or gel permeation chromatography. Widespread application of gel chromatography results from its many advantages: (1) gentleness of the technique permitting separation of labile molecular species, (2) solute recovery approaching 100%, (3) high reproducibility, (4) a broad range of sample sizes from analytical to pilot plant quantities, and (5) relatively short times and inexpensive equipment needed for its performance. Samples differing in molecular size by 25% may be totally separated by this technique on a single gel bed.

MODE OF OPERATION

Although molecular sieve chromatography has been studied in detail (1–9), its mechanism of operation is not yet precisely understood. The various media that can be used for this purpose are spherical beads composed of spongelike matrices containing pores of relatively restricted size distributions. When a mixture of different sized molecules are placed on top of a column containing these beads, the larger molecules cannot easily diffuse into the pores and are eluted from the column with little or no resistance (Figure 5-1). The small molecules diffuse into the pores of the gel beads and are thereby effectively removed from the stream of eluting buffer. Therefore, they stop moving down the column until they diffuse out of the gel matrix. If this process is repeated many times, small molecules are retarded from leaving the column. The degree of retarda-

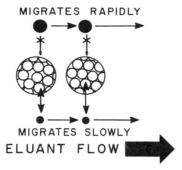

MIGRATES RAPIDLY

MIGRATES SLOWLY

ELUANT FLOW

Figure 5-1. Separation of large and small molecules using gel permeation chromatography. Molecules are represented schematically as large and small filled circles.

tion is determined by how much time a molecule spends inside the gel pores, which is a function of the molecule's size and the pore's diameter. A molecule whose Stokes' radius approaches or exceeds the pore radius does not enter the gel and is said to be excluded. The exclusion limit of a gel, then, is the molecular weight of the smallest molecule incapable of penetrating the gel pores. Molecular weights are used in place of more precise Stokes' radii to indicate exclusion limits because the latter parameter is often unavailable. However, molecular weights do not always accurately reflect the Stokes' radii; thus a given gel has a lower exclusion limit for linear polysaccharides or fibrous proteins than for globular proteins (see ref. 22). The lower limit for effective use of molecular sieve gels is usually about 10% of their exclusion limits.

Although comprehensive mathematical treatment of solute behavior on

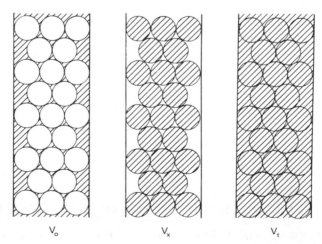

V_o V_x V_t

Figure 5-2. Variables used to characterize the chromatographic bed. (From L. Fischer, *An Introduction to Gel Chromatography*, North-Holland, Amsterdam, 1969).

molecular sieve gels is not necessary for most applications, a few simple and useful relationships are described here. These relationships provide a means of describing the behavior of a solute on a given gel column. As shown in Figure 5-2 the total volume of a gel bed, V_t, is composed of space occupied by the gel beads, V_x, and space occupied by solvent surrounding the gel beads, V_0, which is called the void volume and is usually about 35% of the total bed volume, V_t.

$$V_t = V_0 + V_x \tag{1}$$

The elution volume, V_e, of a solute is the amount of effluent solution exiting from the column between the time that solute first penetrates the surface of the column and it appears in the effluent. In practice V_e is measured from either the half maximum peak height of the leading edge of a solute peak or by an extrapolation of the leading side of a peak to the base line (see Figure 5-13). This is more reliable than the peak crest because the position of that crest depends on the sample volume. These parameters are commonly used in three ways to describe the degree to which a given solute is retarded: (1) the relative elution volume, REV,

$$REV = \frac{V_e}{V_0} \tag{2}$$

(2) the retention constant, R, which is the inverse of REV,

$$R = \left(\frac{V_e}{V_0}\right)^{-1} = \frac{V_0}{V_e} \tag{3}$$

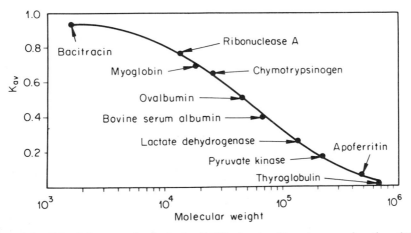

Figure 5-3. Selectivity curve for Sephadex G-200, showing K_{av} values as a function of the molecular weights for several proteins. (Courtesy of I. M. Easterday, Pharmacia Fine Chemicals, Inc.)

and (3) the partition or distribution coefficient, K_d or K_{av},

$$K_{av} = \frac{V_e - V_0}{V_x} \qquad (4)$$

Usefulness of partition coefficients derives from the empirical observation (Figure 5-3) that they may be related to the molecular weight of a solute. Although these observations have been analyzed extensively it must be conceded that as yet they cannot be totally accounted theoretically. When a plot such as that in Figure 5-3 is used to make an estimate of molecular weight, it is advisable to use only the linear region of the plot and to include standard proteins which have molecular weights above and below that of the unknown.

GEL FILTRATION MEDIA

The common use of gel filtration is largely because of technical advances made in production of suitable media for this purpose. The characteristics desired of gel filtration media include (1) chemical inertness of the gel matrix, (2) low content of ionic groups, (3) uniform pore and particle size, (4) wide choice of gel particle and pore sizes, and (5) high mechanical rigidity. Five principal types of media fulfill these criteria to varying degrees. Among the earliest gels developed were the Sephadex® polydextran gels. These gels are produced by permitting a microorganism, *Leuconostoc mesenteroides*, to ferment sucrose into large polymers of glucose. The polyglucose units are purified and cross-linked by treatment with epichlorohydrin into various classes of gel beads with exclusion limits between 1 and 200,000 daltons. Epichlorohydrin introduces glyceryl groups which cross-link the polyglucose units. Pore size is controlled by the molecular weight of the dextran and the amount of epichlorohydrin used in the preparation. These gels are identified by a number such as G-10 or G-200, which refers to the water regain value of the gel multiplied by a factor of 10 (Tables 5-1 and 5-2).

The second type of gel medium is produced by polymerizing acrylamide into bead form. A detailed discussion of the control of the bead and pore size of these gels is deferred to Chapter 6. Polyacrylamide gels (Bio-Rad®) are usually identified by a number such as P-10 or P-100, which indicates the exclusion limit of the gel in thousands of daltons. Large pored polydextran and polyacrylamide gels can be used to separate solutes only up to 300,000 daltons and may be difficult to use owing to their lack of mechanical rigidity. As a result of being pliable, they tend to compress in the column, causing unacceptably slow flow rates.

Table 5-1. Physical Characteristics of Polydextran Gels

Designation	Mesh	Particle Diameter[a] (μ)	Fraction-ation Range (MW)	Water Regain (ml/g dry gel)	Bed Volume (ml/g dry gel)
Sephadex G-10		40–120	700	1.0 ± 0.1	2–3
Sephadex G-15		40–120	1,500	1.5 ± 0.2	2.5–3.5
Sephadex G-25	Coarse	100–300	1,000–5,000	2.5 ± 0.2	4–6
	Medium	50–150			
	Fine	20–80			
	Superfine	10–40			
Sephadex G-50	Coarse	100–300	1,500–30,000	5.0 ± 0.3	9–11
	Medium	50–150			
	Fine	20–80			
	Superfine	10–40			
Sephadex G-75		40–120	3,000–70,000	7.5 ± 0.5	12–15
	Superfine	10–40			
Sephadex G-100		40–120	4,000–150,000	$10. \pm 1.0$	15–20
	Superfine	10–40			
Sephadex G-150		40–120	5,000–400,000	$15. \pm 1.5$	20–30
	Superfine	10–40			
Sephadex G-200		40–120	5,000–800,000	$20. \pm 2.0$	30–40
	Superfine	10–40			

[a] The diameter diven is the dry particle diameter.

These liabilities prompted development of agarose and agarose–acrylamide gels (Table 5-3). Early versions of agarose gels were very poor, because they were prepared from whole or only partially separated agar, which is composed of uncharged agarose and negatively charged agaropectin. The latter compound contains covalently bonded sulfate and carboxylate groups, which impart an unacceptably large amount of charge to the gel. However, total separation of the two components has now been accomplished, making possible production of an uncharged agarose gel. Beads of agarose are not covalently polymerized, but rather are held together by hydrogen bonds. In spite of this, these gels are quite resistant to hydrogen bond breaking agents such as urea and guanidine hydrochloride, raising the possibility that additional unidentified forces are responsible for maintenance of their structural integrity. Although agarose gels differ in bead structure from polydextran gels, their partition coefficients are also a function of molecular size. Recently, gels combining

Table 5-2. Physical Characteristics of Polyacrylamide Gels

Designation	Mesh	Particle Diameter (μ)	Fractionation Range (MW)	Water Regain (ml/g dry gel)	Bed Volume (ml/g dry gel)
Bio-Gel P-2	50–100	150–300	200–2,600	1.5	4
	100–200	75–150			
	200–400	40–75			
	400	40			
Bio-Gel P-4	50–100	150–300	500–4,000	2.4	6
	100–200	75–150			
	200–400	40–75			
	400	40			
Bio-Gel P-6	50–100	150–300	1,000–5,000	3.7	9
	100–200	75–150			
	200–400	40–75			
	400	40			
Bio-Gel P-10	50–100	150–300	5,000–17,000	4.5	12
	100–200	75–150			
	200–400	40–75			
	400	40			
Bio-Gel P-30	50–100	150–300	20,000–50,000	5.7	15
	100–200	75–150			
	400	40			
Bio-Gel P-60	50–100	150–300	30,000–70,000	7.2	20
	100–200	75–150			
	400	40			
Bio-Gel P-100	50–100	150–300	40,000–100,000	7.5	20
	100–200	75–150			
	400	40			
Bio-Gel P-150	50–100	150–300	50,000–150,000	9.2	25
	100–200	75–150			
	400	40			
Bio-Gel P-200	50–100	150–300	80,000–300,000	14.7	35
	100–200	75–150			
	400	40			

agarose and polyacrylamide have become commercially available. These gels are claimed to yield high resolution in the presence of large amounts of hydrogen bond breaking agents and to give substantially increased flow rates over the present preparations. However, they have not been available long enough to critically assess these claims.

The last medium for gel filtration chromatography to be discussed is

Table 5-3. Physical Characteristics of Agarose Gels

Designation	Mesh	Particle Diameter[a] (μ)	Fractionation Range (MW $\times 10^{-6}$)	Agarose Concentration (%)
Sepharose 4B		40–190	0.3–3	4
Sepharose 2B		60–250	2–25	2
Bio-Gel A-0.5m	50–100	150–300	<0.010–0.5	10
	100–200	75–150		
	200–400	40–75		
Bio-Gel A-1.5m	50–100	150–300	<0.010–1.5	8
	100–200	75–150		
	200–400	40–75		
Bio-Gel A-5m	50–100	150–300	0.010–5	6
	100–200	75–150		
	200–400	40–75		
Bio-Gel A-15m	50–100	150–300	0.04–15	4
	100–200	75–150		
	200–400	40–75		
Bio-Gel A-50m	50–100	150–300	0.10–50	2
	100–200	75–150		
Bio-Gel A-150m	50–100	150–300	1->150	1
	100–200	75–150		

[a] Wet particle diameters.

controlled pore glass beads. These fine glass spheres are manufactured to contain large numbers of pores within a very narrow size distribution. Their total rigidity permits high flow rates without seriously compromising resolution. However, the fine glass particles have a tendency to absorb significant amounts of protein to their surfaces. One attempt to circumvent this problem involved covalent bonding of glycerol residues to the glass, but the absorption characteristics of the glass were not totally neutralized by this procedure. A second method which may work somewhat better is treatment of the glass beads with hexamethyldisilazane (19).

In addition to pore size, particle size also influences the resolution attainable. The smaller the bead size, the greater the resolution that can be expected (Figure 5-4). Increased resolution results from an increased surface area of gel being made available to the molecules moving down the column. This provides many more opportunities for a molecule to diffuse into the matrix with ensuing retardation. However, the increase in resolution is acquired at substantially decreased flow rates exhibited by

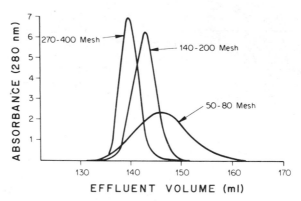

Figure 5-4. Effects of gel particle size upon elution profile of uridylic acid chromatographed on Sephadex G-25. Bed dimensions, 2×65 cm; flow rate, 24 ml/hour. [From P. Flodin, *J. Chromatogr.*, 5:103 (1961).]

the smaller particles. For most analytical applications, 100 to 200 mesh beads are sufficient. For finer separation the gel size may be decreased to 200 to 400 mesh. Conversely, for preparative or batch procedures the gel size may be increased to a mesh of 50 to 100.

PREPARATION OF THE MEDIUM

Polydextran and polyacrylamide gels are provided commercially in dry form and must be hydrated. This may be accomplished by slowly sprinkling the gel onto the surface of an excess amount of eluent solution while gently stirring with a glass rod. Tables 5.1 and 5.2 list water regain values for some of the available gels. The eluent solution should possess a minimum ionic strength of 0.08. This is necessitated by the presence of a few charged carboxyl groups in most permeation gels. Below this ionic strength the gel beads have an ion exchange characteristic that may alter the K_{av} of a solute. The ionic strength of a solution, μ, may be calculated using the equation

$$\mu = \tfrac{1}{2} \sum_i (C_i) \times Z_i^2 \qquad (5)$$

where C_i is the molar concentration of the ion and Z_i is the charge of each ion. Under no circumstances should a magnetic stirrer be used for stirring the gel during hydration because these devices are notorious for fractionating the gel beads, generating large amounts of fines. Gels may be

Table 5-4. Hydration Times of Various Gel Media

Gel	Minimum Hydration Time (hours)	
	At 22°C	At 100°C
Sephadex		
G-10	3	1
G-15	3	1
G-25	3	1
G-50	3	1
Sephadex G-75	24	3
Sephadex		
G-100	72	5
G-150	72	5
G-200	72	5
Bio-Gel		
P-2	4	2
P-4	4	2
P-6	4	2
P-10	4	2
Bio-Gel		
P-30	12	3
P-60	12	3
Bio-Gel		
P-100	24	5
P-150	24	5
Bio-Gel		
P-200	48	5
P-300	48	5

hydrated either at room temperature or in a boiling water bath. Table 5-4 indicates the time required for full hydration of the various gels.

Agarose gels are provided as heavy suspensions and need not be hydrated. Glass bead gels also need not be hydrated prior to use. Unlike polyacrylamide and polydextran gels, agarose gels may not be heated above 36°C. Heating above this temperature results in destruction of the gel structure.

Fines, the small particles of gel that escape removal during the manufacturing process or are generated during hydration by overly vigorous stirring, must be removed from the gel because they will severely decrease column flow rates. This may be done by gently suspending the gel in a two- to fourfold excess of eluent solution. This suspension is poured into a glass cylinder and the gel allowed to settle. When 90 to 95% of the gel has settled out, the remaining gel and supernatant solution are removed rapidly with suction. Decantation of the supernatant is not effective here, as it was for the fining of ion exchange resins, owing to the ease with which the gel is resuspended. This procedure is repeated usually twice or until a rapidly sedimenting preparation is obtained.

The final step in preparation of permeation gel medium is removal of air bubbles that might be present. Their removal is necessary because a bubble that dislodges during chromatography will rise to the column surface, mixing the gel and its contents. Removal of the bubbles is accomplished by transferring the hydrated and fined gel to a vacuum flask and placing it under a vacuum. A properly operating water aspirator provides sufficient vacuum, or a laboratory vacuum pump may be used. This process may be hastened by gently swirling the gel in the flask. Since air bubbles, especially carbon dioxide, are easily accumulated within the gel, it is advisable to degas it just prior to pouring the column.

PREPARATION OF THE COLUMN

Success of gel permeation chromatography requires careful packing of the gel into a column. The design characteristics of a suitable column include (1) a small dead space (the space between the medium support and the outlet of the column), (2) facilities for attachment of capillary tubing to the effluent fitting of the column, (3) a gel bed support that cannot be easily clogged (nylon screening is often used here), and (4) some means of protecting the bed surface. Fine glass beads and fritted glass plates should not be used in columns as gel bed supports, because glass beads tenaciously bind many proteins and fritted glass or glass wool generates fines by cutting the gel beads. The type of gel and column dimensions used depend on the column's anticipated use. Columns used to separate small molecules such as inorganic salts or metabolic intermediates (MW < 1500) from macromolecules (MW > 20,000) should have a volume 4 to 10 times greater than the projected sample volume and a height to diameter ratio of 5:1 to 15:1. This type of column, termed a desalting column, usually employs gels with very small pores (exclusion limits < 25,000). In contrast, columns used to separate one macromolecule from another should have a volume 25- to 100-fold greater than the expected sample volume and a height to diameter ratio of 20:1 to 100:1.

Columns with small diameters (≤ 1 cm) and wettable (glass) walls are subject to "wall effects," that is, diminished solvent flow near the walls. These effects, which are caused by friction between the walls and molecules moving past them, may seriously decrease resolution capabilities of the column. They may be overcome, however, by treating the column walls with dimethyldichlorosilane (20). This is done by coating the walls of a dry glass column with a 1% solution of dimethyldichlorosilane in benzene. After the walls have been in contact with the solution for 5 minutes at room temperature the benzene is permitted to evaporate and excess reagent is removed with a mild detergent. This treatment results in covalent attachment of the dimethyldichlorosilane to the glass and decreases friction encountered by materials as they move down the column walls. A liability incurred with this procedure is the possibility of decreased flow rates when large pore, soft gels are used; decreasing friction between the walls and gel particles permits greater compression of the gel structure by gravity.

Procedures used to pack gel permeation columns differ somewhat for small and large pored gels. However, the initial steps are common to both gel types. Fined and degassed gel and eluent solutions are first brought to the temperature at which the column will be used. Gel beds subjected to temperature changes during or after bed formation acquire severe irregularities caused by nonuniform expansion and contraction of the gel bed. A column whose bed volume has been previously determined is mounted, preferably with clamps at its top and bottom, on a sturdy support. A plumb line or level should be used to assure that it is perfectly vertical. One-half bed volume of eluent is added to the mounted column in order to check for leaks and to flush air bubbles from the space beneath the bed supporting screen. Bubbles are most easily removed by allowing slightly more than one-half of this solution to drain quickly from the column while gently tapping the bottom fittings with your hand. The remaining solution is left in the column to avoid bubbles being trapped beneath or at the screen when gel is added to the column. The amount of gel needed is measured and suspended in sufficient eluent to produce a slurry whose volume is twice that of the desired bed volume. This is poured gently down a glass rod into the column with care to avoid generating bubbles. A column extender or upper reservoir (Figure 5-5) is most useful here because it permits all of the slurry to be added at once.

Subsequent to this point separate procedures must be followed for large and small pored gels. With small pored gels that are quite rigid, the gel particles are permitted to settle 5 to 10 minutes with the column outlet closed. The outlet is then opened, permitting excess eluent to drain away.

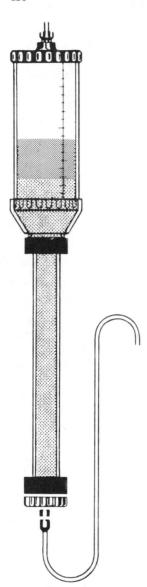

Figure 5-5. Gel permeation column with an attached upper buffer reservoir. (Courtesy of I. M. Easterday, Pharmacia Fine Chemicals, Inc.)

Failure to allow the column to drain during packing results in excessive convection currents (Figure 5-6). These currents will deposit heavy, coarse particles near the column walls and finer particles in the center of the column. Such an unevenly packed column does not flow uniformly and hence yields poor resolution. When the gel bed reaches the depth previously established as the bed volume, excess gel should be removed

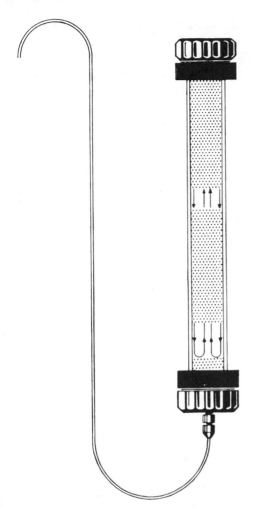

Figure 5-6. Convection currents established during sedimentation of gel particles. Arrows indicate directions of the currents. (Courtesy of I. M. Easterday, Pharmacia Fine Chemicals, Inc.)

from the column by aspiration. If, alternatively, additional gel is needed, it should be added before the gel has fully settled. Gel added to a previously established bed produces visible zones or banding, resulting in poor elution profiles. If gel must be added to an established bed, it is advisable to either repour the column or carefully stir the top 3 to 4 cm of the bed before further gel additions are made. The finished column is stabilized by allowing two or more bed volumes of eluent buffer to pass through it. A small amount of contraction may occur during this process.

An understanding of hydrostatic pressure is required prior to discussing the packing of a column with large pored gels. Hydrostatic pressure is measured as the vertical difference between levels at the top and bottom of a column of solution. Figure 5-7A and B depicts the hydrostatic or operating pressure of two chromatographic assemblies (distance between the broken lines). This figure amply illustrates that hydrostatic pressure is not a function of the length of the column of solution itself, but rather the

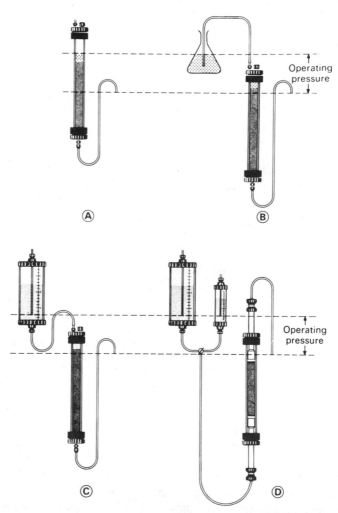

Figure 5-7. Operating or hydrostatic pressure for various column assemblies. A and B: pressure is measured between the free eluent surface in the column or the reservoir and the end of the outlet tubing. C and D: pressure is measured from the bottom of the air inlet tube in the Mariotte flask to the end of the outlet tubing, no matter whether the flow is downward (C) or upward (D).

difference in input and output levels. An easy way of identifying the top and bottom levels of such a column is to find the two points where the column of solution is in direct contact with the atmosphere. In these cases the hydrostatic pressure may be increased by lowering the end of the column outlet tube or raising the level of the effluent solution. Pressure may be decreased in the opposite manner. A disadvantage of the assemblies in Figure 5-7A and B is that the hydrostatic pressure decreases as solution drains from the column. Such pressure changes may be avoided by using a Mariotte flask (Figure 5-7C). This is a vessel with one tube connecting the bottom of the flask with the top of a column and a second tube extending into the solution from an airtight stoppered top. When solution flows out of the flask a partial vacuum is created above the remaining solution (Figure 5-8). As pressure above the solution decreases

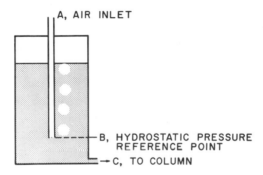

Figure 5-8. Schematic representation of a Mariotte flask in operation.

the level of solution in tube A falls until it reaches the lower end. Further liquid removal causes air to enter the container as bubbles. Pressure at the bottom of tube A is equal to atmospheric pressure; hydrostatic pressure is then measured from the bottom of tube A and is constant regardless of the depth of solution in the flask. Uniform and controlled pressures may also be generated using one of many commercially available peristaltic pumps.

Procedures for pouring a column of large pored gel must account for the compressibility of these soft gels. Fundamental differences between these procedures and those for small pored gels are the precautions taken to ensure that the hydrostatic pressure on the column bed never exceeds that recommended in Table 5-5 during packing and chromatography. Pressures in excess of those indicated will compress the soft gel and seriously reduce effluent flow rates. The freshly poured gel is first allowed

Table 5-5. Maximum Hydrostatic Pressure Permitted for Various Gel Media

Gel	Maximum Recommended Hydrostatic Pressure (cm H_2O)
Sephadex	
G-10	100
G-15	100
G-25	100
G-50	100
Sephadex G-75	50
Sephadex G-100	35
Sephadex G-150	15
Sephadex G-200	10
Bio-Gel	
P-2	100
P-4	100
P-6	100
P-10	100
P-30	100
P-60	100
Bio-Gel P-100	60
Bio-Gel P-150	30
Bio-Gel P-200	20
Bio-Gel P-300	15
Sepharose	
2B	1[a]
4B	1
Bio-Gel	
A-0.5m	100
A-1.5m	100
A-5m	100
Bio-Gel A-15m	90
Bio-Gel A-50m	50
Bio-Gel A-150m	30

[a] Per centimeter of gel depth.

to settle for 10 to 15 minutes with the column outlet closed. After this the outlet is opened, but the outlet tube is raised to the height needed to establish a hydrostatic pressure somewhat below that listed in Table 5-5. As the bed settles the hydrostatic pressure is gradually increased to the final value indicated in the table. Finally the column is stabilized as described for small pored gels.

The use of ascending flow assemblies has somewhat alleviated compression of soft, large pored gels. As shown in Figure 5-7D, an ascending flow column is fitted with adjustable plungers which are sometimes called flow adapters. Sample and eluent enter the bottom of the column rather than the top. The upper flow adapter should not be placed on a column until after the bed has been stabilized. A second precaution taken with ascending flow columns is that the sample and eluent solutions must possess the same densities. Otherwise the sample remains in the bottom of the column. Ascending flow assemblies alleviate compression by arranging for eluent flow and gravity to operate in opposite directions. Upward eluent flow tends to compress the bed in an upward direction whereas gravity exerts an opposite influence. To ensure a balance of these forces, the column should be inverted before each use.

Gel permeation columns may be left unattended during stabilization or chromatographic separations. However, if this is done the column should be provided with a "safety loop" to prevent it from running dry. Two ways of constructing a safety loop are illustrated in Figure 5-9. This device functions as a siphon. When the level of solution from the reservoir reaches the level of the column outlet (dashed line), the siphon is broken and the column flow ceases.

DETERMINATION OF THE VOID VOLUME

Before a freshly prepared column is used for separation of macromolecular species its void volume and uniformity of packing should be determined. This may be done by applying a colored sample material of high molecular weight (above the exclusion limit) to the column. Blue Dextran 2000® (MW approximately 2×10^6) works well for this purpose. The manner in which excluded material migrates through the column is noted, as is the amount of eluent needed to remove it from the column. This amount of eluent is the void volume. If the column has been prepared correctly the material travels in a uniform band and yields a symmetrical elution profile. If, however, a nonuniform resistance is encountered the band of colored material is skewed. In this event the column should be repoured and tested again.

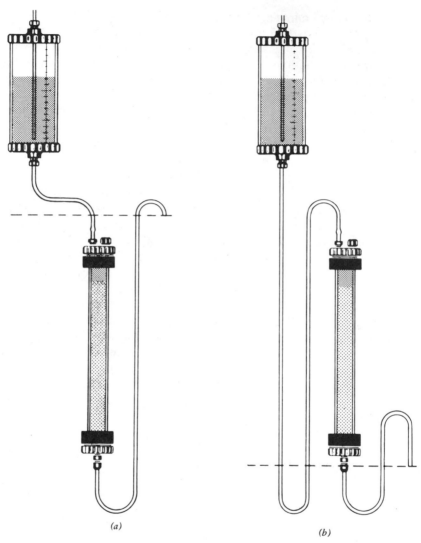

(a)   (b)

Figure 5-9. Safety loop arrangements that prevent a column from running dry. (*A*) The safety loop is placed after the column and the end of the outlet tubing is placed above the column. The flow stops when the eluent in the inlet tubing reaches the level of the outlet tubing. (*B*) The safety loop is placed before the column with the column outlet tubing in any position above the lower loop on the inlet side. The flow stops when the eluent in the inlet tubing reaches the level of the outlet tubing. (Courtesy of I. M. Easterday, Pharmacia Fine Chemicals, Inc.)

SAMPLE APPLICATION AND CHROMATOGRAPHY

Three properties that must be considered with regard to the sample are size, viscosity, and ionic strength. Sample size depends greatly on the sample characteristics and the kind of separation desired. For desalting purposes the sample size may be as large as 10 to 25% of the bed volume, but for separation of macromolecules a sample size of 1 to 5% is advisable. Decreasing the size below the 1% limit, however, does not significantly improve resolution. As shown in Figure 5-10, the upper limit of sample size is largely dictated by the degree of difference between the K_{av} values of the sample components. If components of the mixture require drastically different amounts of eluent to remove them from the column, the sample size may be increased. In fact, such an increase is advisable since under this condition the sample undergoes less dilution during the chromatographic process. A second important sample characteristic is viscosity. As can be seen in Figure 5-11, increasing the viscosity of the sample above that of the eluent results in a broadened and skewed elution profile. At an even higher viscosity differential, total separation is no longer possible. In practice the viscosity of the sample and eluent should never differ by more than twofold. Ionic strength of the sample, as

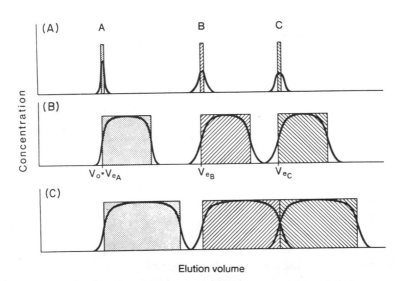

Figure 5-10. Elution profiles for samples of different sizes: (A) a sample much smaller than necessary to obtain separation of components A, B, and C; (B) a sample as large as possible for total separation of the components; and (C) a sample too large for complete resolution. (Courtesy of I. M. Easterday, Pharmacia Fine Chemicals, Inc.)

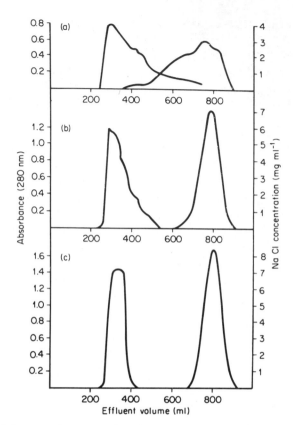

Figure 5-11. Influence of sample viscosity on the elution pattern of a mixture of hemoglobin (0.1%) and sodium chloride (1.0%) chromatographed on a 4×85 cm bed of Sephadex G-25 at a constant flow rate of 180 ml/hour. Dextran 2000 was added to the sample (a) at 5% final concentration (11.8 times greater viscosity than control sample) and (b) at 2.5% final concentration (4.2 times greater viscosity than control sample). (c) Control sample was not altered by addition of dextran. [From P. Flodin, *J. Chromatogr.*, 5:103 (1961).]

indicated above for the eluent, should be greater than 0.08 to overcome any ionic groups that may be present in the gel medium. In most cases sodium or potassium chloride is used at concentrations of 0.2 to $1.0M$ to adjust both the sample and eluent to appropriate ionic strength.

Molecular sieve chromatography depends largely on a mixture of molecules being given an opportunity to diffuse into the gel matrix. The greater the number of opportunities for this to occur, the finer the resolution achieved. Therefore, the flow rate of eluent exerts a profound effect upon the resolution characteristics of a column. As shown in Figure

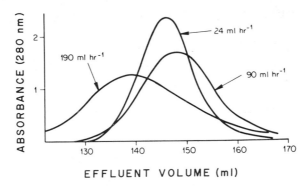

Figure 5-12. Effects of flow rate upon the elution profile of uridylic acid chromatographed on Sephadex G-25 (50 to 80 mesh). Bed dimensions were 2×65 cm and flow rates were those indicated.

5-12, the elution profile of uridylic acid is considerably broadened as the flow rate increases. Clearly, if more than one compound is applied to the column such increases would seriously compromise the column's resolving abilities.

EXPERIMENTAL

Silanization of a Column

5-1. Silanization should be carried out in a fume hood. Disposable rubber gloves should be worn for protection because dimethyldichlorosilane is volatile and extremely toxic. Place a rubber stopper securely in the bottom of a 16×300 mm column.

5-2. Rinse the dry column once or twice with toluene.

5-3. Drain all of the toluene from the column and secure it, in upright position, to a ring stand.

5-4. Prepare 50 ml 5% dimethyldichlorosilane by thoroughly mixing 2.5 ml of this compound with 47.5 ml toluene in a 125 ml Erlenmeyer flask.

5-5. Pour this mixture into the column until the level of solution reaches the very top of the column.

5-6. Allow the column and solution to stand undisturbed for 2 or more hours in the hood.

5-7. Empty the column carefully into the hood sink.

5-8. Thoroughly rinse the column twice with toluene and then repeatedly with water.

5-9. Wash the column using a mild detergent and rinse carefully with glass-distilled water.

Separation of Blue Dextran 2000® and Bromophenol Blue using Sephadex G-25

5-10. Prepare buffer A by dissolving 0.1 moles KCl and 0.01 moles Tris base in 900 ml glass-distilled water. Adjust the pH to 7.0 with concentrated acetic acid. Add sufficient water to give a final volume of 1 liter.

5-11. Slowly sprinkle 20 g of Sephadex G-25 (140 to 200 mesh) onto the surface of 500 ml buffer A while gently stirring the buffer with a glass rod.

5-12. Stir intermittently for 1 hour and then allow the suspension to stand overnight. Alternatively equilibration may be hastened by incubating the suspension at 100°C for 2 hours.

5-13. Remove 400 ml of the supernatant solution from the equilibrated gel. If this gel has been handled gently, fining may not be necessary. If fines are generated they should be removed at this point by transferring the suspension to a 250 ml graduated cylinder and allowing 90 to 95% of the gel to settle. Remove the remaining 5 to 10% of slow sedimenting material with suction. Repeat this operation until a rapidly sedimenting slurry is obtained.

5-14. Transfer the suspension to a 300 ml vacuum filter flask and evacuate the flask for 10 to 15 minutes with intermittent swirling.

5-15. Attach a silanized column to a sturdy support and verify that it is perfectly vertical by holding a small level against the column at three or four different points around it.

5-16. Pour 40 ml buffer A into the column and allow 20 to 30 ml to drain out the bottom before closing the outlet.

5-17. Attach a column extender of reservoir to the column.

5-18. Pour the entire gel slurry gently down the side of the column with the aid of a glass rod and allow it to settle 2 to 5 minutes before opening the outlet.

5-19. With the outlet open allow the column to pack to a bed height of 27 cm.

5-20. Remove any excess gel with suction and close the outlet.

5-21. Fill the remaining volume of the column with buffer A and place an airtight cap on the column.

5-22. Set up a Mariotte flask as described in the text and adjust it to a hydrostatic pressure of 58 cm.

5-23. Allow 100 to 150 ml of buffer A to flow through the column to stabilize it.

5-24. Mark the bed level on the outside of the column with tape or a marking pencil.

5-25. Remove the column cap and allow the column to drain until the solution just reaches the bed surface. Close the outlet.

5-26. Prepare a bromophenol blue solution by dissolving 10 mg dye in 5 ml ethanol. Use a vortex mixer to make certain all of the dye dissolves. Add, dropwise, $1M$ Tris-acetate buffer (pH 7.0) until the solution turns intensely blue.

5-27. Prepare a blue dextran 2000 solution by dissolving 10 mg standard in 2 ml buffer A.

5-28. Prepare the column sample by mixing 0.1 ml bromophenol blue (step 5-26) and 0.5 ml blue dextran solution (step 5-27).

5-29. Transfer the column to a fraction collector adjusted to yield 5.0 ml samples.

5-30. Assemble on the fraction collector a Mariotte flask containing 400 ml buffer A and adjusted to yield a hydrostatic head of 58 cm.

5-31. Transfer very carefully 0.5 ml of the sample solution to the column.

5-32. Open the outlet and allow the sample to enter the gel.

5-33. When the surface of the solution reaches the bed surface, carefully add 1 ml buffer A.

5-34. Repeat step 5-33.

5-35. When this solution reaches the gel surface very gently fill the column with buffer A, replace the column cap, and connect the Mariotte flask to the column. Under these conditions the column will exhibit a flow rate of approximately 5 ml/1.5 minutes. This is much faster than would usually be used and results in considerable dilution of the sample. However, it demonstrates well the technique's potential in a short time. Note also that the two materials completely separate in about 3 to 4 cm of gel. Therefore, a column of only 3 to 5 cm would be needed for this particular sample size and composition. Separations of this type provide a good illustration of the column capacities needed for desalting applications such as those discussed in Chapter 10. Under these conditions a 0.5 ml protein sample could be desalted in less than 10 minutes.

5-36. Collect 60 fractions. Note that the more a sample penetrates the gel pores, the more it spreads out on the column.

5-37. Determine the absorbance of each fraction on a colorimeter or spectrophotometer set at 540 nm.

5-38. The data obtained in this experiment are plotted in Figure 5-13.

5-39. Determine the V_0 and V_e of the blue dextran and bromophenol blue, respectively.

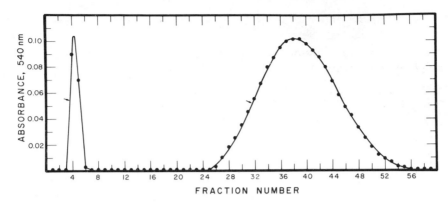

Figure 5-13. Elution profile of bromophenol blue (fractions 24 to 56) and blue dextran 2000 (fractions 4 and 5) chromatographed on Sephadex G-25 (140 to 200 mesh, medium). Arrows indicate the V_0 and V_e values, respectively.

5-40. V_t can be obtained by emptying the column and refilling it with water to the level marked in step 5-24. The volume of the water is then determined by draining the column into a graduated cylinder. If the column is to be reused this procedure may be performed before gel is added to the column and the bed allowed to form to the mark. In this particular experiment V_t was 41 ml.

5-41. Calculate K_{av} for bromophenol blue:

$$K_{av} = \frac{V_e - V_0}{V_t - V_0}$$

$$= \frac{158 \text{ ml} - 18 \text{ ml}}{41 \text{ ml} - 18 \text{ ml}}$$

$$= \frac{140}{23}$$

$$= 6.1$$

REFERENCES

1. D. Rodbard and A. Chrambach, *Proc. Natl. Acad. Sci. US*, **65**:970 (1970). Unified Theory for Gel Electrophoresis and Gel Filtration.

2. L. Fischer, An Introduction to Gel Chromatography in Laboratory Techniques, in *Biochemistry and Molecular Biology* (T. S. Work and E. Work, Eds.), American Elsevier, New York, 1969.

3. J. Porath, *Pure Appl. Chem.*, **6**:233 (1963). Some Recently Developed Fractionation Procedures and their Application to Peptide and Protein Hormones.

4. G. K. Ackers, *Biochemistry*, **3**:723 (1964). Molecular Exclusion and Restricted Diffusion Processes in Molecular-Sieve Chromatography.

5. P. G. Squire, *Arch. Biochem. Biophys.*, **107**:471 (1964). A Relationship Between the Molecular Weights of Macromolecules and Their Elution Volumes Based on a Model for Sephadex Gel Filtration.

6. T. C. Laurent and J. Killander, *J. Chromatogr.*, **14**:317 (1964). A Theory of Gel Filtration and Its Experimental Verification.

7. N. V. B. Marsden, *Ann. N.Y. Acad. Sci.*, **125**:428 (1965). Solute Behaviour in Tightly Cross-linked Dextran Gels.

8. G. A. Gilbert, *Nature*, **210**:299 (1966). Elution Volume versus Reciprocal Elution Volume in the Interpretation of Gel Filtration Experiments.

9. J. C. Giddings and K. L. Mallik, *Anal. Chem.*, **38**:997 (1966). Theory of Gel Filtration (Permeation) Chromatography.

10. J. Reiland, Gel Filtration, in *Methods of Enzymology*, Vol. 22 (W. B. Jakoby, Ed.), Academic Press, New York, p. 287.

11. B. Gelotte and J. Porath, Gel Filtration, in *Chromatography*, 2nd ed. (E. Heftmann, Ed.), Reinhold, New York, 1967.

12. H. Determann, *Gel Chromatography*, Springer, Berlin, 1968.

13. J. M. Curling, The Use of Sephadex® in the Separation, Purification and Characterization of Biological Materials, in *Experiments in Physiology* and *Biochemistry*, Vol. 3 (G. A. Kerkut, Ed.), Academic Press, New York, 1970, p. 417.

14. J. R. Whitaker, *Anal. Chem.*, **35**:1950 (1963). Determination of Molecular Weights of Proteins by Gel Filtration on Sephadex.

15. P. Andrews, *Biochem. J.*, **91**:222 (1964). Estimation of the Molecular Weights of Proteins by Sephadex Gel Filtration.

16. P. Andrews, *Biochem. J.*, **96**:595 (1965). The Gel Filtration Behaviour of Proteins Related to their Molecular Weights over a Wide Range.

17. A. M. Posner, *Nature*, **198**:1161 (1963). Importance of Electrolytes in the Determination of Molecular Weights by Sephadex Gel Filtration, with Especial Reference to Humic Acid.

18. G. A. Locascio, H. A. Tigier, and A. M. del C. Batlle, *J. Chromatogr.*, **40**:453 (1969). Estimation of Molecular Weights of Proteins by Agarose Gel Filtration.

19. A. R. Cooper and J. F. Johnson, *J. Appl. Polym. Sci.*, **13**:1487 (1969). Gel Permeation Chromatography: Effect of Treatment with Hexamethyldisilazane on Porous Glass Packings.

20. E. C. Horning, W. J. A. Vanden Heuvel, and B. G. Creech, *Methods Biochem. Anal.*, **11**:69 (1963). Separation and Determination of Steroids by Gas Chromatography.

21. G. K. Ackers, Molecular Sieve Methods of Analysis, in *The Proteins*, Volume 1 (H. Neurath and R. L. Hill, Eds.) Academic Press, New York, 1975, pp. 2–94.

22. Y. Nozaki, N. M. Schechter, J. A. Reynolds and C. Tanford, *Biochem.*, **15**:3884 (1976). Use of Gel Chromatography for the Determination of the Stokes Radii of Proteins in the Presence and Absence of Detergents. A Reexamination.

Chapter 6

Electrophoresis

Electrophoretic techniques have become principal tools for characterizing macromolecules and for assaying their purity. The method is based on the fact that molecules such as DNA, RNA and proteins possess a charge and therefore are able to move when placed in an electric field. Early applications of electrophoretic methods were performed in sucrose solutions. However, the cumbersome nature of these methods and the expensive equipment needed severely limited their desirability. Use of starch gel in place of sucrose as the supporting medium and anticonvection agent increased the application of electrophoresis to biological problems, but it was the advent of acrylamide gels that projected this method to its present popularity. In the subsequent discussion proteins are used to illustrate the current principles of electrophoresis. However, these principles apply equally to other charged molecules.

ION MOVEMENT IN AN ELECTRIC FIELD

As discussed in Chapter 4, proteins possess charge as a result of the acidic (glutamic and aspartic acids) and basic (lysine and arginine) amino acids they contain. However, here it is only the net charge on the molecule that is significant. If a molecule of net charge q is placed in an electric field, a force F is exerted upon it, which depends on the charge possessed by the molecule and the strength of the field into which it is placed. This is expressed mathematically as

$$F = \frac{E}{d} q \tag{1}$$

where E is the potential difference between the electrodes and d is the distance between them. The quantity E/d is often referred to as the field

194

strength. If this situation were to occur in a vacuum, the molecule would accelerate toward and finally collide with the electrode. In solution, however, this does not occur because the pulling force of the electric field is opposed by the drag or friction occurring between the accelerating molecule and the solution. The extent of drag, as described by Stokes' equation, depends on the size and shape of the molecule and the viscosity of the medium through which it moves:

$$F = 6\pi r\eta v \tag{2}$$

where F is the drag force exerted on a spherical molecule, r is the radius of the spherical molecule, η is the solution viscosity, and v is the velocity at which the molecule is moving. In solution the accelerating force generated by the electric field is opposed by drag and therefore

$$\frac{E}{d}q = 6\pi r\eta v \tag{3}$$

It can be seen by rearranging equation 3,

$$v = \frac{Eq}{d6\pi r\eta} \tag{4}$$

that the velocity (v) at which the molecule moves is proportional to the field strength and charge on the molecule but inversely proportional to its size and the solution viscosity. Since each molecule is expected to possess a unique charge and size, it migrates to a unique position within the electric field in a given length of time. Therefore, if a mixture of proteins is subjected to electrophoresis each of the proteins would be expected to concentrate into a tight migrating band at unique positions in the electric field. Although these considerations describe what occurs when electrophoresis is performed in a sucrose solution, they are not quite adequate to describe electrophoresis in an acrylamide gel. The migration of macromolecules through acrylamide gel is profoundly influenced by the structure of the gel itself.

ACRYLAMIDE GEL ELECTROPHORESIS

Successful performance of electrophoresis requires placing the sample molecule in a stable medium that decreases or eliminates convection and does not react with the sample or in any way retard its movement by binding to it. Such a chemically inert medium was discovered in the production of polyacrylamide gel. The compounds used to construct the polymer matrix are acrylamide, N,N'-methylene-bis(acrylamide), tetra-

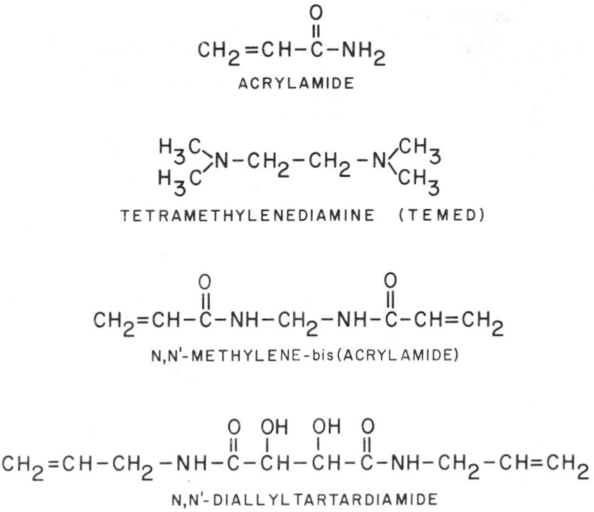

$$CH_2=CH-\overset{\overset{\displaystyle O}{\|}}{C}-NH_2$$

ACRYLAMIDE

$$\begin{array}{c} H_3C \\ H_3C \end{array}\!\!\!> N-CH_2-CH_2-N\!\!<\!\!\begin{array}{c} CH_3 \\ CH_3 \end{array}$$

TETRAMETHYLENEDIAMINE (TEMED)

$$CH_2=CH-\overset{\overset{\displaystyle O}{\|}}{C}-NH-CH_2-NH-\overset{\overset{\displaystyle O}{\|}}{C}-CH=CH_2$$

N,N'-METHYLENE-bis(ACRYLAMIDE)

$$CH_2=CH-CH_2-NH-\overset{\overset{\displaystyle O}{\|}}{C}-\overset{\overset{\displaystyle OH}{|}}{CH}-\overset{\overset{\displaystyle OH}{|}}{CH}-\overset{\overset{\displaystyle O}{\|}}{C}-NH-CH_2-CH=CH_2$$

N,N'-DIALLYLTARTARDIAMIDE

Figure 6-1. Compounds used in the synthesis of acrylamide gels.

methylenediamine, often denoted as TEMED, and ammonium persulfate (Figure 6-1). When ammonium persulfate is dissolved in water it forms free radicals

$$S_2O_8^{2-} \longrightarrow 2SO_4^-\cdot \tag{5}$$

If these free radicals are brought into contact with acrylamide (Figure 6-2) a reaction occurs, with the preservation of the free radical within the acrylamide molecule. This "activated" acrylamide can then react in the same way with successive acrylamide molecules to produce a long polymer chain. A solution of these polymer chains, although viscous, does not form a gel. No gelation occurs because the long chains can slide past one another. Gel formation requires hooking various chains together or cross-linking them to one another. This is done by carrying out polymerization in the presence of N,N'-methylene-bis(acrylamide), a compound that can be thought of as two acrylamide molecules coupled head to head at their nonreactive ends. As shown in Figure 6-3, carrying out polymerization in this manner yields a net of acrylamide chains. The size of the holes or pores in the net is determined by two parameters: (1)

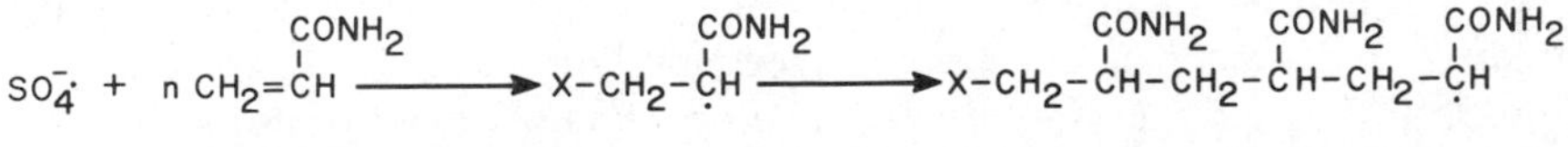

$$SO_4^-\cdot + n\,CH_2{=}\overset{\overset{\displaystyle CONH_2}{|}}{CH} \longrightarrow X-CH_2-\overset{\overset{\displaystyle CONH_2}{|}}{CH} \longrightarrow X-CH_2-\overset{\overset{\displaystyle CONH_2}{|}}{CH}-CH_2-\overset{\overset{\displaystyle CONH_2}{|}}{CH}-CH_2-\overset{\overset{\displaystyle CONH_2}{|}}{CH}\cdot$$

Figure 6-2. Reactions involved in polymerization of acrylamide.

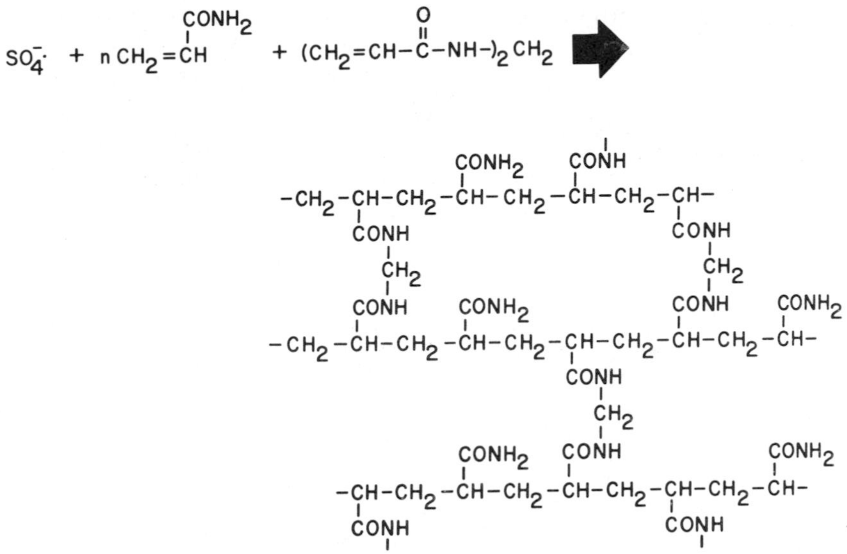

Figure 6-3. Reactions involved in cross-linking acrylamide chains.

the amount of acrylamide used per unit volume of reaction medium and
(2) the degree of cross-linkage. As can be seen in Table 6-1, regardless of
the total amount of acrylamide per unit volume, the average pore size
reaches a minimum when 5% of the total acrylamide used is N,N'-
methylene-bis(acrylamide). Therefore, in many formulations the bis(acrylamide) content is fixed at 5% of the total acrylamide and is not altered

Table 6-1. Effect of Bis(acrylamide) Gel
Concentration on Average Pore Size

Total Acrylamide (%)	Radius (Å)[a]			
	1%	5%	15%	25%[b]
6.5	24	19	28	—
8.0	23	16	24	36
10.0	19	14	20	30
12.0	17	9	—	—
15.0	14	7	—	—

[a] Molecules for which 50% of the gel volume is available. Data taken from reference 27, p. 12.
[b] N,N-methylene-bis(acrylamide).

as a means of controlling pore size. On the other hand, pore size is manipulated by varying the total content of acrylamide. Figure 6-4 indicates an approximation of the pore sizes yielded from various concentrations (w/v) of acrylamide. It must be emphasized that the random nature of pore formation precludes the pores being of one size. The pore size is a distribution, some larger and some smaller. Therefore, great significance should not be accorded the absolute values in Figure 6-4. Rather, this figure illustrates in a gross way the average pore size as

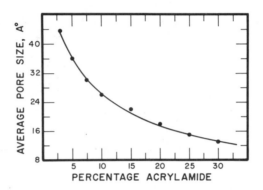

Figure 6-4. Influence of gel concentration (percentage weight/total gel volume) on the average pore size.

a function of acrylamide content. TEMED or β-(dimethylamino) propionitrile is usually added at a concentration of 0.4% to serve as a catalyst of gel formation because of their ability to exist in a free radical form.

A second means of polymerizing acrylamide is through use of riboflavin. In the presence of oxygen and ultraviolet light riboflavin undergoes photodecomposition with resulting production of free radical containing products. These free radicals function just as those described above with regard to ammonium persulfate. Usually an ordinary fluorescent lamp positioned close to the reaction mixture suffices as a source of UV light.

A final consideration in the production of polyacrylamide gel is the nature of the buffer system. The buffer serves (1) to maintain a constant pH within the reservoirs and within the acrylamide gel and (2) as the electrolyte which conducts current across the electric field. If these functions are to be performed acceptably three conditions must be met. First, a buffer must be chosen that does not interact with the macromolecules being separated. Such interactions might change the rate at which a molecule migrates in the electric field. This would occur if the interaction involves the addition to or neutralization of a charge on the macromolecule. Alternatively these interactions might result in the artifactual appearance of two migrating species in place of one. Such an artifact was reported by Cann (1), who observed two protein bands when pure albumin was subjected to electrophoresis in the presence of a borate-containing buffer. He recovered the protein from one of the isolated bands and subjected it to electrophoresis a second time; again two bands were observed. He concluded that the multiple bands were the result of an interaction between borate and some of the albumin molecules. Secondly, the pH of the environment in which electrophoresis occurs must be such that the macromolecules being separated are charged but not denatured. In the case of proteins the usual outside limits are 4.5 and 9.0. However, a broader or more restricted range may be called for, depending on the specific molecules studied. Finally, the ionic strength and concentration of the buffer must be considered. If the electrolyte concentration within the acrylamide gel is too low, the migrating macromolecules conduct a large portion of the current. As a result they are not found in sharp bands but rather spread into diffuse zones and thus greatly decrease the resolution of this method. Alternatively, if the electrolyte concentration within the gel is too high, the amount of current conducted increases while voltage decreases. This usually results in a decreased rate of macromolecular migration and the generation of heat, which can denature the proteins being separated.

As pointed out in the theoretical discussion at the beginning of this chapter, equations 3 and 4 do not adequately describe the electrophoretic movement of an ion through acrylamide gel because they do not account for the existence of gel pores. Intuitively it would seem that the extent to which a macromolecule migrates into the gel depends on both the size of the molecule and the average size of the pores through which it must pass. If most of the pores or holes in the gel have a diameter smaller than that of the protein, then regardless of the protein's charge or the field strength, it will not move into the gel. It is this sieving effect that enhances the resolution obtained with acrylamide gels. As shown in Figure 6-5, logarithms of the relative mobilities for a group of proteins decrease linearly as the total gel concentration is increased. This is expected from the data in Figure 6-4, which indicate that the average pore size of a gel decreases as the gel concentration increases. Rodbard and Chrambach (2) have developed a set of mathematical relationships to describe the effects of gel concentration upon a macromolecule's mobility.

$$\log M = \log M_0 - K_R T \tag{6}$$

where M is the electrophoretic mobility, M_0 is the free mobility in a

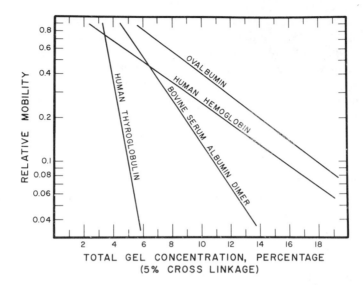

Figure 6-5. Relationship between electrophoretic mobility and total gel concentration observed for various proteins. [From D. Rodbard and A. Chrambach, *Proc. Natl. Acad. Sci. US*, **65**:970 (1970).]

sucrose solution, T is the total gel concentration, and K_R is the retardation coefficient or the slopes of the lines shown in Figure 6-5. The retardation coefficient, K_R, is related to the radius of the macromolecule by

$$K_R = C(R + r) \tag{7}$$

where C is a constant, R is the geometric mean radius of the macromolecule, and r is the radius of the gel fibers, which are assumed to be much longer than the macromolecule itself. Derivation of these equations is necessarily based on statistical considerations because of the random nature of acrylamide gel pore formation. However, the authors present a large collection of data (3–6), indicating that quantitative relationships can be established between the molecular weights of macromolecules and their mobilities within acrylamide gels if sufficient experiments are performed. In spite of this work, acrylamide gel electrophoresis, in the absence of denaturing agents has not been widely used to determine the molecular weights of macromolecules.

ELECTROPHORETIC PROCESS

Only two pieces of equipment are needed for electrophoresis, (1) a d.c. power supply and (2) a buffer reservoir system. Such a reservoir system is shown in Figure 6-6 and is diagrammed in Figure 6-7. This system consists of upper and lower buffer reservoirs with provision for suspending the polyacrylamide gel containing tubes between them. Notice that the only electrical connection between the reservoirs is via the acrylamide gel. Platinum electrodes are positioned in each reservoir (arrows in Figure 6-6) and are connected to terminals extending from the top of the unit. Often the reservoir system is totally enclosed to prevent fatal electrical shock to the operator, who might otherwise accidentally touch one of the reservoirs during operation.

Electrophoretic separation of a group of macromolecules (proteins, for example) is performed by placing a high density solution of the protein mixture on top of the gel. This solution, usually containing glycerol as the high density component, is used to prevent mixing the sample with the upper reservoir buffer which is in contact with the acrylamide gel surface. At pH 9, a commonly used pH for electrophoresis, most proteins are

Figure 6-6. Commercially available chambers for acylamide gel electrophoresis of short and long gels. Arrows in the left apparatus point to the platinum electrode wires. (Courtesy of Savant Instruments, Inc., Hicksville, N.Y.)

negatively charged. Hence the anode is placed in the lower reservoir. When the power is turned on the proteins start migrating toward the positively charged anode. Often a "tracking dye" such as bromophenol blue is included in the sample as a reference. This colored material migrates faster than any of the macromolecules. Therefore, if electrophoresis is continued only until the dye reaches the bottom of the tube one can be reasonably sure that all the macromolecules are still within the gel. However, if the molecules being separated migrate very slowly it may be necessary to determine the length of time required for the dye to traverse the tube and then continue electrophoretic separation for some multiple of this time. The exact multiple necessarily depends on the molecules being separated and can be determined only by trial and error.

The reactions that permit current passage from the cathode to the anode are shown in Figure 6-7. They are essentially the electrolysis of water, producing hydrogen at the cathode and oxygen at the anode. Note that for every mole of hydrogen produced, only one-half mole of oxygen is produced. This affords an easy way of checking the electrodes to make certain that they are operating with the desired polarity. The anode of an operating system always produces about half as many gas bubbles as the

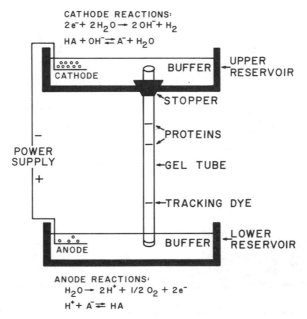

Figure 6-7. Diagram of an electrophoresis apparatus showing the reactions that occur at each electrode.

cathode. The need for a high capacity reservoir buffer system is emphasized by the electrode reactions. As can be seen, one mole of hydroxyl ions and protons are produced at the cathode and anode, respectively, for every mole of electrons that flow through the system.

Occasionally very long time periods are needed for a given separation. When this occurs it may be necessary to recirculate buffer from the lower reservoir to the upper reservoir with a pump in order to prevent exhaustion of the buffer capacity and the ensuing pH changes. However, great care must be exercised when performing this operation. It is necessary to maintain a constant level of solution in both reservoirs and to be certain that a complete electrical circuit is not established between the reservoirs via the recirculation pump. Figure 6-8 illustrates one way in

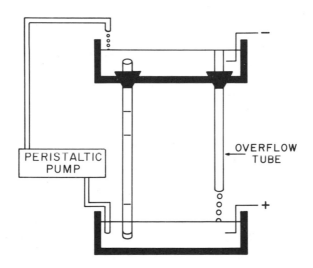

Figure 6-8. Schematic diagram depicting one means of assembling a buffer recycling system for electrophoretic experiments extending over long periods.

which this may be accomplished. Buffer from the lower reservoir is pumped, by means of a peristaltic pump, to the upper reservoir where it is allowed to enter, drop by drop, into the reservoir. If buffer entered the upper reservoir as a stream, clearly a short circuit would result. An overflow tube is positioned in the apparatus in place of one of the gel tubes. As the volume of the upper reservoir buffer increases, it overflows, again drop by drop, back into the lower reservoir.

DISC GEL ELECTROPHORESIS

Discontinuous pH or "disc" electrophoresis is a modification of the zone electrophoretic techniques described above. The significant differences are (1) the use of a two gel system and (2) the unique buffer systems used in the gel matrix and in the buffer reservoirs. The two gel system is illustrated in Figure 6-9. The lower-separating or running gel is prepared using about the same amount of acrylamide (5–10%) as would be used for an analogous zone electrophoretic experiment. It is in this gel that the macromolecules subsequently separate. The buffer used in this gel is usually an amine such as Tris which is adjusted to the proper pH (e.g., 8.7) using hydrochloric acid. After the separating gel has polymerized, a

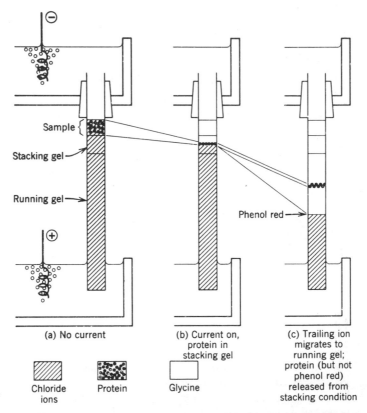

Figure 6-9. Schematic diagram of a two gel system and the movement of the various ionic species during electrophoresis. (From G. Bruening, R. Criddle, J. Preiss, and F. Rudert, *Biochemical Experiments*, Wiley, New York, 1970.)

second, small (1.0 cm) layer of gel is polymerized on top of it. This is called the upper or stacking gel and is prepared using much less acrylamide than was used in preparation of the running gel. A total acrylamide content of 2 to 3% is common. The buffer used in the stacking gel is also an amine such as Tris, but in this case the pH is adjusted with hydrochloric acid to a value about 2 pH units lower than that of the running gel (pH 6.5). The buffer used in the protein sample should be identical to that used in the stacking gel. Buffer used in the lower reservoir is of the same composition and pH as that found in the running gel, although it may be somewhat more dilute. Buffer for the upper reservoir is also an amine. However, it is adjusted to a pH the same as or slightly above that used in the running gel with a weak acid whose pK_a is at the desired pH. Glycine is commonly used for this adjustment.

The electrophoretic behavior observed with this method is best understood by following the events occurring immediately after the power is turned on. Glycine in the upper buffer reservoir exists as both a zwitterion with a net charge of zero and a glycinate ion with a charge of minus one:

$$^+NH_3CH_2COO^- \rightleftharpoons NH_2CH_2COO^- + H^+$$

When the electric field is established, chloride, protein, bromophenol blue, and glycinate anions all begin to migrate toward the anode. However, as glycinate ions enter the sample buffer and stacking gel they encounter a condition of low pH, shifting the equilibrium toward formation of zwitterions, which are immobile. Failure of glycine zwitterions to move into the sample and stacking gel creates a deficiency of mobile ions, which in turn decreases current flow. However, a constant current *must* be maintained throughout the entire electrical system. This is accomplished in the area between the leading chloride ions and the trailing glycinate ions by an increase in voltage. The result is a very high localized voltage gradient occurring between the chloride ions and the glycinate ions. In this condition the relative ion mobilities are glycinate < protein < bromophenol blue < Cl⁻. In this strong local electric field the anionic proteins all migrate rapidly. The stacking gel has large pores so as not to impede their progress. If any of the proteins overtake the leading chloride ions, they slow down because wherever there are chloride ions there is no ion deficiency, and hence the large field strength disappears. Therefore the proteins migrate quickly until they reach the area containing chloride ions and then drastically slow down. The rapid movement of the proteins behind the chloride front and the decreased rate of protein migration as they approach the front result in a piling up or concentration of the protein sample in a tight disc between the glycinate and chloride ions. As the disc of protein encounters the running gel its migration is slowed by the small pores of the gel. This permits the small glycinate ions to catch up with the proteins. Crossing the interface between the stacking and running gels, glycinate ions become fully charged once again, and there is no longer an ion deficiency. Hence from this point on there is a constant field strength throughout the gel, and separation of the proteins proceeds just as with zone electrophoresis. The advantage of this modification is that the protein sample enters the separating gel as a narrow zone. The resulting protein bands are therefore much more compact, and this increases resolution of the technique. A large number of buffer systems have been developed which cover the range of pH values between 3.5 and 9.5 (7, 8, 4, 34).

SDS ACRYLAMIDE GEL ELECTROPHORESIS

Equations 4, 6, and 7 state that the mobility of a protein in acrylamide gels is a function of both its net charge and size. Hypothetically, two proteins of different molecular weights may migrate toward the anode at the same rate if their size differences are balanced by compensating charge differences. For this reason acrylamide gel electrophoresis using gels of only one pore size as discussed above may not be used to gain information about the molecular weight of a molecule. Techniques employing varying pore size are subsequently discussed. A second restriction placed on electrophoretic techniques concerns the number of species observed on the gel. Molecules which are tightly, but not covalently, bound together do not usually separate from one another during electrophoresis. For example, core RNA polymerase (composed of three nonidentical subunits) appears as a single band if subjected to either zone or disc electrophoresis. Therefore, the number of protein or RNA bands observed on a gel represents only the minimum number that may in fact be present.

In an effort to surmount these problems Shapiro et al. (9) attempted to separate a mixture of proteins in the presence of sodium dodecyl sulfate (SDS), an anionic detergent. The results of their preliminary efforts prompted Weber and Osborn (10) to determine the mobilities of about 40 proteins in the presence of SDS. These investigators observed empirically that the mobilities of these proteins were a linear function of the logarithms of their molecular weights (see Figure 6-10). Sodium dodecyl sulfate has been shown to bind to the hydrophobic regions of proteins and to separate most of them into their component subunits. SDS binding also imparts a large negative charge to the denatured, randomly coiled polypeptides. This charge largely masks any charge normally present in

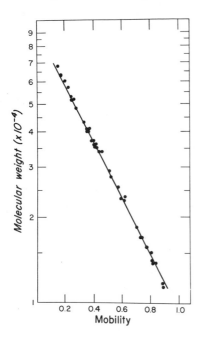

Figure 6-10. Comparisons of the molecular weights of 37 different polypeptide chains in the MW range from 11,000 to 70,000 with their electrophoretic mobilities. [From K. Weber and M. Osborn, *J. Biol. Chem.*, **244**:4406 (1969).]

the absence of SDS. The precise reasons for the success of this technique are obscure, but it is widely used empirically as an assay of the molecular weight and subunit composition of purified proteins. It should not, however, be considered universal because cases have been reported where incorrect information has been obtained with this method. Very large or structural proteins such as collagen are particularly troublesome.

If component subunits of a protein are held together by disulfide bonds, these bonds may be broken before electrophoresis by heating the preparation in the presence of SDS and β-mercaptoethanol, which reduces them to sulfhydryl groups. These groups are then blocked with an appropriate alkylating agent to prevent re-formation of the disulfide bonds. Although the mobilities of proteins with molecular weights between 12,000 and 70,000 daltons behave in the expected manner, some curvature was observed in a plot of mobility versus log molecular weight for proteins in the 40,000 to 200,000 dalton size range. This, however, may be a result of the investigators using a decreased amount of cross-linker [N,N'-methylene-bis(acrylamide)]. Whenever this method is used for molecular weight determination it is advisable to include at least three to four standard proteins with molecular weights spread above and below the unknown. Bracketing the unknown protein in this manner yields a linear plot from the protein standards on which the unknown sample is

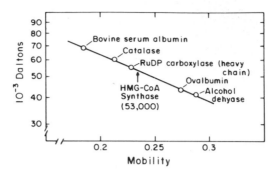

Figure 6-11. Estimation of the subunit mass of mitochondrial HMG–CoA synthase by SDS gel electrophoresis. [From W. D. Reed, K. D. Clinkenbeard, and M. D. Lane, *J. Biol. Chem.*, **250**:3120 (1975).]

appropriately positioned to determine its molecular weight. An example of this approach appears in the molecular weight determination of HMG-CoA synthase shown in Figure 6-11.

A few practical notes should be emphasized regarding SDS electrophoresis. Potassium salts should be avoided and sodium salts used in their places because potassium dodecyl sulfate is quite insoluble. In addition, even sodium dodecyl sulfate is insoluble below about 10°C. Although not specifically cited above, disc gel electrophoresis may also be used in the presence of SDS. These techniques, described by Studier (11) and Ames (12), are of great advantage when the sample volume is in excess of 10 to 20 μliters per gel. A widely used buffer system for SDS acrylamide gel electrophoresis is that developed by Laemmli (41).

A second method for the determination of molecular weights has been described by Hedrick and Smith (8, 13). Their procedures are based on measuring the extent to which a protein migrates into acrylamide gels containing different total amounts of acrylamide and hence possessing different pore sizes. Although this method is more cumbersome than the SDS procedure described above, it has the advantage that molecular weights of "native," undissociated proteins may be obtained. These methods may also be used to distinguish charge isomers of proteins such as lactate dehydrogenase.

VARIATIONS OF ACRYLAMIDE GEL ELECTROPHORESIS

Slab Gel Electrophoresis

There has been a revival of the slab gel techniques originally used when starch was the principal support medium. For this type of electrophoresis, acrylamide is polymerized into a thin square or rectangular slab between two glass plates (11, 12). Sample holes or wells are made at one end of the gel by placing a comb-shaped jig like that shown in Figure 6-12 into the

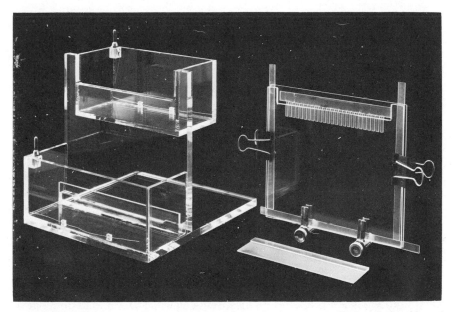

Figure 6-12. Slab gel apparatus. The glass plates are shown assembled with the side and bottom spacers in place and the comb inserted at the notch. [From F. W. Studier, *J. Mol. Biol.*, **79**:237 (1973).]

reaction mixture before it polymerizes. After polymerization the jig is removed leaving sample wells molded into the acrylamide. Any of the electrophoretic techniques discussed thus far can be performed using slab gels in place of the more conventional, tubular gels. The advantage of this technique is the ease with which a number of experimental samples can be compared. Since all of the samples are present in the same gel the conditions of electrophoresis are quite constant from sample to sample. Figure 6-13 depicts the type of data that can be obtained using this technique.

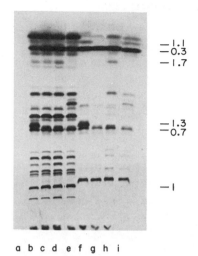

a b c d e f g h i

Figure 6-13. T7 proteins separated by SDS acrylamide slab gel electrophoresis. [From F. W. Studier, *J. Mol. Biol.*, **79**:237 (1973).]

Agarose–Acrylamide Gels

A recent refinement in electrophoretic techniques is the use of complex gels composed of both acrylamide and agarose. Development of these gels was prompted by a need to separate very high molecular weight (in excess of 200,000 daltons) nucleic acids. Acrylamide gels with sufficiently large pores contained such small amounts of acrylamide ($\leq 2.5\%$) that they remained in liquid or near liquid form and as such were unusable. The softness of large pored gels was remedied when Peacock and Dingman (14, 15) added agarose to acrylamide gel to give it physical support. Agarose is a naturally linear polysaccharide. Agarose–acrylamide gels are prepared by dissolving agarose (0.5% final concentration) in boiling water and allowing it to cool to 40°C. At this temperature the components needed for polyacrylamide gel formation are added and the mixture is transferred to the desired mold, that is, tube or slab, which has been prewarmed. On further cooling the gel sets in much the same way as agar. Highly porous yet rigid gels may be formed at acrylamide concentrations as low as 0.5%. However, it is advisable to ensure that the acrylamide polymerizes before the agarose sets. If the agarose sets first an uneven gel surface may result, leading to poor separations. Agarose–acrylamide gels have been especially useful in the electrophoretic separation of molecules up to molecular weights of 1×10^6 daltons for RNA and 3.5×10^6 daltons for DNA. Although little has been said with regard to the electrophoresis of RNA and DNA, these molecules are routinely separated and their molecular weights empirically determined on acrylamide gels (14–17). In the case of RNA the gels may be prepared in 99% formamide in place of water in order to prevent aggregation of the RNA molecules.

Two-Dimensional Gel Electrophoresis

O'Farrell has reported (37) that a mixture containing up to 5000 proteins may be adequately resolved into individual species using two-dimensional electrophoresis. This powerful technique involves first subjecting the mixture to isoelectric focusing on a 1 mm diameter gel in a capillary tube. Isoelectric focusing (38) is carried out in much the same way as electrophoresis with the exception that ampholytes are electrophoresed through the gel prior to sample addition in order to establish a pH gradient within it. As a protein moves down the tube it continuously encounters a higher and higher pH environment. As the protein traverses this pH gradient it reaches, at some point, a pH corresponding to its iselectric point. At this pH the net charge of the protein is zero and it therefore stops moving in the electric field. At the conclusion of isoelectric focusing, the gel is carefully removed from the capillary tube and placed on top of a Studier slab gel apparatus, as shown in Figure 6-14. The

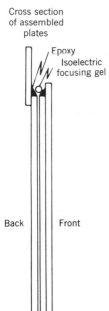

Cross section
of assembled
plates

Epoxy
Isoelectric
focusing gel

Back   Front

Figure 6-14. Cross-section of an isoelectric focusing gel on top of a slab gel in preparation for SDS acrylamide gel electrophoresis of proteins previously separated by isoelectric focusing. [From P. H. O'Farrell, *J. Biol. Chem.*, **250**:4007 (1975).]

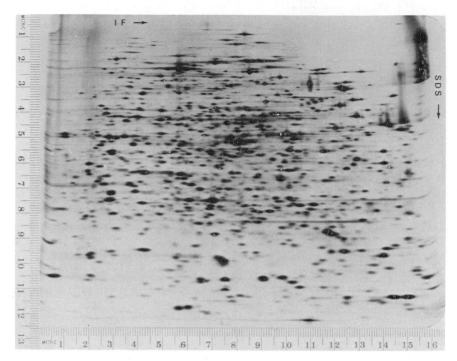

Figure 6-15. Separation of *E. coli* soluble proteins using two-dimensional electrophoresis. The proteins were separated in the horizontal direction by isoelectric focusing and in the vertical direction by SDS acrylamide gel electrophoresis. [From P. H. O'Farrell, *J. Biol. Chem.*, **250**:4007 (1975).]

sample is now subjected to SDS acrylamide gel electrophoresis, which separates the proteins according to their molecular weights. Since the isoelectric point and molecular weight of a protein are unrelated it is possible to obtain an even distribution of proteins using these parameters in two dimensions. This is demonstrated in an autoradiograph (Figure 6-15) of *E. coli* soluble proteins separated in this manner. Recently it has become possible to perform double label experiments with two-dimensional slab gels using the procedures of Nelson and Metzenberg (see also *Anal. Biochem.* **76**:452–457, 1976).

DETECTION OF MACROMOLECULES SEPARATED BY ELECTROPHORESIS

Broad application of electrophoretic techniques has required development of methods that can be used to visualize macromolecules separated on acrylamide gels. A few of these methods are outlined here.

Coomassie Brilliant Blue Staining

The most often used protein stain is coomassie blue. This stain replaced the widely used amido black because of its greater sensitivity, especially in the presence of sodium dodecyl sulfate. The procedure requires fixation of the proteins overnight in at least 10 volumes of 20% sulfa-salicylic acid prior to staining. After fixation the gels are stained in a 0.25% aqueous solution of coomassie blue. The length of staining depends on the gel concentration used; 2 and 4 hours are required for 5 and 10% gels, respectively. Staining in excess of this may result in retention of background staining between the protein bands. Excess stain is removed by repeated washing of the gels in a 7% solution of acetic acid at 37°C.

A widely used variation of this method involves staining and fixing the proteins simultaneously. This is accomplished by soaking gels for 2 to 10 hours in a solution prepared by dissolving 1.25 g of coomassie Blue in a mixture of 454 ml 50% methanol and 46 ml glacial acetic acid. Care must be taken, however, to filter the staining solution (Whatman No. 1 or equivalent) before using it. Destaining is performed by repeated washing of the gels in a solution composed of glacial acetic acid, methanol, and water.

Alternatively the gels may be destained electrophoretically. The gels are replaced in the gel tubes and electrophoresed a second time using a somewhat higher concentration of buffer. There are also instruments available commercially which permit the lateral electrophoresis of gels, thereby shortening the time required for destaining. The advantages of electrophoretic destaining are (1) the speed with which it is accomplished and (2) a reduction in the degree of background staining between the bands compared to that observed after washing the gels. Regardless of the procedure used to destain the gels, they are stored in 7.0 to 7.5% acetic acid.

Fluorescent Staining Techniques

The large amount of time required to visualize proteins by coomassie blue staining resulted in a search for a fast and sensitive means of protein visualization. Two methods may potentially serve this purpose. The first and most sensitive involves fluorescamine (18, 19). This nonfluorescent molecule reacts specifically with primary amines to yield the fluorescent addition product shown in Figure 6-16. Staining may be performed either after the proteins have been separated or before electrophoresis if SDS-containing gels are used. Although there is a direct correlation between the amount of fluorescence observed and primary amino group concentration, the degree of fluorescence cannot be interpreted as a quantitative assay of protein because of the varying amounts of lysine

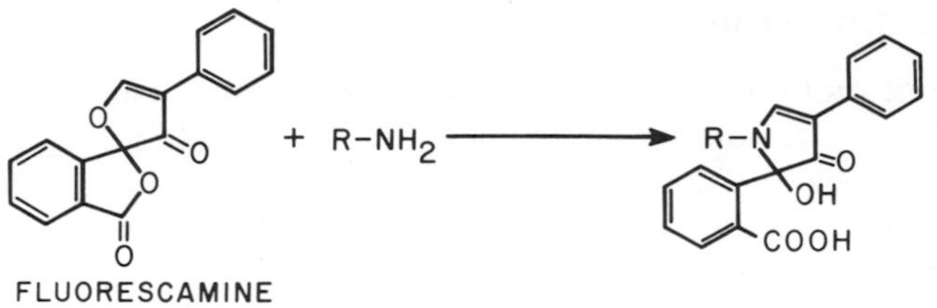

Figure 6-16. Reaction of fluorescamine obtained from Hofmann–LaRoche, Inc. with primary amino groups of amino acids.

found in proteins and because fluorescence fades with time. An additional problem arises if fluorescamine is allowed to react with low molecular weight proteins containing large amounts of lysine prior to SDS electrophoresis. The covalent bonding of fluorescamine to the protein changes its molecular weight. Therefore under these conditions molecular weight determinations may be compromised.

The second fluorescent method, though less sensitive (20 μg), is also convenient. The magnesium salt of anilinonaphthalene sulfonate (ANS) does not fluoresce in water, but does fluoresce brightly when dissolved in organic solvents or when bound to hydrophobic surface regions of proteins. The procedure (20) involves surface denaturation of the protein by incubating the gel from 2 to 5 minutes in $3N$ HCl. Gels are then incubated in a buffered solution of the stain. One advantage of this procedure is that it can be performed without the acid denaturation step. In this manner proteins can be located on the gel without denaturation and hence can be recovered in an active form. However, in the absence of acid denaturation the stain is 10- to 20-fold less sensitive.

Specific Enzyme Visualization

The most specific visualization method is the localization of particular enzymes within an acrylamide gel. Most of these methods involve coupling the enzymatically catalyzed reaction either directly or indirectly to a chemical reaction that generates an intensely colored, insoluble product. Synthesis and availability of tetrazolium salts have greatly increased the variety of enzymes that can be assayed in this way. Reference 21 provides a thorough survey of the enzyme reactions that have been successfully assayed while in an acrylamide gel.

Assay of lactate dehydrogenase illustrates this technique. This enzyme catalyzes the oxidation of lactate, yielding pyruvate and NADH as shown in Figure 6-17. NADH produced in this reaction can be used along with the intermediate electron carrier, phenazine methosulfate (PMS), to

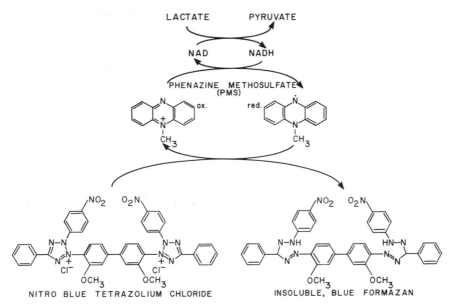

Figure 6-17. Reactions involved in coupling oxidation of lactate with the reduction of *p*-nitroblue tetrazolium chloride.

chemically reduce nitroblue tetrazolium (NBT). Upon reduction the yellow tetrazolium salt is converted to an intensely blue, insoluble formazan. Therefore, when a gel containing lactate dehydrogenase is soaked in a solution containing lactate, NAD, PMS, and NBT, an intense blue precipitate forms at the site or sites of enzyme activity.

The advantages of these techniques are the speed with which the enzymes are visualized and their specificity for given proteins. However, they also suffer from two problems. The first is the possibility of nonspecific staining. For example, any protein that can provide electrons to PMS would yield tetrazolium reduction and perhaps be erroneously designated as lactate dehydrogenase. The most certain means of identifying this problem is to perform the staining procedure on two identical acrylamide gels using one staining mixture that contains lactate and a second staining mixture that contains all the assay reagents except lactate. Any formazan production observed with the second solution is not likely to be due to the presence of lactate dehydrogenase. The second problem is the impermanence of the stained gels. If staining is not terminated once the enzyme bands have stained, the entire gel becomes coated with formazan precipitate. Although nonspecific staining may be reduced by rinsing out unused staining solution, some form of more permanent record such as a photograph or densitometer tracing is advisable.

Miscellaneous Staining Procedures

In addition to the methods described above a battery of other staining procedures are available. These include use of alcian blue (22) to stain glycoproteins, ethidium bromide (23) to stain DNA, and methylene blue (14) and pyronine (16) to stain RNA. A relatively new stain has been nicknamed "stains-all," because of its ability to stain most macromolecules. This dye is a cationic carbocyanine and stains RNA bluish purple, DNA blue, protein red, acid mucopolysaccharides various shades of blue to purple, and phosphoproteins blue (24). It is presently the most widely used stain for RNA.

Detection of Radioactive Macromolecules

A major disadvantage of all the staining techniques is their lack of quantitation. This problem is largely overcome by radioactively labeling the macromolecules being studied. In addition to quantitation of the macromolecules, use of combined radiochemical and electrophoretic techniques has greatly facilitated studies of the pathways of macromolecular synthesis and their control. After being separated electrophoretically radioactive macromolecules may be detected in one of three ways. The oldest method is by means of autoradiography. This method involves first slicing a flat slab longitudinally from the centers of tubular gels with a simple slicing apparatus devised by Fairbanks et al. (25). The flat slabs, or slab gels, produced in this manner are then placed on porous polyethylene or filter paper sheets and dried under vacuum. Gels dried in this manner collapse into a cellophanelike material. Autoradiography of these dried gels is performed by sandwiching the dried gel and a piece of X-ray film between two pressure plates. After an appropriate period of exposure the X-ray film is separated from the gel and developed. Areas in contact with radioactive material are exposed. This method is inexpensive and especially useful for qualitative comparisons in slab gels. However, it can be quantitated only with difficulty and at low accuracy unless special precautions are taken (see Reference 39).

For precise quantitation the gels must be divided into small pieces and the amount of radioactivity per piece determined with a scintillation spectrometer. The most inexpensive apparatus used for this purpose is shown in Figure 6-18. It is simply a large number of razor blades with metal spacers between them; usually only one spacer is placed between blades, yielding slices of 1 mm thickness. Gels to be analyzed are frozen

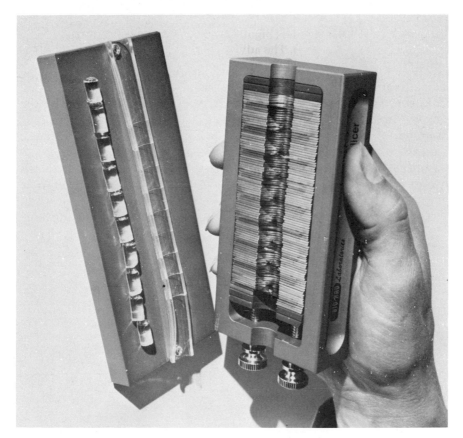

Figure 6-18. Gel slicer used to cut acrylamide gels into slices as small as 1 mm. (Courtesy of Bio-Rad Laboratories, Richmond, Calif.)

and then placed in a jig composed of a piece of Tygon tubing split longitudinally and fixed to a metal block. This holds the gel in place while it is divided into as many as 100 1 mm sections. After being cut the gel slices are removed from between the razor blades with a dissecting needle and transferred to scintillation vials. Before determination of radioactivity the gel must be dissolved. This is accomplished by soaking the acrylamide gels in 30% hydrogen peroxide or a commercially available quaternary base such as Soluene or Protosol at 37°C for 12 to 16 hours. Aqueous scintillation fluid may then be added and radioactivity determined. Usefulness of this method is somewhat compromised by the fact that hydrogen peroxide is a reasonably strong oxidizing and quenching agent. This problem may be circumvented by substituting N,N'-

diallyltartardiamide (DATD) (Figure 6-1) mole for mole in place of N,N'-methylene-bis(acrylamide). The advantage of this variation is that the gels may be dissolved in 2 to 3 hours at room temperature by 2% periodic acid, which does not significantly quench most scintillation fluids or oxidize the sample. Unfortunately, when DATD is used polymerization requires two to three times as long to reach completion and gel concentrations below 10% do not yield linear relationships of mobility to log of molecular weights (26). However, at gel concentrations at or above 10% data identical to those observed using bis(acrylamide) are obtained.

A more elaborate way of dividing the gel is by means of an autogel-divider or gel squasher such as that shown in Figure 6-19. The gel to be analyzed is placed in a tube (left side of the instrument) and then forced by a piston through a fine screen at the end of the tube. Once through the screen the crushed gel is washed away by a stream of buffer pumped past the screen with a syringelike pump located on the right side of the

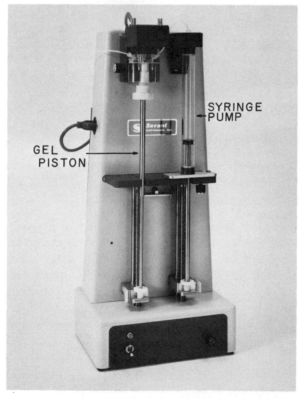

Figure 6-19. Autogeldivider used to crush acrylamide gels and suspend them in a stream of buffer. (Courtesy of Savant Instruments, Inc., Hicksville, N.Y.)

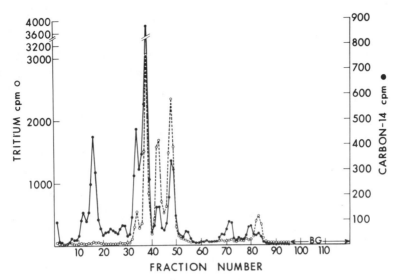

Figure 6-20. Polio virus-induced proteins from the cytoplasm of infected HeLa cells. Proteins were separated by SDS acrylamide gel electrophoresis and the gels fractionated using an autogeldivider. (○) Distribution of tritium-containing proteins from cells infected and then labeled under conditions where only virus-specified proteins are made. (●) Distribution of the label of ^{14}C labeled purified virions. [From Summers et al., *Proc. Natl. Acad. Sci. US*, **54**:505 (1965).]

instrument. The result is a stream of crushed and diluted particles that accurately reflects the sequential arrangement of the macromolecules in the gel. This stream is distributed with a fraction collector into scintillation vials such that 1 to 2 mm of gel is placed in each vial. Water (1 ml) is added to the vials and they are frozen and thawed to facilitate elution of the samples from the gel. Finally an aqueous scintillation fluid is added and the radioactivity of each is determined. The fact that the bits of gel are not dissolved does not interfere with the counting procedures because the radioactive materials are no longer in the gel but rather dissolved in solution. As shown in Figure 6-20, this method, like the gel slicer discribed above, is capable of yielding high quality patterns of radioactivity distribution.

EXPERIMENTAL

Zone Electrophoresis

6-1. Prepare solution A (gel buffer) by dissolving 7.8 g $NaH_2PO_4 \cdot H_2O$, 18.6 g Na_2HPO_4 (or 38.6 g $Na_2HPO_4 \cdot 7H_2O$), and 2.0 g sodium dodecyl sulfate in sufficient distilled water to yield a final volume of 1 liter. The pH of this solution should be 7.2.

6-2.   Prepare solution B (acrylamide) by dissolving 22.2 g recrystallized acrylamide and 0.6 g N,N'-methylene-bis(acrylamide) in sufficient distilled water to yield a final solution volume of 100 ml. This solution should be stored in a brown bottle at 4°C. Solutions more than a month old yield inferior gels and should therefore be replaced. Acrylamide is a strong, cumulative neurotoxin that is easily absorbed through the lungs or directly through the skin. Therefore, extreme care should be taken not to breathe unpolymerized acrylamide or permit it to come in contact with the skin.

Recrystallized preparations of acrylamide are available commercially or may be prepared as follows. Dissolve acrylamide (70 g) in 1 liter chloroform at 50°C. Filter the solution hot and cool it to −20°C to bring about crystallization. Collect the crystalline acrylamide in a chilled Büchner funnel, wash with chilled (−20°C) chloroform and/or heptane, and dry. Additional discussion of the purification of reagents used for electrophoresis may be found elsewhere (35, 36).

6-3.   Prepare solution C (ammonium persulfate) just before it is to be used by dissolving 35 mg ammonium persulfate in 10 ml distilled water.

6-4.   Prepare upper and lower reservoir buffers by diluting solution A of step 6-1 with distilled water in a ratio of 1:1.

6-5.   Prepare a solution of tracking dye (0.05%) by dissolving 5.0 mg bromophenol blue in 10.0 ml distilled water.

6-6.   Prepare staining solution by dissolving 1.25 g coomassie brilliant blue in a mixture of 227 ml distilled water, 227 ml methanol, and 46 ml glacial acetic acid (5:5:1). When the dye is fully dissolved, filter the solution through Whatman No. 1 filter paper or its equivalent.

6-7.   Prepare destaining solution by mixing 875 ml glass-distilled water with 50 ml methanol and 75 ml glacial acetic acid.

6-8.   Prepare gel storage solution by mixing 7.5 ml glacial acetic acid and 92.5 ml water.

6-9.   Before preparing the gels warm all the solutions to room temperature. It is also advisable to remove gas bubbles from the acrylamide solution by placing it in a filter flask, which is then evacuated for 10 to 15 minutes.

6-10.   If glass gel tubes (125 × 5 mm or equivalent) are used, make certain that they have been soaked overnight in dichromate–sulfuric acid cleaning solution and then rinsed thoroughly with distilled water. It is advisable to rinse tubes the last time with a 5% solution of Kodak Photoflo or equivalent wetting agent. This increases the ease of removing the gels from the tubes after electrophoresis.

6-11. Cover the ends of seven gel tubes with Parafilm or serum vial caps, or by some other convenient method. Check to be sure that the tubes do not leak. Plastic tubes may be used in place of glass, but their ends should be covered with dialysis tubing that is held in place with a rubber band to prevent the cast gel from slipping out of the tube. The advantage of using plastic tubes is the ease with which the gels can be removed after electrophoresis. This is especially valuable if very long gels are used.

6-12. Place the covered tubes in a convenient rack such that the tubes are perfectly vertical. A suitable rack is shown in Figure 6-21.

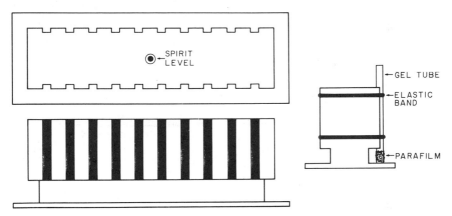

Figure 6-21. Rack for holding gel tubes during the casting of acrylamide gels. Top, front, and end views are shown. The width of the gel tube slots should not permit significant movement of the gel tubes if they are to be held perfectly vertical.

6-13. Gels of 5, 7.5, and 10% acrylamide are prepared by mixing the amounts of solutions shown in the following table:

	Amount (ml)		
Solution	5%	7.5%	10%
A	10	10	10
B	4.5	6.75	9
C	1.0	1.0	1.0
TEMED	0.032	0.032	0.032
Water	4.5	2.25	—

Prepare 7.5% gels for this experiment.

6-14. Mix the above reagents so as not to produce bubbles in the solution. It is also important to note that once ammonium persulfate has been added to the reaction mixture, the gels start to polymerize. Therefore, mix the ingredients and transfer to the casting tubes as quickly as possible. If polymerization occurs too rapidly, it may be slowed down by decreasing the amount of ammonium persulfate that is dissolved per 10 ml solution (step 6-3). Conversely, if polymerization requires greater than 20 minutes the persulfate concentration may be increased slightly.

6-15. Transfer about 2 ml of the reaction mixture prepared in steps 6-13 and 6-14 to each of the tubes prepared in steps 6-11 and 6-12 using a Pasteur pipette. This produces a gel approximately 11.5 cm long. A shorter gel may be prepared merely by decreasing the amount of gel that is placed in the tubes. Make certain that all the gels are of uniform length.

6-16. Gently pipette 0.5 to 1.0 ml of diluted solution A (step 6-4) or water on top of the gel. Take extreme care not to mix the buffer with the polymerizing gel. This may be accomplished by placing a Pasteur pipette with the tip drawn to a fine diameter just below the surface of the liquid as shown in Figure 6-22A. With the pipette in this position allow the buffer to slowly flow. As it does, raise the pipette so that throughout the entire operation the pipette end is just below the surface. If this operation is performed correctly a sharp interface should be observed between the gel and the buffer, as shown in Figure 6-22B. The purpose of this buffer

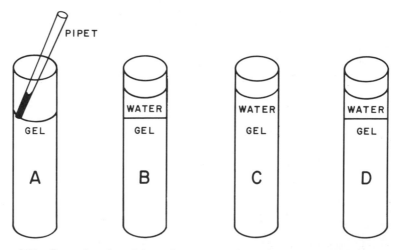

Figure 6-22. Removing the miniscus from acrylamide gels during casting. (A) Correct method of buffer addition to the gel surface. (B) Sharp phase separation visible immediately after buffer addition. (C, D) Disappearance and subsequent reappearance of this phase separation during and after polymerization of the gel, respectively.

addition is to flatten the surface of the gel. If allowed to polymerize in the absence of buffer, the gel would have a concave surface owing to the presence of a miniscus resulting in crescent shaped protein bands after electrophoresis.

6-17. Very shortly after the interface appears between the buffer and the gel surface, it disappears, as shown in Figure 6-22C. This is to be expected and is caused by a small amount of diffusion that occurs at the interface. When the gel has fully polymerized (15–20 minutes) the interface again becomes visible, as shown in Figure 6-22D.

6-18. Leave the buffer on top of the gel to prevent drying until it can be used.

6-19. Prepare 100 ml of a solution composed of 5.0 ml solution A prepared in step 6-1, 2.0 ml 2-mercaptoethanol, 2.0 g of sodium dodecyl sulfate, and 93 ml water.

6-20. Prepare six solutions—one of each protein listed below—by dissolving 10 mg of each protein in 10 ml of the solution prepared in step 6-19. These six solutions will serve as the standards for the following electrophoretic separations.

Protein	MW
Myoglobin	17,200
Carbonic anhydrase	29,000
Glyceraldehyde-3P dehydrogenase	36,000
Aldolase	40,000
Fumarase	48,500
Catalase	57,500

6-21. Six electrophoretic sample solutions are prepared using one of the proteins from step 6-20 for each sample. The final solution is composed of 0.01 ml 50% aqueous glycerol, 0.030 ml of the solution prepared in step 6-19, 0.005 ml of the bromophenol blue solution prepared in step 6-5, and 0.005 ml of the protein solution prepared in step 6-20.

6-22. In addition to the six samples prepared in step 6-21, prepare a seventh sample composed of 0.01 ml 50% aqueous glycerol, 0.005 ml of the solution prepared in step 6-19, 0.005 ml of the bromophenol blue solution prepared in step 6-5, and 0.005 ml of each of the protein solutions prepared in step 6-20.

6-23. The above seven solutions make up the standards to be used in calibration of the gels. If unknown samples are provided, treat them according to steps 6-20 and 6-21. Additional tubes will be needed, however, to accommodate the unknowns.

6-24. Place all the samples in a boiling water bath for 2 minutes.

6-25. If alkylation of the sample is necessary, perform it at this point. In this experiment it is omitted.

6-26. Remove all of the buffer from the top of the gels. This may be done by inverting the tubes and giving each an abrupt snap.

6-27. Carefully pipette 0.01 ml of each sample onto the surface of each gel tube.

6-28. Note that only 1 to 5 μg of protein is placed on each gel. If the solution is more complex, up to 50 μg may be used (note that 30 μg total was used in the sample prepared in step 6-22), but amounts in excess of this yield very poor results owing to overloading.

6-29. Assemble the gel tubes into the upper buffer reservoir and remove the coverings from the tube bottoms. If plastic tubes are being used, pierce the dialysis tubing covering the end of the tube four or five times with a needle. This allows ion exchange and current flow. Do not remove the dialysis tubing, however, or the gels will slip out of the tubes during electrophoresis. Plug any unused holes in the upper reservoir.

6-30. Using the method described in step 6-16, fill the gel tube completely full of the reservoir buffer prepared in step 6-4.

6-31. Fill the lower reservoir with buffer and lower the tubes into it. If bubbles are observed on the bottom interface of the gel remove them. This may be done by flushing the gel bottom with reservoir buffer using a bent Pasteur pipette, as shown in Figure 6-23.

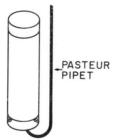

PASTEUR
PIPET

Figure 6-23.   Removal of air bubbles from the bottom of a gel tube using a Pasteur pipette bent into the form of a "J."

6-32. Gently fill the upper reservoir and connect the electrodes to the electrophoretic apparatus. Be sure that the anode is at the bottom.

6-33. Energize the power supply and adjust it to an output of 3 to 4 mA per gel tube (i.e., 21–28 mA total for a total of 7 tubes). DO NOT TOUCH THE SYSTEM BECAUSE A FATAL ELECTRICAL SHOCK MAY RESULT. Allow the system to operate in this manner for 15 to 20 minutes. This provides sufficient time for the protein samples to enter the gels.

6-34. At this time increase the current in the system to 6 to 8 mA per tube (42–56 mA and 56 V total) and operate the power supply at this constant current until the tracking dye has migrated to a position 0.5 to 1.0 cm above the bottom of the tube.

6-35. Turn off the power and unplug the power supply from the wall outlet. Now the electrodes may be safely removed from the apparatus.

6-36. Disassemble the apparatus and pour off the reservoir buffers. These may be used several times before they are discarded if the electrophoretic separations do not require too much time. If the electrophoresis time is extended, however, discard them at the conclusion of each experiment.

6-37. Remove the gels from the gel tubes by gently injecting distilled water between the wall of the glass tube and the gel using a syringe fitted with a 2 in., 24 or 26 gage, blunted needle. Move the syringe in a circular motion around the gel as it is inserted and the water is injected. This is shown in Figure 6-24.

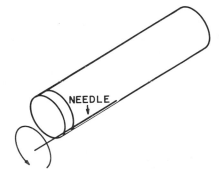

NEEDLE

Figure 6-24. Removal of acrylamide gels from their casting tubes using a 2 in., 26 gage hypodermic needle.

6-38. After removing the gel from the gel tube, determine its length carefully and insert a fine piece of nickel–chromium wire or fine thread (using a fine gage needle) through the dye marker band. This is necessary because during staining the dye will diffuse out of the gel and its position will be lost. Now place the gel gently in a 16×150 mm test tube and cover it with the staining solution prepared in step 6-6, allowing it to stand 10 to 12 hours at room temperature. Staining time may be reduced to a minimum of 2 to 4 hours, but some loss of sensitivity will result.

6-39. Remove the stain from the gels and rinse them several times with water. Transfer the gels to plastic test tubes containing holes like those shown in Figure 6-25. Make holes in the tubes by heating a large nail and

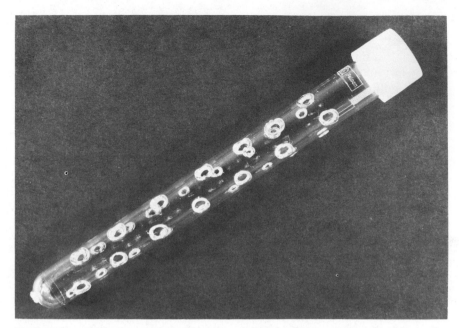

Figure 6-25. Tube used to destain the gels. Note that holes do not extend all the way to the top of the tube. This space is needed to trap an air bubble which keeps the tube afloat. The gel number may be written on the tube cap using water insoluble ink.

gently forcing it through the tube. Cap the tubes trapping a large air bubble in each one. Identify the tubes in some convenient way.

6-40. Place these tubes in a 1 liter Erlenmeyer flask full of destaining solution (step 6-7). Stir gently using a magnetic stirrer. This operation may be performed at room temperature, but destaining will be hastened if it is carried out at 37°C.

6-41. The destaining solution should be changed as soon as it has become highly discolored. Continue destaining with intermittent changes of destaining solution until the background regions between the proteins no longer contain any stain.

6-42. Remove the gels from the destaining tubes at the conclusion of destaining and rinse them with water.

6-43. Gels to be preserved are placed in tubes containing the 7.5% acetic acid solution (step 6-8). Keep stained gels out of direct sunlight or other strong light sources because they cause the stained bands to fade.

6-44. Measure the distance that the dye marker has migrated from the origin.

6-45. Measure the distance that each of the standard and unknown proteins have migrated from the origin.

6-46. Calculate the relative mobility of each protein using the equation:

$$\text{mobility} = \frac{(\text{distance of protein migration})(\text{gel length before staining})}{(\text{distance of dye migration})(\text{gel length after destaining})}$$

The gel length before and after destaining must be considered in this equation because the gels expand to different degrees during the staining process.

6-47. Plot the mobilities obtained from step 6-46 as a function of the molecular weights given in step 6-20 on semilogarithmic coordinates. Such a plot is shown in Figure 6-26.

Zone Electrophoresis of Fluorescamine Labeled Proteins

6-48. Repeat steps 6-1 to 6-4 inclusively.

6-49. Repeat steps 6-9 to 6-18 inclusively.

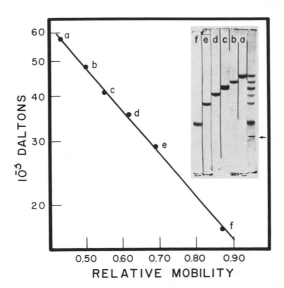

Figure 6-26. Separation of standard proteins using SDS acrylamide gel electrophoresis. The standards are *a*, catalase; *b*, fumarase; *c*, aldolase; *d*, glyceraldehyde-3P dehydrogenase; *e*, carbonic anhydrase; and *f*, myoglobin. The inset shows Coomassie Brilliant Blue staining of the acrylamide gels from which these data were derived. Arrow indicates the location of the dye marker. In this case it has been marked by inserting a small piece of nickel–chromium wire through the gel at the appropriate level.

6-50. Prepare fluorescamine staining solution by dissolving 5.0 mg fluorescamine in 1.0 ml dimethyl sulfoxide. This solution is very sensitive to light. Therefore, the container used for its storage should be wrapped in aluminum foil.

6-51. Repeat step 6-20 with the exception that the 10 mg samples should be dissolved in 3 ml 0.1M phosphate or borate buffer adjusted to pH 9.0.

6-52. For each of the protein solutions prepared in step 6-51, prepare an electrophoretic sample solution composed of 0.01 ml 50% aqueous glycerol, 0.020 ml of the solution prepared in step 6-19, and 0.005 ml of the protein solution prepared in step 6-51.

6-53. Prepare unknown samples, if they exist, in the same manner.

6-54. Repeat step 6-24. Cool the samples back to room temperature.

6-55. Add 0.030 ml of the fluorescamine solution prepared in step 6-50 to each of the cooled samples from step 6-54. Mix the samples as vigorously and quickly as possible after adding fluorescamine so that reaction between the stain and proteins occurs before the reagent decomposes.

6-56. Add 0.005 ml tracking dye (step 6-5) to the solution from step 6-55.

6-57. Repeat step 6-26.

6-58. Carefully pipette 0.01 ml of each of the samples prepared in step 6-56 onto the surface of each of the prepared gel tubes.

6-59. Repeat steps 6-29 to 6-37, inclusively. Darken the room during electrophoresis and shine an ultraviolet lamp on the tubes.

6-60. Rinse and acrylamide gels briefly with water and place them under a long wavelength ultraviolet lamp.

6-61. Repeat steps 6-44 and 6-45.

6-62. Calculate the relative mobility of each protein using the equation:

$$\text{mobility} = \frac{\text{distance of protein migration}}{\text{distance of dye migration}}$$

6-63. Repeat step 6-47. Compare these data with those obtained in step 6-47.

Disc Gel Electrophoresis of Lactate Dehydrogenase Using Nitroblue Tetrazolium for Enzyme Visualization

6-64. Prepare solution A (lower gel buffer) by dissolving 56.75 g Tris buffer in 200 ml water. Adjust the pH to 8.9 using concentrated HCl. Add sufficient water to bring the volume to 250 ml.

6-65. Prepare solution B (acrylamide) by dissolving 93.75 g acrylamide and 2.5 g N,N'-methylene-bis(acrylamide) in sufficient distilled water to yield a final volume of 250 ml.

6-66. Repeat step 6-5.

6-67. Prepare the upper reservoir buffer by dissolving 6 g Tris buffer and 28.8 g glycine in sufficient distilled water to yield a final volume of 1 liter. Check the pH of the solution. If it is not 8.3, adjust the pH with a small amount of Tris or glycine.

6-68. Prepare the lower reservoir buffer by dissolving 908 g Tris buffer in 2 liters water. Add sufficient concentrated HCl to yield pH 8.9. Add sufficient water to yield a final volume of 4 liters. Recheck the pH and adjust if necessary.

6-69. Prepare solution C (ammonium persulfate) by dissolving 17.5 mg ammonium persulfate in 10.0 ml distilled water.

6-70. Prepare solution D (upper gel buffer) by dissolving 8.9 g Tris buffer and 40.0 ml $1.0M$ H_3PO_4 in sufficient distilled water to yield a final volume of 250 ml. The final pH of the solution should be 6.5 to 6.7.

6-71. Prepare solution E (acrylamide) by dissolving 31.25 g acrylamide and 7.8 g N,N'-methylene-bis(acrylamide) in sufficient distilled water to yield a final volume of 250 ml. Store all acrylamide solutions at 4°C in brown bottles.

6-72. Prepare solution F (riboflavin) by dissolving 2.5 mg riboflavin in 100 ml distilled water. Store at 4°C in a brown bottle.

6-73. Repeat steps 6-9 to 6-12 inclusively.

6-74. Gels of 5, 7.5, and 10% are prepared by mixing the following amounts of solutions:

	Amount (ml)		
Solution	5%	7.5%	10%
A	0.4	0.4	0.4
B	1.33	2.0	2.67
C	1.0	1.0	1.0
TEMED	0.016	0.016	0.016
Water	7.27	6.60	5.93

Prepare a 7.5% gel for this experiment.

6-75. Repeat steps 6-14 and 6-15 using the reagents prepared for this experiment with the exception that the depth of the gel in the present case should be 10.5 cm instead of the 11.5 cm used in step 6-15.

6-76. Repeat steps 6-16 and 6-17 but replace solution A called for in step 6-16 with an equivalent amount of solution A prepared in step 6-64.

6-77. Prepare the upper, stacking gel by mixing the amounts of solutions shown in the following table:

Solution	Amount (ml)
D	1.0
E	1.0
F	1.0
Water	4.0

6-78. Mixing of the above reagents should be performed so as not to produce bubbles in the solution.

6-79. Remove buffer A and rinse the top of each gel with a small amount of the solution prepared in steps 6-77 and 6-78.

6-80. Transfer a sufficient amount of the solution prepared in steps 6-77 and 6-78 onto the top of each gel to yield a depth of 1.0 to 1.3 cm.

6-81. Repeat steps 6-16 and 6-17 but replace solution A called for in step 6-16 with an equivalent amount of water.

6-82. Place the gel tubes in close proximity to a strong fluorescent light source (within 2–5 cm). Allow the gels to polymerize until the upper gel becomes opalescent.

6-83. Carefully remove the water from the top of the gel.

6-84. Prepare a sample of freshly drawn human or rabbit serum by mixing 1.0 ml serum, 0.25 ml 50% aqueous glycerol, and 0.10 ml tracing dye (step 6-5).

6-85. Gently pipette 0.02 to 0.05 ml of this solution onto the top of the gel (step 6-83). Other amounts may be placed on the remaining gels. Place the electrophoresis apparatus and power supply in a cold room (4°C).

6-86. Repeat steps 6-29 to 6-32 inclusively but replace the buffers called for in these steps with the following solutions: upper reservoir buffer— the solution prepared in step 6-67 diluted 1 part buffer with 9 parts water; lower reservoir buffer—the solution prepared in step 6-68 diluted 1 part buffer with 4 parts water.

6-87. Energize the power supply and adjust it to an output of 1 to 2 mA/gel. The voltage should be held constant at no more than 75 V. DO NOT TOUCH THE SYSTEM BECAUSE FATAL ELECTRICAL SHOCK MAY RESULT.

6-88. Continue in this manner until the tracking dye has migrated to a position 0.5 to 1.0 cm above the bottom of the tube.

6-89. Repeat steps 6-35 to 6-37 inclusively.

6-90. Rinse the gels with water and place them in 13 × 100 mm test tubes.

6-91. Fill each tube with a solution composed of the following components at the final concentrations indicated: sodium lactate, 20 μliters 60% syrup/ml; NAD, 0.7 mg/ml; phenazine methosulfate, 0.023 mg/ml; p-nitroblue tetrazolium, 0.4 mg/ml; and 100mM Tris–HCl buffer adjusted to pH 8.0. The components of this solution may be prepared ahead of time and kept frozen. Even in this condition the phenazine methosulfate and nitroblue tetrazolium containers should be wrapped with aluminum foil because these compounds are very sensitive to light. Once the components are combined, they should be used immediately. Incubate one of the gel tubes known to have a large amount of protein in a reaction mixture identical to that just described above but omit lactate from the mixture.

6-92. Cover the tubes with Parafilm and gently invert them several times to ensure that any water adhering to their surfaces is replaced with the enzyme reaction mixture.

6-93. Incubate the tubes at 37°C until the three to five blue bands of insoluble formazan are clearly visible.

6-94. When the gels have stained to the desired intensity remove them from the reaction mixture and wash them thoroughly with water. This will slow the reaction but these gels should not be considered as permanent; the results should be recorded by photographing them, as shown in Figure 6-27.

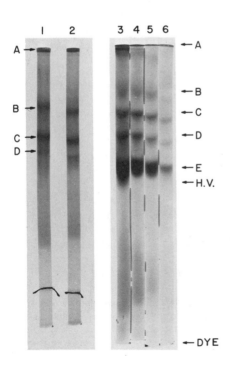

Figure 6-27. Acrylamide gel electrophoresis of lactate dehydrogenase from human (gels 1 and 2) and rabbit (gels 3–6) sera. The sample sizes used on gels 1 and 2 were 0·035 and 0·020 ml, respectively. A piece of thread marks the dye front. The smear extending below band D of gels 1 and 2 results from association of the bromophenol blue dye with some of the serum proteins. The sample sizes of gels 3, 4, 5, and 6 were 0.05, 0.04, 0.03, and 0.02 ml, respectively. Staining the rabbit sera was carried out overnight at room temperature in the dark. In the case of human serum, however, staining is complete in 40 to 60 minutes at 37°C. The voltage used during electrophoresis of the rabbit serum was intentionally increased to twice (150 V) the recommended value (75 V) to demonstrate the effects of this error. Note that it caused a diffuse band to run ahead of the main protein bands. This is indicated by the letters $H.V.$ in the figure.

REFERENCES

1. J. R. Cann, *Biochemistry*, **5**:1108–1112 (1966). Multiple Electrophoretic Zones Arising from Protein–Buffer Interaction.
2. D. Rodbard and A. Chrambach, *Proc. Natl. Acad. Sci., US*, **65**:970–977 (1970). Unified Theory for Gel Electrophoresis and Gel Filtration.
3. A. Chrambach and D. Rodbard, *Science*, **172**:440–451 (1971). Polyacrylamide Gel Electrophoresis.
4. D. Rodbard and A. Chrambach, *Anal. Biochem.*, **40**:95–134 (1971). Estimation of Molecular Radius, Free Mobility and Valence Using Polyacrylamide Gel Electrophoresis.
5. D. Rodbard, G. Kapadia, and A. Chrambach, *Anal. Biochem.*, **40**:135–157 (1971). Pore Gradient Electrophoresis.
6. J. Lunney, A. Chrambach, and D. Rodbard, *Anal. Biochem.*, **40**:158–173 (1971). Factors Affecting Resolution, Bandwidth, Number of Theoretical Plates and Apparent Diffusion Coefficients in Polyacrylamide Gel Electrophoresis.
7. O. Gabriel, Analytical Disc Gel Electrophoresis, in *Methods in Enzymology*, Vol. 22 (W. B. Jakoby, Ed.), Academic Press, New York, 1971, pp. 565–577.
8. J. L. Hedrick and A. J. Smith, *Arch. Biochem. Biophys.*, **126**:155–164 (1968). Size and Charge Isomer Separation and Estimation of Molecular Weights of Proteins by Disc Gel Electrophoresis.
9. A. L. Shapiro, E. Vinuela, and J. V. Maizel, *Biochem. Biophys. Res. Commun.*, **28**:815–820 (1967). Molecular Weight Estimation of Polypeptide Chains by Electrophoresis in SDS–Polyacrylamide Gels.
10. K. Weber and M. Osborn, *J. Biol. Chem.*, **244**:4406–4412 (1969). The Reliability of Molecular Weight Determinations by Dodecyl Sulfate–Polyacrylamide Gel Electrophoresis.
11. W. F. Studier, *J. Mol. Biol.*, **79**:237–248 (1973). Analysis of Bacteriophage T₇ Early RNAs and Proteins on Slab Geis.
12. G. F. L. Ames, *J. Biol. Chem.*, **249**:634–644 (1974). Resolution of Bacterial Proteins by Polyacrylamide Gel Electrophoresis on Slabs.
13. T. D. Kempe, D. M. Gee, G. M. Hathaway, and E. A. Noltmann, *J. Biol. Chem.*, **249**:4625–4633 (1974). Subunit and Peptide Compositions of Yeast Phosphoglucose Isomerase Isoenzymes.
14. A. C. Peacock and C. W. Dingman, *Biochemistry*, **7**:688–674 (1968). Molecular Weight Estimation and Separation of Ribonucleic Acid by Electrophoresis in Agarose–Acrylamide Composite Gels.
15. A. E. Dahlberg, C. W. Dingman, and A. C. Peacock, *J. Mol. Biol.*, **41**:139–147 (1969). Electrophoretic Characterization of Bacterial Polyribosomes in Agarose–Acrylamide Composite Gels.
16. E. G. Richards and R. Lecanniduo, *Anal. Biochem.*, **40**:43–71 (1971). Quantitative Aspects of the Electrophoresis of RNA in Polyacrylamide Gels.
17. A. S. Lee and R. L. Sinsheimer, *Proc. Natl. Acad. Sci. US*, **71**:2882–2886 (1974). A Cleavage Map of Bacteriophage $\phi\chi$ 174 Genome.
18. S. Udenfriend, S. Stein, P. Bohlen, W. Dairman, W. Leimgruber, and M. Weigele, *Science*, **178**:871–872 (1972). Fluorescamine: A Reagent for Assay of Amino Acids, Peptides, Proteins and Primary Amines in the PicoMole Range.
19. W. L. Ragland, J. L. Pace, and D. L. Kemper, *Anal. Biochem.*, **59**:24–33 (1974).

Fluorometric Scanning of Fluorescamine-labeled Proteins in Polyacrylamide Gels.

20. S. Udenfriend and B. K. Hartman, *Anal. Biochem.*, **30**:391–394 (1969). A Method for Immediate Visualization of Proteins in Acrylamide Gels and its Use for Preparation of Antibodies to Enzymes.

21. O. Gabriel, Locating Enzymes on Gels, in *Methods in Enzymology*, Vol. 22 (W. B. Jakoby, Ed.), Academic Press, New York, 1971, pp. 578–604.

22. A. H. Wardi and G. A. Michos, *Anal. Biochem.*, **49**:607–609 (1972). Alcian Blue Staining of Glycoproteins in Acrylamide Disc Electrophoresis.

23. K. Timmis, F. Cabello, and S. N. Cohen, *Proc. Natl. Acad. Sci. US*, **74**:4556–4560 (1974). Utilization of Two Distinct Modes of Replication by a Hybrid Plasmid Constructed *in vitro* from Separate Replicons.

24. M. R. Green, J. V. Pastewka, and A. C. Peacock, *Anal. Biochem.*, **56**:43–51 (1973). Differential Staining of Phosphoproteins on Polyacrylamide Gels with a Cationic Carbocyanine Dye.

25. G. Fairbanks, C. Levinthal, and R. H. Reeder, *Biochem. Biophys. Res. Commun.*, **29**:393–399 (1965). Analysis of ^{14}C-labeled Proteins by Disc Electrophoresis.

26. H. S. Anker, *FEBS Lett.*, **7**:293 (1970). A Solubilizable Acrylamide Gel for Electrophoresis.

27. A. H. Gordon, Electrophoresis of Proteins in Polyacrylamide and Starch Gels, in *Laboratory Techniques in Biochemistry and Molecular Biology* (T. S. Work and E. Work, Eds.), Vol. 1, Part I, North-Holland, Amsterdam, 1971.

28. M. Bier, *Electrophoresis: Theory, Methods, and Applications*, Vol. I, Academic Press, New York, 1959.

29. M. Bier, *Electrophoresis: Theory, Methods, and Applications*, Vol. II, Academic Press, New York, 1967.

30. D. J. Shaw, *Electrophoresis*, Academic Press, New York, 1969.

31. H. R. Maurer, *Disc Electrophoresis*, 2nd ed., DeGruyter, New York, 1971.

32. E. DeVito and J. A. Santome, *Experientia*, **22**:124 (1966). Disc Electrophoresis in the Presence of Sodium Dodecyl Sulfate.

33. E. F. Ambrose, *Cell Electrophoresis*, Little, Brown, Boston, 1965.

34. B. Paterson and R. C. Strohman, *Biochemistry*, **9**:4094–4105 (1970). Myosin Structure as Revealed by Simultaneous Electrophoresis of Heavy and Light Subunits.

35. U. E. Loening, *Biochem. J.*, **102**:251 (1967). Fractionation of High-Molecular Weight Ribonucleic Acid by Polyacrylamide-Gel Electrophoresis.

36. D. H. Grant, J. N. Miller, and O. T. Burns, *J. Chromatogr.*, **79**:267–273 (1973). The Fluorescence Properties of Polyacrylamide Gels.

37. P. H. O'Farrell, *J. Biol. Chem.*, **250**:4007–4021 (1975). High Resolution Two-Dimensional Electrophoresis of Proteins.

38. J. S. Fawcett, *FEBS Lett.*, **1**:81–82 (1968). Isoelectric Fractionation of Proteins on Polyacrylamide Gels.

39. R. A. Laskey and A. D. Mills, *Eur. J. Biochem.*, **56**:335–341 (1975). Quantitative Film Detection of ^{3}H and ^{14}C in Polyacrylamide Gels by Fluorography.

40. K. Weber and M. Osborn, Proteins and Sodium Dodecyl Sulfate: Molecular Weight Determination on Polyacrylamide Gels and Related Procedures, in *The Proteins*, Vol. 1 (H. Neurath and R. L. Hill, Eds.), Academic Press, New York, 1975.

41. U. K. Laemmli, *Nature*, **227**:680–685 (1970). Cleavage of Structural Proteins during the Assembly of the Head of Bacteriophage T4.

Chapter 7

Affinity Chromatography

Conventional purification procedures such as those discussed in Chapters 4, 6, and 10 all depend heavily on relatively small physiochemical differences of the proteins in a mixture. With such low specificity a large number of methods are usually required to attain reasonable levels of purity. The time consumed by such laborious approaches and the unsatisfactory yields obtained prompted development of a powerful isolation procedure called affinity chromatography. The unequaled success of this technique derives from its ability to exploit the most unique and specific property of macromolecules, their biological function. The in vivo task of most significant macromolecules may be divided into two parts, recognition of other molecules and a biochemical response following that recognition. The response may take the form of catalysis if the macromolecule is an enzyme, or prevention of transcription if the macromolecule is a control protein. The responses in fact are as varied as the nucleic acids and proteins that carry them out, but in all cases appropriate recognition must occur first. Recognition almost always involves binding one or more molecules to the macromolecule. Enzyme proteins are used as examples here, but the same principles apply to the other macromolecules as well. In simple terms enzymatic reactions occur according to the reaction sequence

$$E + S \underset{K_{-1}}{\overset{K_1}{\rightleftharpoons}} ES \xrightarrow{K_2} P \tag{1}$$

The first reaction in which substrate (S) binds to the enzyme (E) to yield an enzyme–substrate complex (ES) represents the recognition needed for efficient biological function and is the step of interest here; P is the normal product of the reaction. The basis of affinity chromatographic techniques

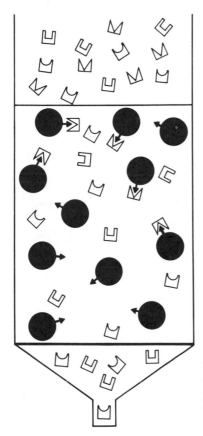

Figure 7-1. Separation of macromolecules by affinity chromatography. Cutout squares, semicircles, and triangles are schematic representations of ligand binding sites on the macromolecules.

is to covalently affix the molecules being recognized (S) to an immobile solid matrix such as an agarose bead (see Figure 7-1). A mixture containing the desired macromolecule is then permitted to percolate through this matrix. The vast majority of molecules have no affinity for the bound molecule or ligand and flow through the matrix unretarded. The desired macromolecule, however, recognizes the bound molecule, binds to it, and thus is retarded. After all the undesired components have been flushed from the column the conditions of the wash solution are altered to bring about dissociation of macromolecule and bound ligand. The desired macromolecule consequently appears in the effluent largely purified from the original mixture. Purification capabilities such as these are demonstrated by the data in Figure 7-2. A crude extract of *E. coli* was allowed to flow through an appropriately constructed column designed to isolate β-galactosidase. Acrylamide gel electrophoresis of fractions 40 (unad-

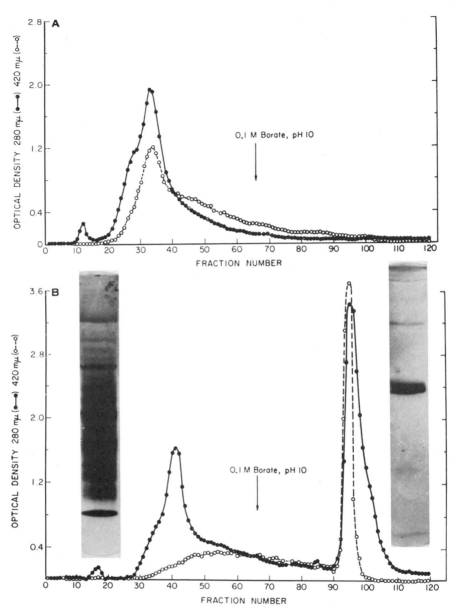

Figure 7-2. Affinity chromatographic patterns of extracts from *E. coli* on unsubstituted Sepharose 4B (*A*) and on Sepharose 4B substituted with *p*-aminophenyl-β-D-galactopyranoside (*B*). The pyranoside is a substrate analogue for β-galactosidase. The column was equilibrated and chromatographed with 0.05*M* Tris–HCl buffer, pH 7.5. Elution of adsorbed protein was carried out with 0.1*M* sodium borate, pH 10.05. The extract (20 ml) was applied to a column, 1.5 × 22 cm. The column was run at room temperature (23°C) with a flow rate of 80 ml/hour. Fractions containing 0.8 ml were collected. Protein was measured spectrophotometrically at 280 nm (●————●), enzymatic activity at 420 nm (○————○). Inset photographs depict polyacrylamide gel electrophoresis of samples derived from fractions 40 (left inset) and 90 (right inset).

sorbed fraction) and 96 (specifically bound proteins) amply demonstrate the large degree of purification obtained.

In principle, these methods can be used to purify almost any macromolecule including enzymes, antibodies, specific and general classes of nucleic acids, vitamin-binding proteins, repressors and other control elements, transport proteins, and drug and hormone receptors. Table 7-1 lists a few of the large number of diverse molecules purified in this manner. Reference 1 has a much more extensive compilation but even that list is incomplete owing to the ever-increasing application of these methods. Less apparent advantages of affinity chromatography include (1) rapid separation of the macromolecules being purified from destructive contaminants such as proteases or nucleases, (2) separation of active (capable of ligand recognition) and inactive forms of the protein, and (3) protection from denaturation, resulting from ligand-active site associations.

Careful planning must be done before specific laboratory procedures are performed. Major considerations include (1) the type of matrix used, (2) the nature of ligand and means of covalently binding it to the matrix, and (3) the conditions used to absorb and elute the macromolecule from the column. Although some general guidelines are available, it must be emphasized that conditions needed to purify a given macromolecule must be specifically tailored to meet the unique biological properties of that molecule.

Table 7-1. Purification of Various Macromolecules Using Affinity Chromatography

Enzyme	Immobilized Ligand
Adenosine deaminase	Adenosine
Amino peptidase	Hexamethylenediamine
APO-aspartate aminotransferase	Pyridoxal-5'-phosphate
Avidin	Biocytin
Carbonic anhydrase	Sulfanilamide
Chorismate mutase	Tryptophan
α-Chymotrypsin	Tryptophan
Glycerol-3P dehydrogenase	Glycerol-3-phosphate
Isoleucyl-tRNA synthetase	Aminoacyl-tRNA
Thrombin	Benzamidine
Xanthine oxidase	Allopurinol
Coagulation factor	Heparin
Follicle-stimulating hormone	Concanavalin A
Gal repressor	p-aminophenyl-α-thiogalactoside
Interferon	Antibody
Thyroxine-binding protein	Thyroxine

CHROMATOGRAPHIC MATRIX

Characteristics desired of a matrix for affinity chromatography resemble in many ways those sought for a molecular sieve medium. They are as follows:

1. Low nonspecific absorption.
2. Good flow characteristics.
3. Chemical and mechanical stability over a broad range of pH, ionic strengths, and denaturant concentrations.
4. Availability, in large numbers, of chemical groups capable of being activated.
5. High effective porosity.

Desirability of the first three characteristics is self-evident in view of their discussion in Chapter 6. The remaining two characteristics may require brief explanation. As is discussed later, the degree of binding observed for a given macromolecule is a function of the available ligand concentration. This is especially true if the affinity between the ligand and macromolecule is low. Therefore, many matrix sites must be available for ligand attachment under mild conditions. These sites once substituted must also be easily approachable by the macromolecule and hence the medium must exhibit high porosity as well.

There are two instances when the degree of porosity is not a crucial factor. The first arises when very high affinity exists between the macromolecule and bound ligand. In this case the limited number of sites available on the matrix surface are quite sufficient for total binding of the macromolecule. The second arises when the binding species is much too large to penetrate even the most porous support media. Such circumstances are encountered when large particles such as polysomes, membrane fragments, or even whole cells are being isolated by these techniques. For example, insulin–agarose is used for isolation of fat cell ghosts containing insulin acceptors and hapten–polyacrylamide is used for isolation of lymphocytes with appropriate acceptors. However, for separations of this type, steps must be taken to ensure that the matrix does not form a mesh that physically denies cells or subcellular organelles free movement through the medium.

Presently three media can be considered as potentially useful in affinity chromatographic techniques: agarose, polyacrylamide, and controlled porosity glass beads. Of these three, agarose beads are by far the most often used because they possess all the desirable features cited above. The only major disadvantage of agarose is its susceptibility to contraction when denaturant solutions are used for elution.

Beaded polyacrylamide meets many of the desired criteria. Most notably it possesses an extremely large number of modifiable carboxamide groups, permitting preparation of a highly substituted matrix. Unfortunately, the medium suffers from a lack of porosity, which is decreased further on substitution. During evaluation of this medium for possible use in the purification of β-galactoside it was found that tenfold more ligand could be covalently bound to polyacrylamide than could be affixed to an equivalent amount of agarose. No binding of β-galactosidase to the substituted polyacrylamide could be demonstrated, whereas in the case of substituted agarose, binding occurred readily. Although it could be argued that the large size of β-galactosidase ($\sim 200,000$ daltons) made this a poor test case, similar difficulties were observed with staphylococcal nuclease, a small protein of only $\sim 17,000$ daltons. Such problems have limited use of this medium in the past, but are now beginning to be circumvented.

A medium that may find increased use in the future is controlled pore glass beads. This medium, briefly mentioned in Chapter 6, is produced by heating borosilicate glass to 700 to 800°C. At these temperatures the borate and silicate phases separate. After cooling, borate phases are dissolved in acid leaving a porous silica glass. Further etching produces pore sizes ranging from 45 to 2500 Å in diameter. The glass is ground and the resulting beads are sized with fine mesh screens. The greatest advantages of this medium are its mechanical and chemical stability, which permit good flow rates under a wide variety of conditions. The most serious problems encountered with glass beads as a medium are a high degree of nonspecific protein adsorption and the lack of large numbers of easily activated functional groups. Some progress is being made to circumvent these problems by first coating the glass surfaces with dextran or other substances, but considerable advancement in these technologies will be necessary before this becomes the medium of choice.

LIGAND SELECTION

Selection of the ligand to be used in construction of an affinity column requires careful consideration. Possible candidates for this include substrate analogues, effectors, enzyme cofactors, and under certain circumstances the enzyme substrate. When a substrate is used, conditions must be arranged such that the enzyme does not function catalytically. This may be accomplished by omission of required metal ions, a change in pH if the K_m and K_{cat} pH profiles are different or low temperatures. For those enzymes which catalyze biomolecular reactions, one substrate may be easily used if (1) the other needed substrate of the reaction is scrupulously excluded from the reaction mixture and (2) binding of one substrate occurs in the absence of the other.

Two criteria must be met in selection of the best ligand. First there must be a strong interaction between the ligand and desired protein. In practice ligands with dissociation constants at or above $5mM$ in solution are poor candidates. On the other hand, too high an affinity between ligand and protein can also be detrimental because the conditions needed for dissociation of the complex may also irreversibly damage the protein. A good example of this is use of avidin as the ligand to purify biotin-containing carboxylases. The dissociation constant for the biotin–avidin complex is approximately $10^{-15}M$ which requires $6M$ guanidine–HCl at pH 1.5 for dissociation. These conditions will irreversibly damage most, if not all, of the labile carboxylases. Second, the small molecule to be bound must have functional groups that can be modified to form the covalent linkage with the supporting matrix. Equally important, the linkage modification should not seriously impair binding of the ligand to the desired protein. If the affinity observed in solution is very great (a dissociation constant in the micromolar range) a decrease in affinity of 1000-fold on covalent bonding may be tolerable. The important parameter is the effective affinity observed between the insolubilized ligand and desired protein. Insight regarding this parameter may be gained from preliminary experiments which measure in solution the apparent K_i values of the chosen ligand modified in a manner similar to that planned for its insolubilization.

LINKAGE OF LIGAND AND SUPPORTING MATRIX

Procedures used to covalently couple the ligand to the supporting matrix involve (1) activation of the matrix functional groups and (2) coupling of the ligand to these activated groups. The chemical reactions used to bring about linkage must be mild enough to be tolerated well by both ligand and matrix. Following linkage the support matrix must be exhaustively washed to remove any remaining unbound ligand, and the amount of ligand bound must be determined. The latter measurement is most usefully expressed in terms of capacity per milliliter of packed matrix rather than in terms of its dry weight. These measurements are easily made if radioactive ligand of low but known specific activity is used. Otherwise, ligand must be removed from a sample of the prepared absorbent and its concentration determined in some other way.

The number of methods for linking ligands to insoluble supports has grown enormously in the past several years. Only a few prominent examples are discussed in detail here to provide an appreciation of the important parameters. Variation of the conditions under which linkage is carried out profoundly affects both the resulting affinity absorbent and its use.

Sepharose 4B is one of the most often used beaded agarose derivatives. It is more porous than the 6B derivative and has higher substitution capacity than the 2B medium. The following procedures are necessary to activate this medium and covalently attach primary aliphatic or aromatic amines to the active form. The activation is usually carried out using cyanogen halide treatment. Any procedures involving cyanogen bromide must be carried out in a well ventilated fume hood (even weighing the material) because of its toxicity. Cyanogen bromide is best stored in a dry double container at 4°C. A given volume of agarose gel is gently suspended in an equivalent amount of water and pH electrodes are placed in the solution for constant monitoring. The suspension is stirred using an overhead stirring assembly such as that shown in Figure 7-3. Use of magnetic stirring bars should be avoided because it generates fines in the agarose. An appropriate amount (50–300 mg/ml of packed gel) of cyanogen bromide is weighed and ground into a fine powder using a cold mortar and pestle. The amount of cyanogen bromide added dictates the amount of agarose activated and hence the degree of substitution obtained. Therefore, if a very high degree of substitution is desired, amounts near 300 mg/ml may be used. Conversely, as the desired degree of substitution decreases, so also does the amount of cyanogen bromide needed. The pH of the suspension is raised to 11 using sodium hydroxide and all of the finely divided cyanogen bromide is added at one time. Under

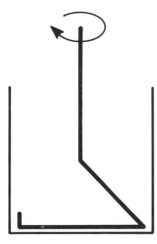

Figure 7-3. Overhead stirring assembly used during substitution of agarose. The stirrer is made from a piece of glass rod bent into the shape shown here. The rod is affixed to the chuck of a variable speed stirring motor.

these conditions activation occurs. It must be emphasized, however, that the precise structure of the activated intermediate is not known. The pH must be constantly maintained at 11 by manual addition of sodium hydroxide (2M or 8M solutions for 1–3 or 20–30 g of cyanogen bromide, respectively). Since this reaction is also exothermic the temperature must be regulated to about 20°C by addition of ice. The reaction usually requires 10 to 15 minutes for completion, which is signaled by cessation of proton generation. At this point a large amount of ice is added to cool the mixture to 4°C or below. From this point on the procedures must be carried out as rapidly as possible because the activated agarose is destroyed with a half-life of about 15 minutes at 4°C. The suspension is transferred to a coarse Büchner funnel and washed under suction with 10 to 20 times the gel volume of buffer. An efficient vacuum is necessary if washing is to proceed rapidly. The wash buffer should be at a pH of 9.5 to 10 and should be precooled at 4°C. Buffers containing ammonia or amino groups (glycine, ammonium acetate, Tris, or ammonium bicarbonate) should be avoided because they will compete with the ligand being coupled. Sodium bicarbonate or borate buffers work well. The activated and washed gel is resuspended in an equal volume of the same buffer containing the desired ligand. This mixture is gently stirred from overhead for 16 to 20 hours. During this time the ligand is covalently affixed to the supporting medium as postulated in Figure 7-4 (2). The absence of adjacent hydroxyl groups in agarose makes it likely that ligands are bound via isourea linkage. It is important that both the iminocarbonate and isourea derivatives retain the positive charge on the amino nitrogen since it may play a role in the binding of ligand to protein.

The final step in production of an affinity absorbent is removal of unbound ligand. Although the activated groups of agarose would not be expected to survive more than 12 hours it is advisable to wash the absorbent at room temperature with a 0.1M solution of pH 9.0 glycine buffer. This ensures that any possible surviving activated groups are

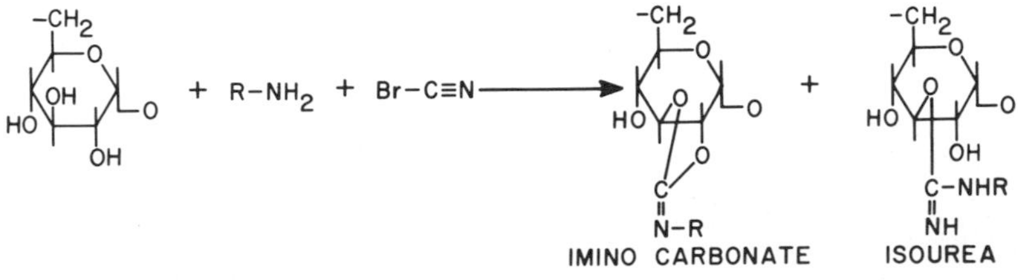

Figure 7-4. Reaction of a primary amino group with a carbohydrate polymer in the presence of cyanogen bromide.

destroyed. This is followed by extensive washing to remove unbound ligand.

A number of parameters may be varied to modify this procedure for specific applications. As stated earlier, the amount of cyanogen bromide used influences the degree of substitution. It also has an obvious effect on the amount of heat and acid generated during the reaction. Two parameters may be varied at the coupling step. The first is the concentration of ligand added. In general, the concentration of ligand in the coupling reaction mixture should be 20- to 30-fold higher than that desired in the final product. However, if the concentration is raised in excess of this it may be necessary to also raise the concentration of cyanogen bromide to a value of about 200 mg/ml of gel. If less ligand is desired in the final product its initial addition may be correspondingly decreased. A second parameter of the coupling reaction is the pH. Above a pH of 9.5 to 10 the activated agarose intermediate is very labile (see Figure 7-5). Since the

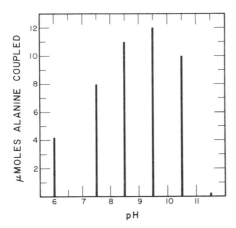

Figure 7-5. Effect of pH on the amount of alanine covalently bound to cyanogen bromide activated agarose. [From P. Cuatrecasas, *J. Biol. Chem.*, **245**:3059 (1970).]

unprotonated form of the ligand amino group participates in the coupling reaction, a decrease in pH reduces the reactive concentration of ligand. As shown in Figure 7-5, this also decreases the amount of ligand bound to the absorbent. The lower pK values of aromatic amines reduce somewhat the hydrogen ion concentrations that still yield efficient coupling. When proteins are bound to insoluble support media it is desirable to have them attached by as few covalent bonds as possible. This allows greater flexibility of the bound molecule and reduces the possibility that attachment will interfere with its function. One of the best methods of accomplishing this is to carry out the coupling at low pH, that is, 6.5 to 7.0.

ABSORBENT DERIVATIVES

Construction of a successful affinity chromatography absorbent requires
the ligand to be sufficiently distant from the solid matrix as not to interfere
with its binding to the desired macromolecule. If the ligand is bonded
directly to the matrix, as in Figure 7-6, macromolecules cannot approach
the ligand owing to steric hindrance of the matrix. This problem may be
conveniently resolved by placing an "arm" between the matrix and
ligand. Effectiveness of this solution is demonstrated by the data shown in
Table 7-2 which indicate the amount of staphylococcal nuclease bound to
absorbents with arms of increasing length. The most versatile way of
constructing "arms" is by ω-aminoalkylation of agarose. First an amine of
appropriate length is covalently bound to agarose, using the cyanogen
bromide procedure described above. Two amines often used for this are
hexamethylenediamine and 3,3'-diaminodipropylamine, shown in Figure
7-7. The resulting derivatives possess arms and are quite stable. However,
their greatest advantage is the ease with which further substitution may
be performed. A number of the more common substitutions are shown in
Figure 7-8. Any carboxylic acid may be bonded directly to the primary
amine at pH 4.7 in the presence of a water soluble carbodiimide. The
compound most often used for this is 1-ethyl-3-(3-dimethyl-
aminopropyl)carbodiimide, or EDAC. The reaction sequence proposed
for this substitution is shown in Figure 7-9. It involves reaction of the
carboxylic acid and carbodiimide to yield an O-acylisourea which then
reacts with a good nucleophile, that is, primary amino group of the
derivatized agarose, to yield the desired product (3). The bromoacetyl
derivative of agarose (Figure 7-8) may be prepared by reacting the

Figure 7-6. The approach of a macromolecule to an affinity absorbent for which the
ligand is bound directly to the matrix (right) or bound indirectly to the matrix through an
"arm" (left). Note that in the right figure no interaction can occur between the
macromolecule and the bound ligand.

Table 7-2. Specific Staphylococcal Nuclease Affinity Chromatographic Adsorbents[a]

ADSORBENT DERIVATIVE	CAPACITY

A — ξ—NH—⟨benzene⟩—PO_4^-—T—$PO_4^=$ 2

B — ξ—NHCH$_2$CH$_2$NHCCH$_2$NH—⟨benzene⟩—PO_4^-—T—$PO_4^=$ 8

C — ξ—NHCH$_2$CNHCH$_2$CNHCHCH$_2$—⟨phenol ring⟩—N=N—⟨benzene⟩—PO_4^-—T—$PO_4^=$ 8

D — ξ—NHCH$_2$CH$_2$CH$_2$NHCH$_2$CH$_2$CH$_2$NHCCH$_2$CH$_2$CNH—⟨benzene⟩—PO_4^-—T—$PO_4^=$ 10

[a] Prepared by attaching the competitive inhibitor, pdTp-aminophenyl, to various derivatives of Sepharose-4B or Bio-Gel P-300. (A) The inhibitor was attached directly to agarose, after activation of the gel with cyanogen bromide, or to polyacrylamide via the acyl azide step. (B) Ethylenediamine was reacted with cyanogen bromide-activated Sepharose. This amino gel derivative was then reacted with the N-hydroxysuccinimide ester or bromoacetic acid to form the bromoacetyl derivative; the latter was then treated with the inhibitor. (C) The tripeptide Gly-Gly-Tyr was attached by the α-amino group to agarose by the cyanogen bromide or acyl azide procedure, respectively; this gel was then reacted with the diazonium derivative of the inhibitor. (D) 3,3'-Diaminodipropylamine was attached to the gel matrix by the cyanogen bromide or acyl azide step. The succinyl derivative, obtained after treating the gel with succinic anhydride in aqueous media, was then coupled to the inhibitor with a water-soluble carbodiimide. The jagged vertical lines represent the agarose backbone. Capacity is reported as milligrams of staphylococcal nuclease bound per milliliter of gel. The various substituted gels were diluted with unsubstituted gel to obtain a ligand concentration of 2 μmoles/ml of packed gel. Therefore, the ligand concentrations for all the derivatives are the same when nuclease is added to the column.

From P. Cuatrecasas, *J. Biol. Chem.*, **245**:3059 (1970).

ω-aminoalkyl agarose with *O*-bromoacetyl-*N*-hydroxysuccinimide. The resulting bromoacetyl derivative effectively reacts with sulfhydryl, amino, phenolic, or imidazole groups. A carboxylic acid agarose may be generated by reacting (Figure 7-8) ω-aminoalkyl agarose with succinic anhydride. Then in the presence of a soluble carbodiimide, aliphatic or aromatic amines may be covalently bound to the resulting derivative. By using these various reactions in tandem, it is possible to build up "arms"

$$NH_2 - (CH_2)_6 - NH_2$$

HEXAMETHYLENE DIAMINE

$$NH_2 - (CH_2)_3 - NH - (CH_2)_3 - NH_2$$

3,3'- DIAMINODIPROPYLAMINE

Figure 7-7. Two compounds used as intermediate "arms" in the attachment of ligands to agarose.

of considerable length for subsequent binding of most functional groups. In addition to these procedures, a large variety of others are also available for specifically tailoring the coupling reaction to the characteristics of the ligand and the requirements of the desired affinity absorbent. Recently many agarose derivatives have become available commercially.

A number of practical considerations should be noted. Some of the derivatizations can be carried out in organic solvents such as dioxane, 50% aqueous dimethylformamide, or ethylene glycol. Such conditions are most advantageous if water-insoluble ligands such as fatty acids or steroid hormones are involved. Secondly, the stability at 4°C of affinity absorbents is limited only by stability of the covalently bound ligand. Antibacterial and antifungal agents, however, should be included in the storage medium.

CHROMATOGRAPHY

The concept of "progressively perpetuating effectiveness" enunciated by Cautrecasas (4) as an explanation of the functioning of affinity chromatography should be considered before discussing experimental procedures for absorption and elution of macromolecules from affinity columns. This concept has been diagrammed in Figure 7-10. At the onset of passing a protein mixture through an affinity column the concentration of bound ligand is known and constant. At this point the macromolecule concentration is zero. As the sample begins to enter the column, chance collisions occur between the ligand and desired protein, forming complexes. Some of these complexes, however, dissociate. Additional sample moving through the column adds to the amount of enzyme already present. Therefore, the enzyme concentration constantly increases as more sample passes through the column and is retarded. The increasing enzyme concentration results in a shift of equation (1) toward the right in favor of complex (ES) formation, leading to an apparently more tightly bound protein.

The concept of a self-perpetuating or increasing effectiveness during sample absorption explains why affinity chromatography can be used successfully even when relatively low affinity ($K_i = 1 \times 10^{-3} M$) systems are being processed. It is also for this reason that conditions needed to elute complexed proteins from the column are far more drastic than would normally be expected. An important experimental implication of this principle involves the decision of whether to use batch or column procedures. If the observed affinity between ligand and macromolecule is very high, batch procedures may be used. These are desirable because of their increased efficiency and ability to handle large volumes of material. However, if the observed affinities are more limited the column procedure with its self-perpetuating effectiveness is advisable. Conversely, if overly tight binding occurs, one possible solution is to resuspend the chromatographic material in a 10- to 20-fold excess of solution. This effectively decreases the concentration of free enzyme and favors dissociation.

The conditions used to absorb material on the column must be chosen to promote binding. Very high protein concentrations (>20–$30\,mg/ml$) should be avoided because at these concentrations the proteins behave more as aggregates and may decrease interaction of the desired protein with the ligand. Table 7-3 lists the conditions used for adsorption of a variety of materials. It may be useful to partially purify the protein before performing affinity chromatographic procedures. This is especially true for large preparations because decreasing the amount of protein being processed decreases the time required. If, however, the affinities are quite strong it may be possible to proceed directly to affinity dependent procedures.

Table 7-3. Conditions for Binding and Elution of Proteins to Affinity Columns

Enzyme	Binding Solution	Eluting Solution
Adenosine deaminase	$0.1M$ KCl and $0.1M$ phosphate, pH 7.0	$2mM$ mercaptopurine riboside (substrate analogue)
Aspartate aminotransferase	$5mM$ phosphate, pH 5.5	$100mM$ phosphate, pH 5.5 or 1 mg/ml pyridoxal phosphate
Carbonic anhydrase	$0.01M$ Tris, pH 8.0	0–$10^{-4}M$ gradient of acetazolamide (enzyme inhibitor)
Xanthine oxidase	$0.01M$ $Na_2S_2O_4$	O_2-saturated $1mM$ salicylate
Coagulation factor	$0.05M$ Tris, pH 7.5	0.1–$0.4M$ NaCl
Gal repressor	$0.05M$ KCl, pH 7.5	$0.1M$ borate, pH 10.5

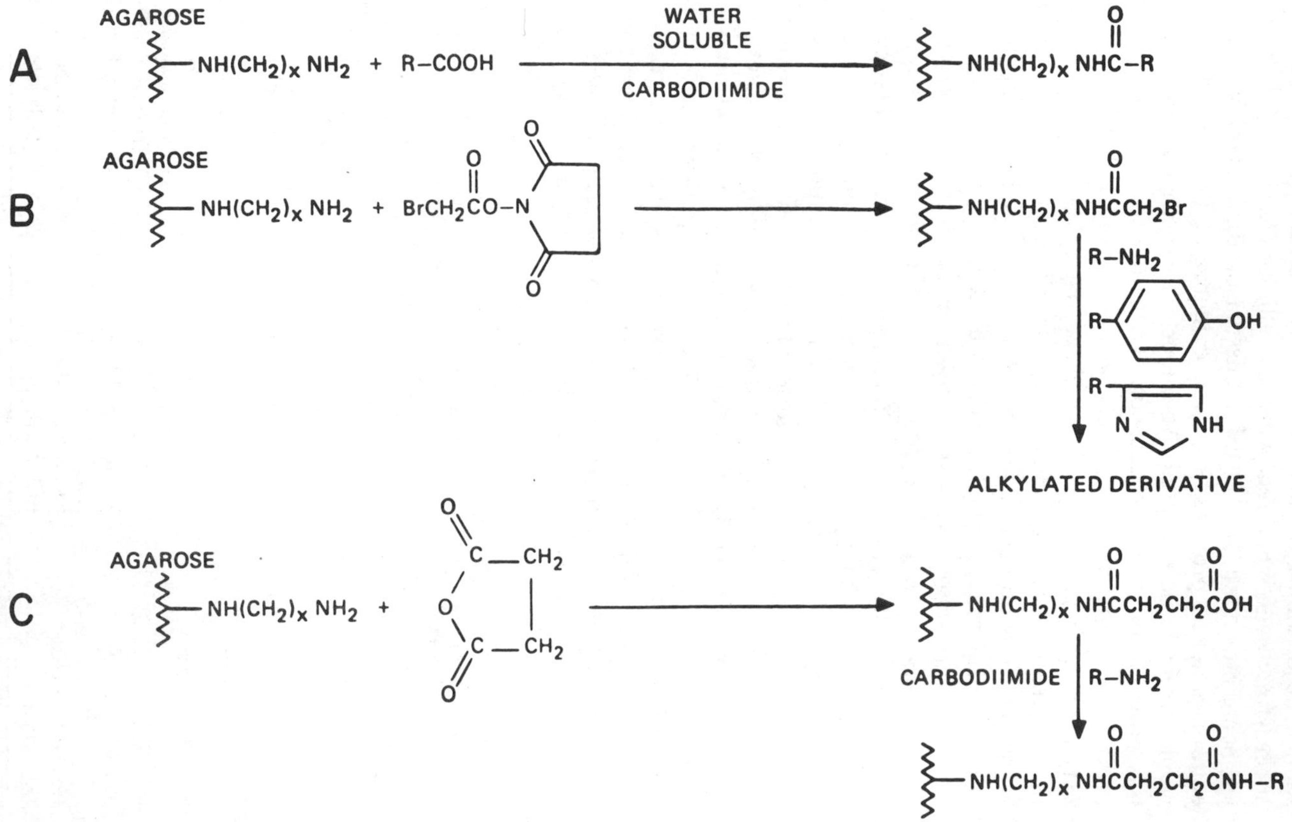
A
AGAROSE
NH(CH2)x NH2 + R–COOH
WATER SOLUBLE CARBODIIMIDE
NH(CH2)x NHC–R
B
AGAROSE
NH(CH2)x NH2 + BrCH2CO–N
NH(CH2)x NHCCH2Br
R–NH2
R–OH
ALKYLATED DERIVATIVE
C
AGAROSE
NH(CH2)x NH2 +
NH(CH2)x NHCCH2CH2COH
CARBODIIMIDE
R–NH2
NH(CH2)x NHCCH2CH2CNH–R

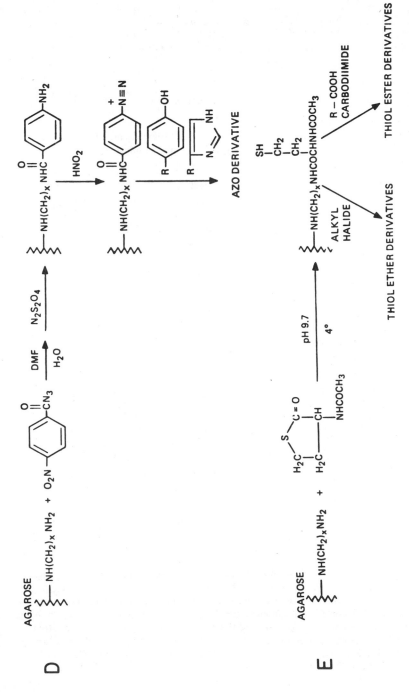

Figure 7-8. Derivatizations of ω-aminoalkyl agarose that can be used for preparation of selective adsorbents for affinity chromatography. Detailed procedures for the preparation of these derivatives are described in reference 9. (From P. Cuatrecasas, Affinity Chromatography of Macromolecules, in *Advances in Enzymology* (A. Meister, Ed.), Vol. 36, Wiley, New York, 1972, p. 29.)

$$RCOOH + {}^{'}R-N=C=N-R'' + H^{+} \longrightarrow R-\overset{\overset{\displaystyle O}{\|}}{C}-O-\overset{\diagup NHR'}{\underset{\overset{+}{\diagdown} NHR''}{C}}$$

CARBODIIMIDE

O–ACYLISOUREA

$$\downarrow X-NH_2$$

$$R\overset{\overset{\displaystyle O}{\|}}{C}-NH-X \ + \ O=C\overset{\diagup NHR'}{\diagdown NHR''} \ + \ H^{+}$$

Figure 7-9. Activation of carboxylic acids with water-soluble carbodiimide and subsequent reaction of the *O*-acylisourea formed with a primary amine.

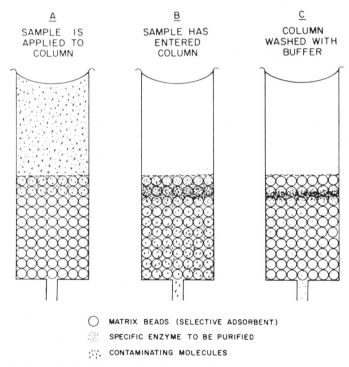

A

SAMPLE IS
APPLIED TO
COLUMN

B

SAMPLE HAS
ENTERED
COLUMN

C

COLUMN
WASHED WITH
BUFFER

◯ MATRIX BEADS (SELECTIVE ADSORBENT)

SPECIFIC ENZYME TO BE PURIFIED

CONTAMINATING MOLECULES

Figure 7-10. Steps involved in the adsorption of enzyme on a column containing a specific adsorbent during affinity chromatography. The formation of a sharp highly concentrated enzyme band on the column is demonstrated. The enzyme is progressively concentrated over a narrow band during application of the sample to the column. The progressively increasing concentration of enzyme during the procedure greatly increases the strength with which the enzyme adsorbs to the column. (From P. Cuatrecasas, Affinity Chromatography of Macromolecules, in *Advances in Enzymology* (A. Meister, Ed.), Vol. 36, Wiley, New York, 1972, p. 29.)

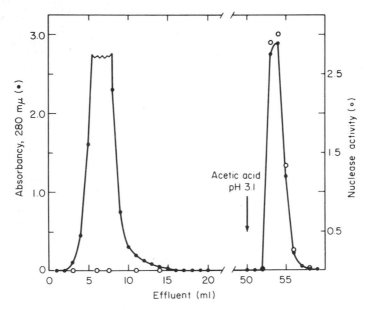

Figure 7-11.   Purification of staphylococcal nuclease by affinity adsorption chromatography on a nuclease-specific agarose column (0.8 × 5 cm). The column was equilibrated with 50mM borate buffer, pH 8.0, containing 10mM CaCl$_2$. Approximately 50 mg of partially purified material containing about 8 mg nuclease was applied in 3.2 ml of the same buffer. After 50 ml of buffer had passed through the column, 0.1M acetic acid was added to elute the enzyme. 8.2 mg nuclease and all the original activity was recovered. The flow rate was about 70 ml/hour. [From P. Cuatrecasas, M. Wilchek, and C. B. Anfinsen, *Proc. Natl. Acad. Sci. US*, **61**:636 (1968).]

As indicated above, the conditions needed to elute bound macro-molecules are usually much more drastic than those used for binding. This is also evident from the data presented in Table 7-3. There are essentially two approaches to eluting a bound protein. First, it may be removed from the adsorbent by washing the column with a solution of a compound which competes with the bound ligand and has higher affinity. This approach usually results in the need for large volumes of eluent and, unfortunately, increasing the competitor concentration may not improve the situation. This derives from the fact that elution is governed by the rate of dissociation of the complex and very stable complexes may have dissociation half-lives of 5 to 15 minutes. Complexes with half-lives of this order may require as much as 1 hour or more to be 95% eluted. Increasing the competitor concentration may improve the degree to which reassociation is prevented, but it does not alter the time required to

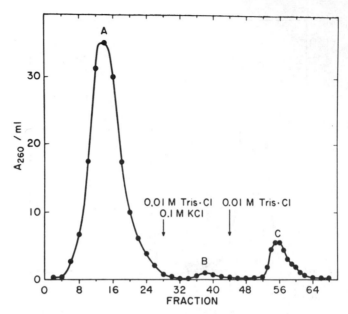

Figure 7-12. Isolation of poly A containing RNA using oligo (dT)-cellulose as the affinity absorbent. Crude rabbit reticulocyte polysomal RNA was applied to the column in a solution containing 0.01M Tris–HCl (pH 7.5) and 0.5M KCl. Nonabsorbed RNA (peak A) was eluted from the column by continued washing with the application buffer. The column was washed with a buffer of intermediate ionic strength (peak B) and finally with a buffer of low ionic strength (peak C) which removed poly A containing RNA from the column. [From H. Aviv and P. Leder, *Proc. Natl. Acad. Sci. US*, **69**:1408 (1972).]

remove the desired macromolecule. If unacceptably long times are required, one possible solution is to initiate elution with inhibitor and then stop the elution process and allow the column to incubate in the presence of competitor for a while before resuming column flow. The second approach is to drastically change the environment of the complex to a point where the macromolecule–ligand complex can no longer be maintained. Methods that have been used, listed in Table 7-3, include changes of pH (Figure 7-11), temperature, and ionic strength (Figure 7-12). Although the alteration should destroy the complex, it must not totally unfold the protein because this usually results in irreversible damage to the protein. If large changes of pH are required for elution, care should be taken to reestablish optimum pH conditions either by neutralization or rapid dialysis. When drastic procedures are used it is advisable to carefully determine whether or not the protein was damaged in the process. This may be done by assay of its catalytic and regulatory

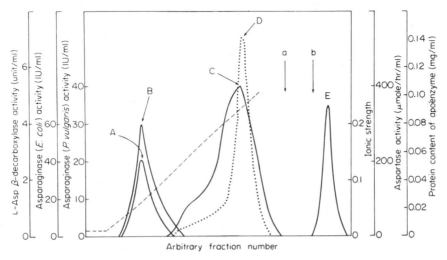

Figure 7-13. Group-specific adsorption of immobilized L-Asp column with the elution profile of each enzyme superimposed. Asparaginase from *Proteus vulgaris*, asparaginase from *Escherichia coli*, holoenzyme of aspartate β-decarboxylase, and apoenzyme of the enzyme were each dissolved in sodium acetate buffer (pH 7.0, ionic strength = 0.01, 1.0 ml) and separately charged to the column. Elution was carried out with linear gradients. In the case of aspartase, the enzyme preparation was dialyzed against sodium acetate buffer (pH 7.0, ionic strength = 0.01), and dialyzed sample was charged to the column. Stepwise elution was carried out using 0.1M sodium acetate buffer containing the following concentrations of NaCl: (*a*) 0.2M (ionic strength = 0.275); (*b*) 0.3M (ionic strength = 0.379). Fraction volume was 3.0 ml in all cases except for aspartate β-decarboxylase (2.5 ml). ———, activity; · · · ·, protein; – – –, ionic strength. A, Asparaginase from *P. vulgaris*; B, asparaginase from *E. coli*; C, holoenzyme of aspartate β-decarboxylase; D, apoenzyme of aspartate β-decarboxylase; E, aspartase. [From I. Chibata, T. Tosa, T. Sato, R. Sno, K. Yamamoto, and Y. Matus, in *Methods in Enzymology* (W. B. Jakoby and M. Wilchek, Eds.), Vol. 34, Academic Press, New York, 1974, p. 405.]

characteristics or by rechromatographing a small sample and determining whether or not its ability to bind the bound ligand has been destroyed.

Throughout this chapter it has been assumed that only one macromolecule complexes with the bound ligand. This, however, may not be the case. An affinity column with aspartate as the ligand retarded asparaginase, aspartate decarboxylase, and aspartase (see Figure 7-13). In such a case it is advisable to attempt sequential elution of the complexed proteins using less drastic procedures. One method used with good success is elution with gradients of increasing ionic strength or competitor (Figure 7-14). This procedure was used in a concave gradient to elute the various forms of LDH using NADH as the eluting competitor.

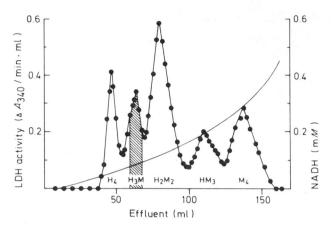

Figure 7-14. Elution of lactate dehydrogenase (LDH) isozymes with a concave gradient of NADH. Protein, 0.2 mg, in 0.2 ml 0.1M sodium phosphate buffer, pH 7.0, 1mM β-mercaptoethanol, and 1M NaCl was applied to an AMP analogue–sepharose column (140 × 6 mm, containing 2.5 g wet gel) equilibrated with 0.1M sodium phosphate buffer, pH 7.5. The column was washed with 10 ml of the latter buffer, then the isozymes were eluted with a concave gradient of 0.0 to 0.5mM NADH in the same buffer, containing 1mM β-mercaptoethanol. Fractions of 1 ml were collected at a rate of 3.4 ml/hour. The hatched area indicates the pooled fractions that were rechromatographed. [From P. Brodelius and K. Mosbach, *FEBS Lett.*, **35**:223 (1973).]

REFERENCES

1. M. Wilchek and W. B. Jakoby, The Literature on Affinity Chromatography, in *Methods of Enzymology*, Vol. 34 (W. B. Jakoby and M. Wilchek, Eds.), Academic Press, New York, 1974, p. 3.

2. J. Porath, *Nature*, **218**:834 (1968). Molecular Sieving and Adsorption.

3. D. G. Hoare and D. E. Koshland, *J. Biol. Chem.*, **242**:2447 (1967). A Method for the Quantitative Modification and Estimation of Carboxylic Acid Groups in Proteins.

4. P. Cuatrecasas, Affinity Chromatography of Macromolecules, in *Advances in Enzymology*, Vol. 36 (A. Meister, Ed.), Wiley, New York, 1972, p. 29.

5. E. Steers, P. Cuatrecasas, and H. B. Pollard, *J. Biol. Chem.* **246**:196 (1971). The Purification of β-Galactosidase from *Escherichia coli* by Affinity Chromatography.

6. P. Cuatrecasas and C. B. Anfinsen, Affinity Chromatography, in *Methods in Enzymology*, Vol. 22 (W. B. Jakoby, Ed.), Academic Press, New York, 1971, p. 345.

7. P. Cuatrecasas and C. B. Anfinsen, *Ann. Rev. Biochem.*, **40**:259 (1971). Affinity Chromatography.

8. P. Cuatrecasas, Selective Adsorbents Based on Biochemistry Specificity, in *Biochemical Aspects of Reactions on Solid Supports* (G. R. Stark, Ed.), Academic Press, New York, 1971, p. 79.

9. P. Cuatrecasas, *J. Biol. Chem.* **245**:3059 (1970). Protein Purification by Affinity Chromatography. Derivatizations of Agarose and Polyacrylamide Beads.

10. J. Porath and T. Kristiansen, Biospecific Affinity Chromatography and Related Methods, The Proteins Vol. I (H. Neurath and R. L. Hill, Eds.), Academic Press, New York, 1975, Pg 95–178.

Chapter 8

Immunochemical Techniques

The enormous specificity and resolution of immunochemical techniques are responsible for their increasing application to problems in biochemistry and molecular biology. The ability to specifically measure picogram quantities of one molecule or to isolate milligram quantities of another is indeed awesome. However, as with most powerful techniques, these potentials are attainable only by discriminating application and careful use. Major aspects of immunochemical methodology are described along with some of their inherent problems. The discussion, however, is by no means complete. Some specialized immunochemical procedures are dealt with only briefly and the reader is referred to other relevant sources for a more complete discussion.

ANTIBODY STRUCTURE

Tiselius and Kabat showed very early that antibodies were part of the γ-globulin fraction of serum. In their classic experiment, they compared electrophoretic profiles of two samples of serum obtained from a rabbit heavily immunized with ovalbumin. One profile was obtained for whole serum (solid line of Figure 8-1) and the other for an identical serum sample from which ovalbumin-specific antibodies had been precipitated by addition of purified ovalbumin (dashed line of Figure 8-1). This treatment markedly decreased only the γ-globulin portion of the serum profile, indicating the presence of antibodies. Since these early studies, the γ-globulin fraction has been carefully resolved into a set of structurally similar but immensely diversified proteins called the immunoglobulins (Igs). On the basis of structural differences, immunoglobulins are

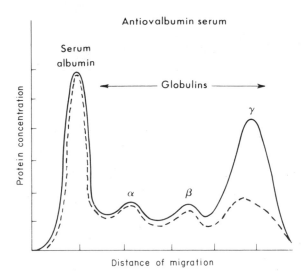

Figure 8-1. Electrophoretic separation of serum taken from a rabbit immunized with ovalbumin. The solid line indicates the result obtained before addition of ovalbumin to the serum and the broken line indicates the result obtained when ovalbumin antibodies are specifically precipitated prior to electrophoresis by addition of ovalbumin. [From A. Tiselius and E. A. Kabat, *J. Exp. Med.*, **69**:119 (1939) as redrawn by B. D. Davis, R. Dulbecco, H. N. Eisen, H. S. Ginsberg, and W. B. Wood, Jr., *Microbiology*, 2nd ed., Harper and Row, New York, 1973.]

classified at three successively finer levels of resolution. The first subdivision separates these proteins on the basis of structural and immunochemical differences into five groups called isotypes (Table 8-1). Each of these isotypes is divided into subgroups called allotypes. Allotypic differences are minor amino acid variations observed from one individual to another; that is, they are allelic forms within an isotype. The third level of classification, idiotypes, depends on differences that are clone specific for clones of immunoglobulin-producing cells. Idiotypes are variations of structure occurring close to the ligand binding site of the γ-globulin.

The most detailed structural studies have been performed using IgG because this isotype accounts for 85% or more of the serum immunoglobulins. These molecules are Y-shaped (see Figure 8-2) with a molecular weight of 150,000 daltons. They contain two heavy chains (MW 50,000 daltons) and two light chains (MW 25,000 daltons) held together by three to seven of the 20 to 25 disulfide bonds occurring per molecule. The heavy chains are bent at their midpoints to permit formation of the Y shape. Wide variation of the bend angle has been observed, but it is unclear whether a single molecule can bend to multiple angles. Cleavage of

Table 8-1. Characteristics of Immunoglobulin Isotypes

Immunoglobulin Isotype	Molecular Weight	Carbohydrate Content (%)	Serum Concentration[a]	Biological Properties
IgG	160,000	2.9	750–2200	Main serum immunoglobulin
IgA	160,000[b]	7.5	51–380	Main immunoglobulin of secretions
IgM	1,000,000[b]	10.9	21–279	Increases in initial immunological response
IgD	160,000		1–56	
IgE	160,000		$1–14 \times 10^{-4}$	Allergic antibodies

[a] Concentration in normal human serum expressed as mg/100 ml of serum.
[b] These immunoglobulins polymerize.

From J. Clausen, Immunochemical Techniques for the Identification and Estimation of Macromolecules, in *Laboratory Techniques in Biochemistry and Molecular Biology* (T. S. Work and E. Work, Eds.), American Elsevier, New York, 1969, p. 413.

purified antibodies with papain yields three fragments as shown in Figure 8-2. Two fragments forming the arms of the Y (denoted F_{ab}, antigen-binding fragments) retain the specific ligand binding sites but are univalent, that is, contain only one binding site per fragment. The remaining fragment (denoted F_c, crystallizable fragment) appears to be uniform in all rabbit IgG molecules.

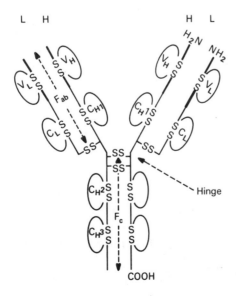

Figure 8-2. Schematic representation of polypeptide chains in the basic immunoglobulin structure. [From D. R. Davies, E. A. Padlan, and D. M. Segal, *Ann. Rev. Biochem.*, **44**:639 (1975).]

ANTIBODY FORMATION

A detailed discussion of antibody production and its control is beyond our purpose. However, minimal appreciation of the events leading to antibody production and a few definitions are necessary if immunochemical techniques are to be used knowledgeably. An antigen may be defined as any compound that (1) can stimulate production of antibodies when injected into a test animal and (2) reacts specifically with the antibodies produced. Both parts of this definition are necessary to distinguish antigens from haptens (see Figure 8-3), which are small molecules that

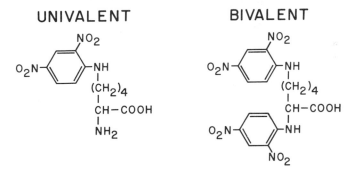

Figure 8-3. Examples of a univalent hapten, ϵ-DNP-lysine, and a bivalent hapten, α,ϵ-bis-DNP-lysine.

can react with specific antibodies, but do not elicit specific antibody production unless injected in a conjugated form, that is, bound to an antigen. Figure 8-4 depicts a conjugated protein with the hapten molecules represented as open circles.

The most important parts of the antigen are its antigenic determinants, which are those specific regions of the macromolecule directly involved in determining specificity of the immunochemical reaction. Each antigen may have many different determinants, each of which elicits a specific immune response and reacts with specific antibodies produced as a part of that response. Antibodies, like enzymes, possess very specific three-dimensional requirements for binding to an antigenic determinant. Departure from the required shape of determinant or the antibody binding site markedly decreases or totally prevents binding. Figure 8-5 illustrates the complementary nature of the antigenic determinant with the antibody binding site. Although maximum binding requires a perfect fit, it must be emphasized that significant binding may be possible even in the absence

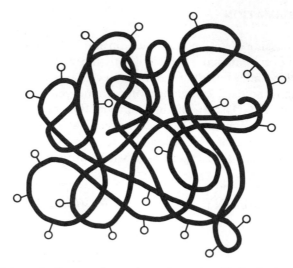

Figure 8-4. Schematic diagram of a conjugated protein with the haptens represented as "lollipop" structures.

of total complementarity; this is discussed below. Figure 8-6 shows a representative haptenic group, 2,4-dinitrophenol, bound to an ϵ-amino group of a lysine residue (outlined by a solid line) along with an antigenic determinant (outlined by a broken line). Note that the amino acids forming the determinant need not be contiguous; only their three-dimensional proximity is necessary.

An antigen injected into a test animal elicits an immune response by binding to determinant specific receptor sites on two types of small lymphocytes denoted B and T. These receptor sites have been shown to be immunoglobulins integrated into the lymphocyte cell membrane, which accounts for their determinant specificity. A few characteristics of B and T lymphocytes are listed in Table 8-2, but their ontogeny, function, and interactions are immensely more complex than such a casual description

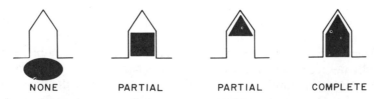

NONE   PARTIAL   PARTIAL   COMPLETE

Figure 8-5. Schematic diagram depicting various degrees of complementarity between the determinant of an antigen and the determinant binding site of an antibody.

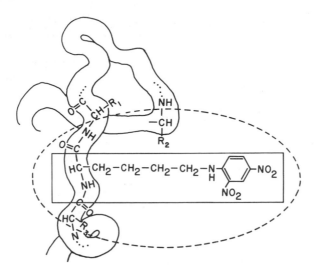

Figure 8-6. A haptenic group, DNP covalently bound to the ε-amino group of a lysine residue. The solid line encloses the hapten and the broken line encloses what might represent a determinant. Note that residues R_1 and R_2 are found in very different areas of the polypeptide chain even though both residues are part of the antigenic determinant.

indicates (see references 1 and 2 for more detailed descriptions). In response to the binding of an antigen, both lymphocyte varieties divide and increase in size, becoming large lymphocytes (Figure 8-7). Large B lymphocytes begin producing limited quantities of antibody, but more importantly continue differentiating into plasma cells containing an extensive rough endoplasmic reticulum (Figure 8-8). It is the specialized plasma cells that secrete most of the immunoglobulins. Although T lymphocytes

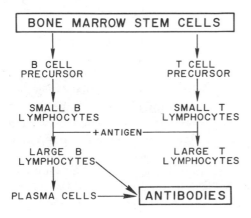

Figure 8-7. Development routes of the principal components involved in the immune response. Macrophages, which play an important accessory role in the immune response have been omitted.

Table 8-2. Properties of Mouse B and T Lymphocytes

Properties	B Cells	T Cells
Site of differentiation	Bursa of Fabricius (in birds) as yet unknown equivalent in mammals	Thymus
Antigen-binding receptors on the cell surface	Abundant Igs, (restricted to 1 isotype, 1 allotype, 1 idiotype per cell)	Nature of specific receptors is uncertain; Igs are sparse
Approximate frequency (%)		
Blood	15	85
Lymph (thoracic duct)	10	90
Lymph node	15	85
Spleen	35	65
Bone marrow	Abundant	Few
Thymus	Rare	Abundant
Distribution in lymph nodes and spleen	Clustered in follicles around germinal centers	In interfollicular areas
Functions		
Secretion of antibody molecules	Yes (large lymphocytes and plasma cells)	No
Helper function (react with "carrier; moieties of the immunogen)	No	Yes
Effector cell for cell-mediated immunity	No	Yes

Modified from B. D. Davis, R. Dulbecco, H. N. Eisen, H. S. Ginsburg, W. B. Wood, and M. McCarty, in *Microbiology*, 2nd ed., Harper and Row, New York, 1973.

do not secrete immunoglobulins, they do play a critical accessory role in stimulating and regulating proliferation and differentiation of B lymphocytes.

One of the puzzles concerning antibody formation is the observation that a wide variety of antibody molecules are produced against each type of antigenic determinant injected into the test animal. Of the several attempts to explain these observations, Burnet's clonal selection hypothesis currently receives greatest support. This view suggests that there are many individual, diversified B lymphocytes which may each respond to only one or a few antigenic determinants. An antigen injected into an animal interacts with only those cells carrying a receptor site that is complementary or nearly complementary to the injected determinant (Figure 8-9). On dividing, each of the responding lymphocytes gives rise to a clone of identical progeny, and these in turn give rise to clones of

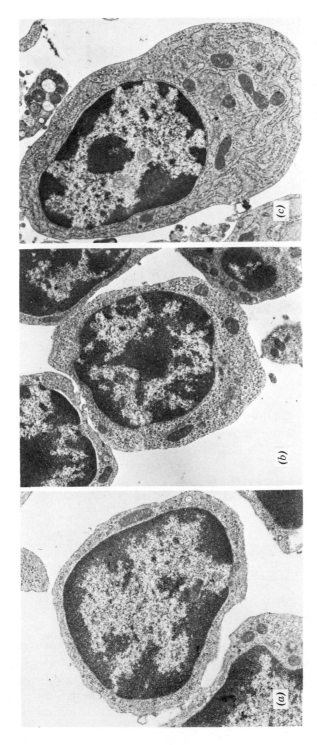

Figure 8-8. Principal cells involved in antibody formation. (*a*) Small lymphocyte, (*b*) large ("transitional") lymphocyte, and (*c*) plasma cell. Note the secretory vesicles and extensive rough endoplasmic reticulum in the plasma cell. Magnifications are approximately (*a*) ×4125; (*b*) ×3630; (*c*) ×3630. (Courtesy of Dr. R. G. Lynch.)

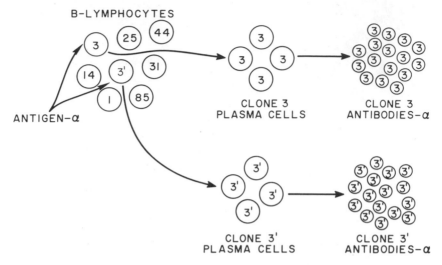

Figure 8-9. Schematic diagram representing the salient features of the clonal selection theory. (Based on F. M. Burnet, *The Clonal Selection Theory of Acquired Immunity*, Vanderbilt University Press, Nashville, Tenn., 1959.)

plasma cells, each producing their own specific type of antibody. All the antibodies produced by the member cells of one clone are identical, but those produced by different clones are each different, reflecting the structure of the immunoglobulin receptor to which the antigen was originally bound.

PRACTICAL ASPECTS OF ANTIBODY PRODUCTION

The Antigen

Success or failure of immunochemical methods is largely determined at the initial step, preparation of the antigen. The immune systems of most animals are intensely sensitive, detecting even the smallest traces of contaminant antigens. B lymphocyte proliferation leading to plasma cell secretion of antibody is primarily responsible for the amplified sensitivity. Therefore, every effort should be made to ensure that the antigen to be injected is homogeneous. At minimum the preparation should yield only one band, with no visible contamination, when subjected to acrylamide gel electrophoresis. At best, however, this ensures only that antibody is produced against one protein. If the protein is made of subunits, each subunit is likely to have multiple antigenic determinants and result in production of a set of antibodies in varying proportions. In view of this, the specificity of antibody preparations may be increased by first separating the protein into subunits and injecting only one type of polypeptide. This is not absolutely necessary, however, unless maximum specificity is required. It is also important to note that injecting only one subunit of a protein requires that the isolated subunit possesses a set of determinants similar to those of the subunit in its native position within the complete protein.

Adjuvants

The immunogenic potency of soluble antigens may be considerably increased if they persist within tissues of the test animal. This may be accomplished by giving multiple small injections of antigen over a period of time. It has been shown with diphtheria toxoid, for example, that several small doses given over a period evoke a much greater response than an identical quantity of toxoid given all at once. Another method of maintaining antigen persistence is through use of an adjuvant. This term may be used broadly to denote any substance that increases the immune response toward an injected immunogen. Inorganic gels such as aluminum hydroxide or aluminum phosphate served as the earliest adjuvants. Immunogens were adsorbed on their surfaces and the mixture injected. These compounds, however, have been largely replaced by the oil–water emulsion adjuvants developed by Freund (Table 8-3). Immunogen in aqueous solution is mixed with one of Freund's adjuvants and a stable emulsion produced. This may be done by very rapid mixing, sonification, or drawing the mixture repeatedly into a syringe (20–24 gage needle) and squirting it out again forcibly. Stability of the emulsion may be ascertained by placing a single drop in a beaker of cold water. If it is stable the first droplet spreads evenly across the liquid surface but subsequent drops remain on or below the surface and do not break up. Injecting an unstable emulsion into a test animal results in loss of most

Table 8-3. Composition of Freund's Complete and Incomplete Adjuvants

Components	Complete	Incomplete
Mannide monooleate, 1.5 ml	+	+
Paraffin oil, 8.5 ml	+	+
Mycobacterium butyricum, 5 mg (killed and dried)	+	−

advantages incurred by using adjuvants. After either subcutaneous or intramolecular injection, the stable emulsion forms small droplets which spread widely from the site of injection. The stability of emulsion results in a very slow release of immunogen, lasting sometimes up to several months or longer.

The presence of heat-killed mycobacteria in Freund's complete adjuvant serves to generally stimulate the entire reticuloendothelial system of the test animal. In addition, there is some evidence these dead cells also stimulate T lymphocyte proliferation and their production of B cell-stimulating factors. Use of the complete adjuvant is recommended for the first dose of immunogen but incomplete adjuvant is adequate for subsequent injections.

Animals, Dose, and Route of Inoculation

Animals commonly used for antibody production are rabbits, goats, guinea pigs, chickens, rats, donkeys, and horses. Of these sources, young New Zealand White rabbits are most often employed when limited quantities of serum are needed and goats for larger quantities. If antibody preparations are to be used for low or medium specificity studies, sera may be collected from a number of rabbits immunized with the same preparation of immunogen. However, if fine specificity is required it is advisable to use a larger animal such as a goat, so that the entire study can be performed using well characterized antibody preparations obtained from a single or a few closely spaced bleedings. This is because, as discussed below, antibody preparations differ from animal to animal and even within the same animal over a period of time.

The correct dose of immunogen depends on its nature, whether adjuvant is used, and whether the animal has received previous injections. For example, 100 μg to 1 mg serum albumin injected into a rabbit in the absence of adjuvant evokes only a minimal response, whereas in combination with Freund's adjuvant it is highly effective. A second injection at a later time would require only 10 to 50 μg for a threshold response. After the threshold dose is exceeded, increasing quantities of immunogen lead to increasing responses. The degree of increase, however, is not proportional to dose. In fact, there is a very real danger of using an excessive amount of immunogen and establishing a state of unresponsiveness or immunological tolerance. Such tolerance is also evoked when certain immunogens are given in amounts just below threshold levels. Some trial and error is necessary, therefore, to establish the best dosage for a specific antigen and the conditions of injection.

For most primary injections using Freund's complete adjuvant, intramuscular injection provides rapid access to the lymphatic circulatory system. This route is much better than using foot pad injections, which usually produce severe swelling, ulceration, and necrosis. Secondary injections in the presence of incomplete adjuvant may be given either intramuscularly or subcutaneously. The latter route is advantageous in that the antigen is absorbed slowly, minimizing the possibility of anaphylactic shock. This route should not be used with complete adjuvant because an abscess will form.

Response to Inoculation

The rate of antibody production following inoculation depends on whether it is the first (primary) or a subsequent (secondary) injection. The two patterns of response are shown in Figure 8-10. After the first injection of immunogen there is a lag period of 1 to 30 days before appearance of serum antibodies; 5 to 7 days is quite common for many soluble proteins. After this the concentration of serum antibodies increases exponentially, reaching a maximum usually at around 9 to 11 days when soluble proteins

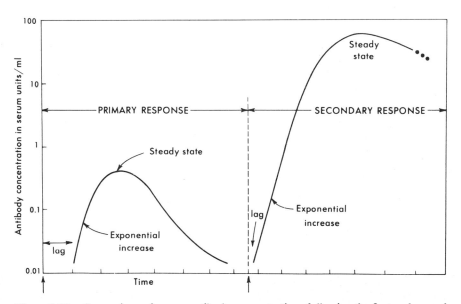

Figure 8-10. Comparison of serum antibody concentrations following the first and second injections of immunogen (indicated as arrows beneath the abscissa). Note the logarithmic scale for antibody concentration. Time units are unspecified to indicate the great variability encountered with different immunogens. (From B. D. Davis, R. Dulbecco, H. N. Eisen, H. S. Ginsberg, and W. B. Wood, Jr., *Microbiology*, 2nd ed., Harper and Row, New York, 1973.)

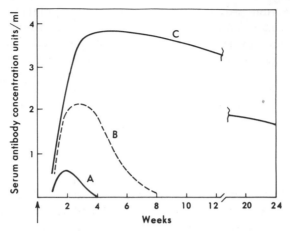

Figure 8-11. Influence of adjuvants on the production of antibodies following injection of an immunogen. Schematic view of amounts of antibody produced by rabbits in response to one injection (arrow) of a soluble protein (*A*) in a dilute salt solution, (*B*) adsorbed on precipitated alum, or (*C*) incorporated into a stable water-in-oil emulsion containing mycobacteria (Freund's complete adjuvant). (From B. D. Davis et al., *Microbiology*, 2nd ed., Harper and Row, New York, 1973.)

are used. The duration of maximum antibody production and the rate of its decline are largely influenced by the effectiveness of adjuvant and stability of the adjuvant–antigen emulsion. These facts are illustrated in Figure 8-11. In the absence of adjuvant, serum antibody concentration increases only a small amount and falls rapidly again. Alum adjuvant increases the maximum level and slows the rate of decline, but not nearly to the degree observed with Freund's complete adjuvant.

As pointed out earlier in this chapter, antibody preparations obtained from an immunized animal change with time. The changes involve both avidity of the antiserum (i.e., affinity between immunogen and antibody) and its specificity or cross-reactivity. As shown in Table 8-4 the average intrinsic association constant between anti-DNP antibodies and DNP-lysine is very low shortly after injection of immunogen and increases dramatically with time. These effects probably derive from the low affinity for the antigen of many of the B lymphocytes of a freshly inoculated animal. The relatively large quantities of antigen present immediately after injection are sufficient to activate these cells. This results in very early appearance of antibodies of low affinity. With time the concentration of antigen decreases, dropping below the level needed to trigger antibody production in low affinity lymphocytes. At these low levels only higher affinity lymphocytes are activated resulting in produc-

Table 8-4. Affinities of Anti-DNP Antibodies Isolated at Varying Times After Injection of DNP–Bovine γ-Globulin

	Average Intrinsic Association Constants for Binding of ϵ-DNP–Lysine to Anti-DNP Antibodies		
Antigen Injected (mg)	2 weeks[a]	5 weeks[a]	8 weeks[a]
5	0.86	14	117
50	0.38	0.73	9
100	0.38	0.44	0.51
250	0.22	0.17	0.19

[a] Times elapsing between initial injection of DNP–bovine γ-globulin and isolation of immune serum used in the experiments.

From H. N. Eisen and G. W. Siskind, *Biochemistry*, 3:996 (1964).

tion of antibodies with higher antigen affinities. Similarly, as the dose of antigen is increased the observed average affinity decreases and remains low even after extended periods.

Time dependent broadening of antibody specificity likely arises as a results in very early appearance of antibodies of low affinity. With time determinants, each provoking production of specific antibodies with different lag times. Therefore, shortly after inoculation there are comparatively fewer diverse types of antibodies but this diversity increases with time. This would also explain the observation of Heidelberger and Kendall that an antigen behaves as though it has a greater number of antibody binding sites per molecule when incubated with immune sera obtained long after primary inoculation than with sera obtained very soon thereafter.

Increases in cross-reactivity with time are also due to a combination of the above factors. First, an antibody with high affinity for one determinant would be expected to have a lower but still measurable affinity for structurally similar determinants located on other proteins. Therefore the high affinity antibodies produced later exhibit broader specificities. Second, as a greater variety of antibodies appear in response to the multiple determinants of an antigen the probability of similarity between a given antibody type and a structurally related determinant of some other antigen increases.

If a second injection of immunogen is made after the primary response has declined, a secondary (anamnestic or memory) response ensues. The nature of this response is shown in Figure 8-10. By comparison the secondary response (1) occurs after a much shorter lag time, (2) results in significantly higher levels of circulating antibodies, and (3) persists longer. Of practical significance is the much smaller amount of antigen needed to evoke a secondary response than for the primary inoculation. Therefore, care should be taken to decrease the dose rate. In addition, complete adjuvant is no longer needed since the reticuloendothelial system is still stimulated from the primary injection; incomplete adjuvant is usually used for booster injections.

When large quantities of immune serum are collected over extended periods an important consideration is the booster regimen. Only a very slight advantage is gained by making repeated primary inoculations at multiple sites on the animal or repeated secondary inoculations at too frequent intervals. As a matter of fact, this greatly increases the possibility of establishing immunological tolerance. Therefore, "patience is often more valuable than the ever-ready syringe." Booster inoculations should be given only after the level of circulating antibodies (titer) has decreased significantly. An increase in titer usually reaches its maximum 5 to 7 days after the booster injection and remains high for many weeks. Subsequent to a booster injection the test animal should not be inoculated again for weeks to months.

Serum Collection and Preparation

Since rabbits are by far the most often used animals for antibody generation, collection of their blood will be discussed here. Procedures for bleeding other small animals may be found in other references (3–5). Blood is collected from larger animals such as goats, donkeys, and horses via the jugular vein. In these cases an inexperienced investigator would be well advised to obtain the assistance of a veterinarian or other qualified individual to demonstrate proper handling of the animal and withdrawal procedures.

Rabbits should always be restrained when being inoculated or bled. The rabbit may be transferred from its cage to the restrainer by seizing it at the scruff of the neck. The animal should never be picked up by its ears or by placing one hand under its belly. The hind legs of a rabbit are quite powerful and their nails are long and sharp; placing one's hand anywhere on the underside of the animal almost always results in being scratched. If this occurs the wound should be immediately washed with 70% ethanol and left uncovered; a tetanus shot may also be advisable. The animal should be restrained quite firmly so that no movement is possible in order to prevent sudden jerking and self-inflicted injury. Devices such as that shown in Figure 8-12 are commercially available and work quite well. Bleeding is more efficient if the hair is first shaved from the animal's ear

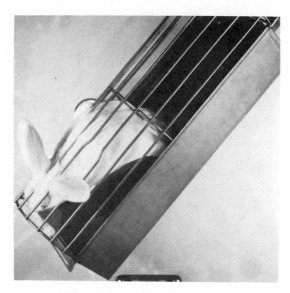

Figure 8-12.   A rabbit in a commercially available stainless steel restrainer.

using electric hair clippers. The edge of the ear should not be shaved because the flesh is so thin that invariably the ear gets caught in the clipper teeth. The best procedure is to lay the ear on one's hand and shave at an angle as shown in Figure 8-13. The ear is then swabbed with toluene, taking care not to use so much that it runs into the animal's ear. In 30 seconds or so all the blood vessels dramatically dilate and become easily visible. The vein should be cut longitudinally 1 to 2 cm from the head. This cut (0.25–0.3 cm long) should be made in one bold, quick, smooth movement. If it is done squeamishly the animal may bolt, resulting in a cut that either is too long or extends through the ear. I have found the best device for making this incision to be the corner of a new, single-edged razor blade. With a good incision, the blood should flow freely or pulsate and 40 to 50 ml can be collected in a matter of a few minutes. At times during collection the flow may slow and then increase again; this is quite normal. The key to obtaining a good yield of blood is to make a good incision (one of proper length and of sufficient depth to enter the vein, rather than just cutting the skin superficially) while the vein is fully delated, because after application of toluene the vessels gradually return to their normal size and blood flow decreases, aiding the clotting process. Flow should stop by itself as a clot forms. With these procedures the vein will heal and may be bled again after a week or so. If healing is slow it is possible to use both ears alternately for weekly bleedings. In the event

Figure 8-13. The correct position for shaving a rabbit's ear without catching the ear margin in the clipper teeth.

that the vein is irreversibly damaged a new site of incision may be chosen closer to the artery (see Figure 8-14). It is for this reason that the first incision is made close to the head; otherwise the length of vein available for subsequent incisions is more limited. When clotting is complete the ear should be washed thoroughly with water to remove any remaining toluene. Constant exposure to toluene will irritate the skin. If this occurs a mild hand cream will alleviate the problem.

If the collected blood is cooled and processed immediately the red blood cells may be removed before clotting. Alternatively the blood may be allowed to fully clot. In the latter situation the clot should be broken up with disposable wooden applicator sticks. In either case red blood cells are removed by centrifuging the whole blood at $5000 \times g$ (4000–7000 rpm in a Sorvall ss-34 rotor). If the cells are subjected to too strong a centrifugal force hemolysis occurs. Therefore moderate speeds should be used for centrifugation even if some decreases in yield are observed. If cell lysis (hemolysis) is minimal the serum supernatant should be yellow to slightly pink. Following centrifugation the serum is carefully decanted into a Corvex tube. A slight cloudiness at the top of the serum layer is caused by fat. This may be prevented by starving the animal for 24 hours or more before bleeding. In certain cases the serum clots after the red blood cells have been removed and does not flow. If this occurs, place two wooden

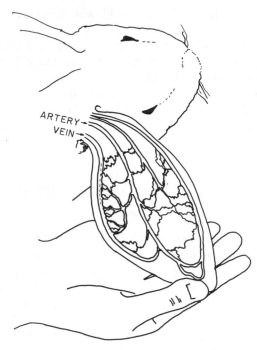

Figure 8-14. Circulatory system for the ear of a rabbit. The area of the vein to be longitudinally cut is denoted by a dashed line.

applicator sticks in the gelled serum, avoiding the precipitated red blood cells, and roll the sticks rapidly back and forth between the thumb and index finger. This usually winds the fibrin and stroma of the clot onto the sticks and permits the serum to be decanted. The material can be seen as a thin film when the sticks are pulled apart. If some red blood cells are carried over in the decantation, the supernatant should be recentrifuged before proceeding further.

Untreated serum contains a set of 11 proteins called complement. These proteins are not immunoglobulins and do not increase quantitatively after immunization. However, they do bind to antigen–antibody complexes and can cause problems for certain immunochemical applications. These proteins may be inactivated by heating the serum to 56°C for 10 to 20 minutes. This treatment will not harm the immunoglobulin fraction if the temperature is carefully regulated.

The decomplemented serum is now ready for either storage or further purification. Antibodies are quite stable and may be stored for many years frozen (-20 to -70°C) or lyophilized. Storage at 4°C requires addition of an antimicrobial agent such as cyanide or azide and is only recommended for short periods.

The immunoglobulin fraction may be isolated and concentrated by ammonium sulfate fractionation or column chomatography (6). Specific antibodies may be isolated using affinity chromatography with the original antigen as the bound ligand. In early purifications of this type antibody was eluted with dilute acid or base, often resulting in antibody inactivation. These procedures are now replaced by elution with $4M$ magnesium chloride, with much less inactivation.

REACTION OF MACROMOLECULAR ANTIGENS AND ANTIBODIES IN SOLUTION

Most, if not all, contemporary applications of immunochemical techniques are based on the reaction of antibodies with the antigen used to induce their production, to yield a very stable complex. If both antigen and antibody are present in solution, a precipitate forms as long as the antibody is in molar excess. This is known as a precipitin reaction and may be quantitated as shown in Figure 8-15. For this example a highly purified preparation of avidin was injected into a rabbit which then produced avidin-specific antibodies. Since biotin forms a very stable complex (dissociation constant $\sim 10^{-15}$) with avidin, complexation of avidin with ^{14}C-biotin provides a convenient means of detecting the protein's whereabouts. The data shown in Figure 8-15 were generated by adding increasing amounts of ^{14}C-biotin–avidin complex to a constant amount of either avidin-immune serum (Figure 8-15A) or nonimmune control serum (Figure 8-15B). Following 24 hours of incubation at 4°C the immunoprecipitate was separated from soluble components by centrifugation and the amount of radioactive biotin in the precipitate and supernatant was determined. The curve describing the amount of radioactivity found in the precipitate (closed squares of Figure 8-15A) may be divided into three portions: an ascending limb termed the region of antibody excess, a region of equivalence, and a descending limb termed the region of antigen excess. In the regions of antibody excess and equivalence all of the antigen is found in the precipitate. In the region of antigen excess, however, less and less ^{14}C-biotin precipitates and more can be found in solution (closed circles in Figure 8-15A). Note that none of the ^{14}C-biotin is precipitated when nonimmune control serum is used in place of immune serum (Figure 8-15B). This is essentially a titration of the serum's capacity to precipitate avidin. Heidelberger and Kendall performed a similar experiment but also isolated the precipitates and determined the

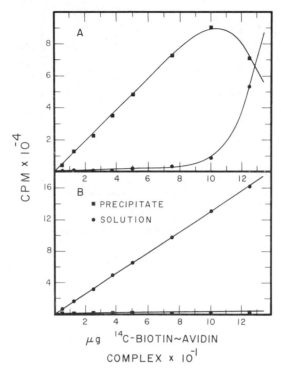

Figure 8-15. Immunoprecipitation of ^{14}C-biotin–avidin complex with avidin-immune serum. (*A*) A constant amount of serum obtained from a rabbit immunized with avidin was added to increasing quantities of ^{14}C-biotin–avidin complex. (*B*) An identical amount of control serum, obtained before immunization, was used in place of the immune serum.

antibody–antigen mole ratios at various antigen concentrations. Their data, depicted in Figure 8-16, indicate that this ratio declines continuously as more antigen is added. From this observation they concluded that immunoprecipitation resulted from growth of an antigen–antibody aggregate as shown in Figure 8-17. When these aggregates attain sufficiently large volumes they settle out of solution forming the precipitates observed experimentally. The lattice theory, as this is now known, requires that both the antigen and antibody be multivalent; that is, each possesses more than one binding site for the other. Earlier in this chapter antibodies were described to have two ligand binding sites per molecule and the data in Table 8-5 indicate that antigens also have a varying number of determinant sites, which increases with increasing molecular weight. Antibody bivalency and the lattice theory explain the failure of immunoprecipitates to form at low antibody–antigen molar ratios; there is an

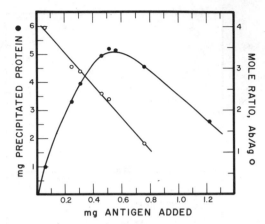

Figure 8-16. Mole ratio of antibody to antigen in immunoprecipitates obtained by mixing increasing quantities of antigen with a constant amount of immune serum. [From M. Heidelberger and F. E. Kendall, *J. Exp. Med.*, **62**:697 (1935).]

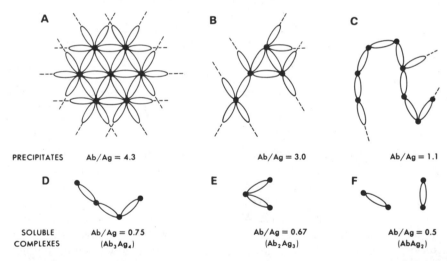

Figure 8-17. Hypothetical structure of immune precipitates and soluble complexes according to the lattice theory. Numbers refer to mole ratios of antibody to antigen. Dotted lines associated with precipitates indicate that the complexes extend as shown. The precipitates may be visualized as those found in the antibody (Ab) excess zone (*A*), the equivalence zone (*B*), and the antigen (Ag) excess zone (*C*). The soluble complexes correspond to those in supernatants having moderate (*D*), large (*E*), and extreme (*F*) Ag excess. (From B. D. Davis et al., *Microbiology*, 2nd ed., Harper and Row, New York, 1973.)

Table 8-5. Composition of Immunoprecipitates Obtained with Various Sets of Antigens and Corresponding Antibodies

Antigen	Molecular Weight	Mole Ratio of Antibody–Antigen[a]
Bovine ribonuclease	13,400	3
Egg albumin	42,000	5
Horse serum albumin	67,000	6
Human γ-globulin	160,000	7
Horse apoferritin	465,000	26
Thyroglobulin	700,000	40
Tomato bushy stunt virus	8,000,000	90
Tobacco mosaic virus	40,700,000	650

[a] Precipitates were prepared under conditions of extreme antibody excess. These values represent a minimal estimate of the antigen valence.

From E. A. Kabat and M. M. Mayer, *Experimental Immuno-chemistry*, 2nd ed., Charles C. Thomas, Springfield, Ill., 1961.

insufficient number of binding sites on the antibodies compared to the number of available determinant sites to form the required lattice.

ANTIGEN–ANTIBODY REACTIONS IN GELS

The most often used procedures for evaluating the specificity of antibody preparations are double diffusion methods developed by Ouchterlony and immunoelectrophoretic methods developed by Grabor and Williams. For a discussion of variations of these techniques as well as other useful procedures the reader is referred elsewhere (7–9).

The two-dimensional double diffusion or Ouchterlony method depends on the formation of a micellar structure during gelation of agar and agarose. Molecules or aggregates with molecular weights less than 200,000 daltons diffuse freely through channels between the micelles whereas those with greater molecular weights are retarded. Therefore, individual antibody and antigen molecules diffuse through the gel at a rate dependent only on their initial concentrations and Stokes' radii, whereas large macromolecular aggregates formed by their interaction are immobile. Small wells are cut in agar or agarose plates and filled with antigen and antibody, respectively. Many geometric arrangements of the wells are possible; one of those available commercially is shown in Figure 8-18. As the two reactants diffuse radially from their respective wells (Figure 8-19) a concentration gradient is established. When the concentrations of overlapping antigen and antibody are sufficiently great to form a visible aggregate, a precipitin band appears (see Figure 8-21). The quantitative aspects of these events are described in Figure 8-19, where it is assumed that a precipitate will form at an antibody–antigen molar ratio of 4 : 1. The precipitate formed between antigen and antibody acts as an immuno-specific barrier preventing further diffusion of that particular antigen–antibody pair but failing to hinder movement of other antigen and antibody molecules. Consequently in systems containing appropriate concentrations of both antigen and antibody the precipitate grows peripherally, forming lines of increasing length or arcs, when the same antigen or antibody is placed in adjacent wells (see Figure 8-21 for an example of this). If, as shown by the dashed line in Figure 8-19, the concentration of only one reactant is increased, the position of the precipitin band moves away from that well and closer to that of the complementary reactant. This occurs as a result of a greater diffusion rate of the more concentrated reagent. If the antigen and antibody concentrations are grossly out of proportion, however, the precipitate may migrate

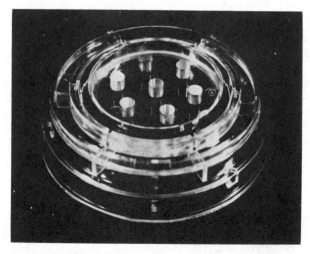

Figure 8-18. Intact (*a*) and exploded (*b*) views of a disposable micro-Ouchterlony plate. The components include, from bottom to top, (1) a base unit containing a small circular channel into which water is placed to preserve a moist atmosphere during incubation (arrow); (2) agarose through which the antigen and antibody diffuse; (3) a plastic center-piece containing the sample reservoirs; (4) a moisture seal to prevent the cells from drying out during storage; and (5) a cap for the entire cell. (Courtesy of Cordis Laboratories, Miami, Fla.)

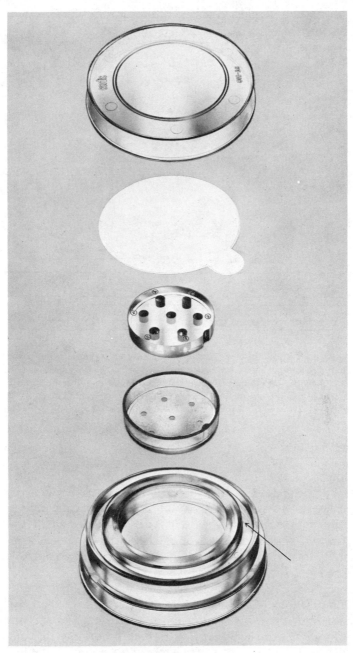

Figure 8-18. (*Continued*)

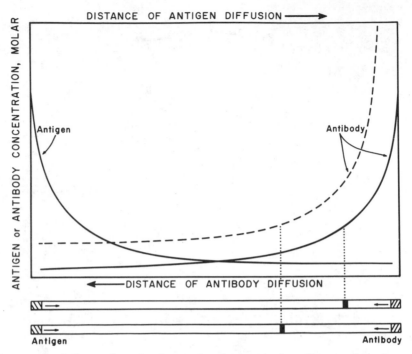

Figure 8-19. Concentration of antigen and antibody at various distances from the wells containing these components. The dashed line describes the concentrations observed when a greater concentration of antibody is used. The lower panel depicts two Ouchterlony plates in cross-section showing the location of the immunoprecipitate at the two antibody concentrations used above. Vertical dotted lines mark the point where the antibody–antigen ratio is 4:1.

or multiple precipitin bands become visible owing to alternate solubilization and reprecipitation of the initial immunoprecipitate.

So far we have discussed only systems containing one pure antigen and one antibody made in response to it. The Ouchterlony pattern formed in response to a system of this type is shown in Figures 8-20A and 8-21. Patterns of this type are said to indicate reactions of complete identity. It should be emphasized that the above result does not suggest that either (1) there is only one type of determinant in the two antigen wells, or (2) the total compositions of the two wells are identical. On the contrary, for most situations there are very many determinants and perhaps different types of antigens in each well. The data indicate only that antibodies in the center well are reacting to the same antigen in both the left and right antigen wells.

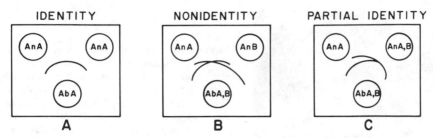

Figure 8-20. Schematic diagram of the precipitin bands expected from different homologous and heterologous combinations of antigens (An) and antibodies (Ab).

The second possible experimental situation is one in which the center well contains antibodies against two antigens, and each of the two antigen wells contains one of the two different antigens that will react with these antibodies (Figures 8-20B and 8-22). Here two totally separate precipitin bands form independently of one another. There is no immunospecific barrier formed and hence there is no "fusion" of the two precipitin bands to form a smooth arc as in the preceding case. This behavior is called a reaction of nonidentity.

A third experimental possibility is to have an antibody preparation containing antibodies against two different forms of an antigen. This

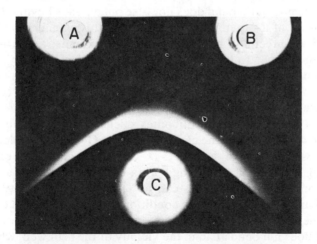

Figure 8-21. Precipitin band observed when two concentrations of avidin and avidin-immune serum are permitted to diffuse toward one another. Wells A and B contained 15 and 30 μg of avidin, respectively, and well C contained the avidin-immune serum. This photograph was taken $2\frac{1}{2}$ days after the components were added to the Ouchterlony plate. Incubation was at 4°C.

Figure 8-22. Precipitin bands observed when avidin and catalase are placed in adjacent wells and allowed to diffuse toward a center well containing antibodies against both antigens. Wells *A* and *B* contained 20 μg of avidin and 25 μg of catalase, respectively. Well *C* contained a mixture of avidin- and catalase-immune sera. This photograph was taken after 4 days of incubation at 4°C.

serum is placed in the center well of an Ouchterlony plate in which one antigen well contains both forms of the antigen and the second adjacent well contains only one of the two forms. The significant features of this situation are that the two forms of antigen possess some antigenic determinants in common and that one form has a number of determinants that are not found in the second form. Such an experimental condition results in a pattern similar to that shown in Figure 8-20*C* and is called a reaction of partial identity. There is total fusion of the precipitin bands formed with antigenic determinants that are identical in both forms of the antigen. However, there is no immunospecific barrier against antigen and antibody diffusion for the form of antigen that is present in only one well. Therefore, its precipitin band does not fuse but extends beyond the fusion arc in the form of a "spur." A spur formed in this manner is usually less dense than the precipitin band from which it extends, requiring careful examination at times to be detected. Figure 8-23 depicts such an experimental result obtained with *E. coli* RNA polymerase. This enzyme is composed of five subunits, α_2, β, β', σ, and may be found in two forms: (1) complete enzyme containing all five subunits and (2) core enzyme, which lacks the σ subunit. Antibodies elicited against complete RNA polymerase were placed in the center well of an Ouchterlony plate (well *G* in Figure 8-23). Antigen wells *A*–*D* contained complete enzyme at various stages of purity beginning with a crude cell extract, and well *E* contained only core enzyme. As can be seen in Figure 8-23 two sets of precipitin bands form under these conditions: a heavy inner set containing core enzyme and an outer set containing complete enzyme. Note that the outer band does not bend at all in the vicinity of the fifth well, which lacks this form of the protein. Spurs (indicated by arrows) may be seen at the fusion points between wells *C*–*D* and *D*–*E*, indicating that wells *C* and *E* contain forms of the antigen that are not present in well *D*. The procedures used to obtain the data shown in Figure 8-23 are quite useful

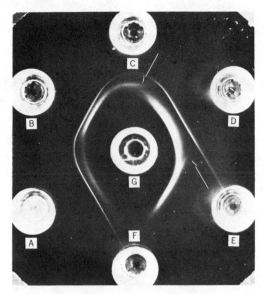

Figure 8-23. Precipitin bands observed when holo-RNA polymerase (complete enzyme including the σ subunit) immune serum is allowed to diffuse toward wells containing RNA polymerase at various stages of purity. The wells contained protein samples derived from the following purification steps: (A) crude extract, (B) 33 to 42% ammonia sulfate fractionation, (C) DEAE cellulose eluent, (D) RNA polymerase following sedimentation in a glycerol gradient, and (E) phosphocellulose eluent which contained only core enzyme devoid of σ subunit. The final well (F) was empty. The purification procedures used here were similar to those reported by R. R. Burgess, *J. Biol. Chem.*, **244**:6160 (1969). This photograph was taken after 12 days of incubation at 4°C.

for determining the immunospecificity of an antibody preparation and for detecting any changes that might have occurred in an antigen during its purification.

Immunoelectrophoresis

A powerful modification of the double diffusion procedures discussed above is immunoelectrophoresis. This technique is performed using a double diffusion chamber similar to the one depicted in Figure 8-24A. Such chambers are available commercially or may be prepared by coating a glass microscope slide with purified agar or agarose and punching appropriate wells into the agar with a cookie-cutter type of dye. The antigen preparation is placed in the small round well and subjected to electrophoresis (Figure 8-24B). At the conclusion of electrophoresis the rectangular well is filled with immune serum and the gel incubated to

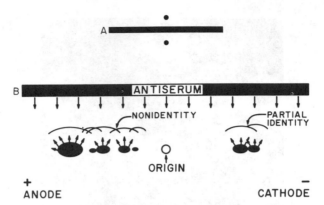

Figure 8-24. Immunoelectrophoresis: mode of operation (*B*) and Ouchterlony plate well pattern (*A*).

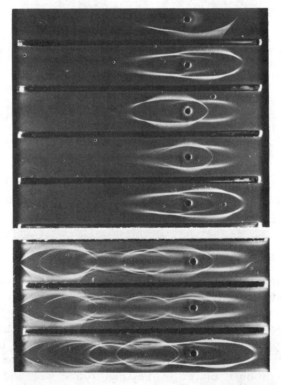

Figure 8-25. Precipitin bands observed following immunoelectrophoresis of normal and pathological mammalian sera. (Courtesy of Hyland Laboratories, Costa Mesa, Calif.)

permit diffusion of antigen and antibody molecules toward one another. This results in formation of precipitin bands such as those shown in Figure 8-25. The major advantage of this method is its increased resolving ability gained by the combined electrophoretic and double diffusion methods.

USE OF ANTIBODIES FOR SPECIFIC, HIGH RESOLUTION ASSAY OF PROTEINS

Past studies concerning enzyme production and its control depended heavily on measurement of enzymatic activity to detect the presence of a protein. Although major advances in our understanding of molecular biology attest to the success of such approaches, they are nonetheless limited. Their principal deficiency is the requirement for activity before a given enzyme may be detected. It is becoming clear, however, that many interesting and important events occur prior to the point at which an enzyme becomes active. For example, some enzymes are synthesized in an inactive form and are subsequently activated by proteolytic cleavage or other chemical modification. Protein aggregates such as phage structural proteins, ribosomes, and multienzyme complexes undergo elaborate assembly procedures prior to becoming functional. An increasing number of cases have been found in which an enzyme is inactivated by chemical modification rather than being proteolytically degraded. In all these situations, enzyme activity is absolutely inadequate as an experimental probe.

Reliance on enzyme activity as the monitoring device also limits study to enzyme-type proteins, a most unfortunate situation since many interesting proteins do not have catalytically assayable functions. Some proteins are regulatory elements which modulate synthesis or activity of enzymatically active proteins; others are structural elements or proteins that can perform their function only when integrated into a membrane, that is, permeases. The functions and modes of operation for many of these proteins are being elucidated with the help of immunochemical assay procedures.

Immunochemical assay procedures may be divided into two classes, direct and indirect. Direct methods involve synthesis or modification of the molecule to be measured in the presence of a radioactively labeled precursor followed by its isolation using immunochemical precipitation. Indirect methods involve competition of a nonradioactive protein obtained from an experimental sample with a radioactively labeled standard protein for a limited amount of antibody; the quantity of desired protein in the experimental sample is determined by the degree to which it successfully competes with radioactive standard for binding to antibody. The latter method is known as radioimmunoassay or RIA.

Safety

Both direct and indirect immunochemical assay methods often involve high levels of radioactivity. It is imperative, therefore, that proper precautions be taken in the handling and disposal of isotopes used in such experiments. Safety procedures should always be approved by radiation health officials before experiments involving high levels of ^{14}C, ^{3}H, and ^{35}S are undertaken. The same is true for experiments involving ^{32}P, ^{125}I, and ^{131}I at any level. Disintegration of ^{32}P molecules results in emission of an energetic β particle which requires minimal amounts of lead shielding. Radioactive iodine, on the other hand, is a strong γ-ray emitter and requires much greater shielding and special handling procedures. An additional danger involved in the use of either phosphorus or iodine is their efficient uptake and sequestration by mammals; radioactive phosphorus becomes stably incorporated into bones and nucleic acids, and iodine is concentrated in the thyroid gland, covalently bound to the hormone thyroxine.

Choice of Radioactive Label for the Antigen

The specific activity of proteins assayed by direct immunochemical methods or those used as standards in radioimmunoassays profoundly affects the resolution attainable by these techniques. Therefore, the nature of the radioactive label to be used in such experiments must be considered carefully. Table 8-6 lists the isotopes available for this purpose along with the number of atoms of each isotope that must be incorporated to produce an arbitrary counting rate. As can be seen here, 557 atoms of ^{3}H and 261,672 atoms of ^{14}C must be incorporated into every molecule of protein to yield the same number of disintegrations per minute as only one ^{131}I or 11 ^{35}S molecules. ^{35}S-methionine is often the isotope of choice for many direct immunochemical procedures since it is relatively inexpensive to prepare at high specific activity. On the other hand, the relative ease with which radioactive iodine may be incorporated into a purified antigen makes it the isotope of choice for radioimmunoassay methods. Of the two iodine isotopes available, ^{125}I is most often used because of its longer half-life. This is an important consideration since it usually takes more than 1 week to prepare and test a labeled antigen prior to its experimental use.

Table 8-6. Number of Various Isotopic Atoms Needed to Produce a Given Counting Rate[a]

Isotope	Number of Isotopic Atoms Incorporated per Macromolecule
^{131}I	1
^{32}P	1.8
^{125}I	7.5
^{35}S	10.9
^{3}H	557.0
^{14}C	261,672.0

[a] The values in this table were calculated through use of the equation

$$\lambda = \frac{0.693}{T_{1/2}}$$

where λ is the rate of isotope disintegration and $T_{1/2}$ is the half-life of the isotope in seconds. Disintegration rates may be converted to curies using the conversion factor $1\ c = 3.7 \times 10^{10}$ dps.

The expected specific activity of a labeled antigen may be calculated using the equation

$$\lambda = \frac{0.693N}{t_{1/2}} \tag{1}$$

where λ is the rate of particle emission (disintegrations per second), N is the number of atoms present, and $t_{1/2}$ is the half-life of the isotope in seconds. This equation is a rearrangement of equations 8 and 10 from Chapter 3. The following example demonstrates this type of calculation.

Example. Calculate the specific activity ($\mu c/\mu g$) of a protein whose molecular weight is 250,000 and contains one molecule of iodine-125 per molecule of protein.

$t_{1/2}$ for ^{125}I is 5.184×10^6 seconds and N for 1 mole of iodine is Avogadro's number, 6.0228×10^{23}. Therefore,

$$\lambda = \frac{(0.693)(6.0228 \times 10^{23})}{5.184 \times 10^6}$$

$$= 8.051 \times 10^{16}\ \text{dps/mole}$$

One mole of protein weighs 250,000 g or 2.5×10^{11} μg. Therefore,

$$\text{specific activity} = \frac{8.051 \times 10^{16} \text{ dps/mole}}{2.5 \times 10^{11} \text{ } \mu\text{g/mole}}$$

$$= 3.221 \times 10^{5} \text{ dps}/\mu\text{g}$$

Since 1 μc is equivalent to 3.7×10^{4} dps,

$$\text{specific activity} = \frac{3.221 \times 10^{5} \text{ dps}/\mu\text{g}}{3.7 \times 10^{4} \text{ dps}/\mu\text{c}}$$

$$= 8.704 \text{ } \mu\text{c}/\mu\text{g}$$

DIRECT IMMUNOPRECIPITATION OF ANTIGENS

Direct immunoprecipitation of an antigen, for example, a protein, is performed by mixing a radioactively labeled experimental preparation with an excess of suitable antibody. This may be done successfully over a pH range of 6 to 9, salt concentrations up to approximately $0.25M$, and temperatures between 0 and 37°C. Kabat and Mayer's work (10) may be consulted for more complete discussions of these and other pertinent parameters. After a suitable period of incubation, the length of which must be determined by trial and error (usually 40 minutes to several days), the precipitate is collected by centrifugation. Following removal of adsorbed proteins the amounts of protein and radioactivity in the precipitate are determined.

Direct precipitation procedures may be used for a variety of applications, each requiring a specific set of control experiments and precautions if useful and accurate information is to be gained. Three such applications serve to illustrate the kinds of problems that may arise. They are (1) demonstration of de novo protein synthesis, (2) determination of structural identity between two antigens from different preparations, and (3) visualization of an inactive antigen.

Demonstration of de novo Protein Synthesis

One of the earliest applications of immunochemical techniques in the field of molecular biology was the immunoprecipitation of an enzyme as a means of demonstrating its de novo synthesis. Such a demonstration in bacteria might involve growing the cells under two sets of physiological conditions, one in which the enzyme is expected to be present and a second in which it is not. A radioactive amino acid is added to each culture as a means of labeling newly synthesized proteins. The cells are then killed, broken open, and the resulting cell-free extract is incubated in the presence of excess antibody prepared against the enzyme to be studied. A small amount of pure, carrier enzyme can be added to ensure the presence of an immunoprecipitate in both cases. However, if carrier is needed to form a precipitate it is usually better to use a modification of the indirect RIA methods discussed below. The precipitates are finally recovered by centrifugation and assayed for the amount of radioactivity each contains. Only one precipitate is expected to contain substantial amounts of radioactivity; in practice, however, both precipitates are likely to contain radioactive material, for example, 1500 and 300 cpm, respectively.

Interpretation of these results depends heavily on the homogeneity of the precipitated material. It is possible that the 300 cpm observed in the one precipitate does not arise from the enzyme being studied, but rather from nonselective contamination of the immunoprecipitate by a wide variety of other cellular proteins. This is a very common occurrence which may seriously compromise this type of experiment, especially when small amounts of an enzyme are to be measured. Problems of nonspecific adsorption or precipitation may be eliminated in several ways. First the antibody preparation may be purified by ammonium sulfate fractionation and affinity chromatography. If affinity chromatographic techniques are used $4.5M$ magnesium chloride may be more gentle as an eluting solution than the acidic or basic solutions reported in the past. Second, the immunoprecipitation may be carried out in the presence of high salt concentration ($\sim 150mM$) and detergents (Triton X-100 and deoxycholate) (6). These conditions do not inhibit formation of legitimate immunoprecipitates but do retard nonspecific precipitation and adsorption. Finally the immunoprecipitates may be sedimented through a stepped sucrose gradient (11) in very small (1 ml) plastic tubes rather than merely being pelleted in centrifuge tubes. After centrifugation the tubes may be placed in an acetone–dry ice bath to freeze their contents. The bottom tip of the tube containing all of the immunoprecipitate may then be cut off with a razor blade and transferred to a scintillation vial for precipitate digestion and radioactivity determination. A superspeed centrifuge (20,000 rpm capability) equipped with a swinging bucket rotor may be conveniently used for these procedures in place of an ultracentrifuge. An adapter such as that shown in Figure 8-26 permits simultaneous centrifugation of 24 samples in the standard four well rotors.

Radioactive protein present in the precipitate after these precautions have been taken likely represents that which has been legitimately immunoprecipitated. This protein may represent the basal level of enzyme present in the cells or existence of an inactive precursor. That the

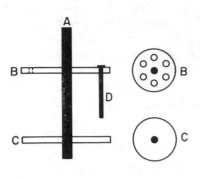

Figure 8-26. Adapter for the centrifugation of small capped plastic vials (Beckman Corp.). This adapter is for use with the Sorvall swinging bucket rotor or its equivalent. The centerpiece (*A*) extends to the bottom of the 50 ml tube well. Disk *C* is used as a means of centering the adapter in conjunction with the upper disk (*B*) of identical diameter. The test tubes are suspended by a lip from the upper disk. Centrifuging speed should not exceed 9000 rpm when this apparatus is used. Similar small plastic tubes (without caps) and four-place adapters are available commericially from the Sorvall Corp. These tubes may be centrifuged at significantly higher speeds without damage.

precipitated protein arises from the specific gene whose expression is being studied may be verified by conducting a similar experiment with a strain in which that particular gene has been genetically deleted. One and two-dimensional SDS electrophoretic techniques, described in Chapter 5, could also be used to distinguish a basal level of active enzyme from the existence of an electrophoretically distinct inactive precursor. These methods are often performed on the immunoprecipitate as supporting evidence to certify that the precipitated material is the desired protein. What would be the experimental result in each of these experiments?

Demonstration of an Inactive Form of an Enzyme

The techniques described above are often used to demonstrate de novo synthesis of specific proteins, but may not be used alone to argue that the protein synthesized is exactly the same as the purified antigen. On the contrary, all that is known with certainty is that the newly produced protein has determinants in common with the pure antigen used to elicit antibody formation. More precise knowledge requires rigorous structural identification of the precipitated material. Considerations such as these are encountered when immunochemical procedures are used to demonstrate the existence of an inactive form of an antigen. The data shown in Figure 8-27 provide one example of this. The amount of thiogalactoside transacetylase activity and material precipitable with transacetylase specific antibody were assayed in a set of extracts derived from a series of bacterial strains carrying nonsense mutations at various points within the Z gene of the *lac* operon. As shown in the figure, there are a number of strains that possess low levels of enzyme activity (the strain carrying a mutation at position 6), yet contain considerable amounts of immuno-

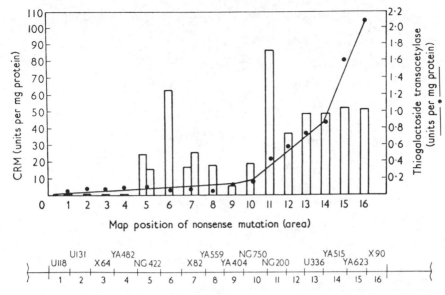

Figure 8-27. Cross-reacting material (bars) and thiogalactoside transacetylase activity (filled circles) in extracts of nonsense mutants. A genetic map of the *lac* Z gene, indicating the locations of some of the mutations used in this experiment, is shown below. [From A. V. Fowler and I. Zabin, *J. Mol. Biol.*, **33**:35 (1968).]

precipitable material called CRM (cross-reaching material). Such observations are usually presented as evidence for the presence of an inactive form of a protein.

Similar examples of immunochemical isolation of inactive proteins may be found in the isolation of a precursor protein. Such situations are well documented for the zymogen systems such as trypsinogen–trypsin, chymotrypsinogen–chymotrypsin, and proinsulin–insulin.

In all the above examples the precipitated material must be shown to be structurally related or identical to the purified antigen. The most convincing demonstration is to prepare cyanogen bromide fragments of the pure antigen and the immunoprecipitated material and show their coincidence of elution from ion exchange resins or congruence on two-dimensional silica gel or paper chromatograms. As shown in Figure 8-28 cyanogen bromide reacts specifically with methionine residues of a protein, yielding a cyanosulfonium bromide intermediate. Decomposition of this intermediate cleaves the peptide, leaving a peptide fragment, peptidyl homoserine lactone, and methyl thiocyanate as products. This method is particularly good for "fingerprinting" a protein because the fragments generated are usually of reasonable size and the specificity of the reaction

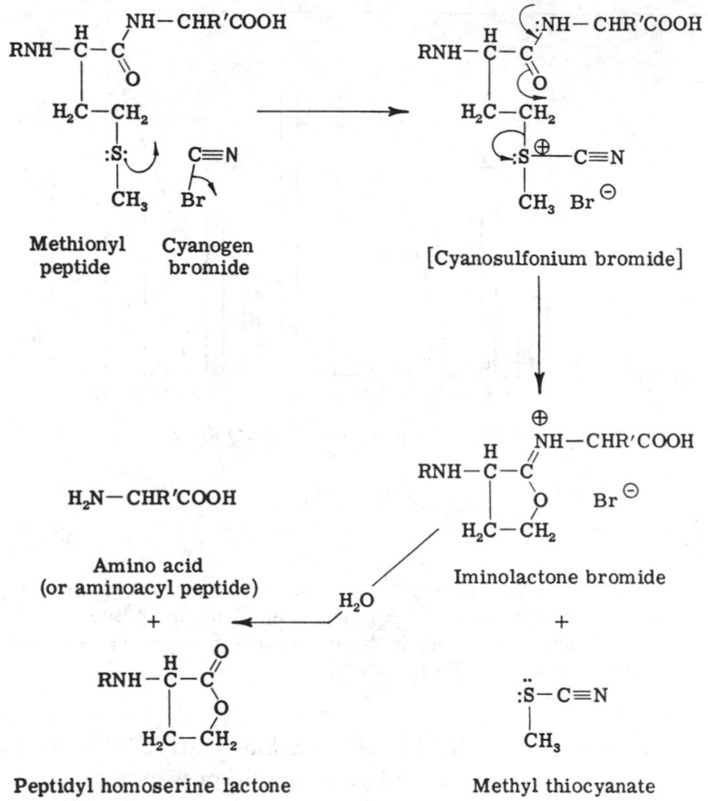

Figure 8-28. Reaction of cyanogen bromide with methionine residues, followed by cleavage of the peptide chain containing those residues. [From E. Cross, in *Methods of Enzymology*, Vol. II (C. H. W. Heis, Ed.), Academic Press, New York, 1967, p. 238.]

is high under acidic conditions (under basic conditions the specificity broadens to include most basic residues).

Alternatively, the standard and experimental samples may be subjected to limited proteolytic digestion with one of many available proteases. In the case of precursor studies all the fragments derived by cleavage of the purified antigen (active protein for the zymogen systems) would be expected to have identical counterparts in the collection of fragments generated by cleavage of the immunoprecipitated precursor (zymogens in the above situation). The cleaved precursor, however, might contain additional fragments that are not observed in the purified antigen. These would correspond to portions of the molecule that are lost during conversion of precursor to active protein.

[35]S-Methionine Synthesis

Earlier in this chapter ^{35}S-methionine was argued to be the isotope of choice for use in direct immunochemical precipitation procedures. However, the value of such a choice may be compromised by excessive expense if the radioactive amino acid is purchased commercially. This problem may be solved by preparing ^{35}S-methionine from inexpensive ^{35}S-sodium sulfate. One of the most convenient methods of preparation involves permitting a culture of *Saccharomyces cerevisiae* (baker's yeast) to incorporate carrier free ^{35}S-sulfate into cellular protein. The highly radioactive proteins are then proteolytically digested with papain and ^{35}S-methionine is isolated from the hydrolysate by paper chromatography. Purified ^{35}S-methionine may then be eluted from the paper as shown in Figure 8-29. A number of precautions should be observed during the preparation of radioactive methionine in this manner. First it is necessary to trap ^{35}S-hydrogen sulfide gas that may be given off by the growing yeast. As shown in Figure 8-30 this may be done by culturing the organisms in a closed vessel. Air entering the culture vessel (*B*) via an asparger is prewarmed and prehumidified (*A*); any hydrogen sulfide that is given off is trapped by bubbling effluent gases through $0.5N$ NaOH (*C*). The second precaution involves preventing oxidation of methionine to methionine sulfoxide. As shown in Figure 8-31, the sulfoxide may be easily separated from methionine chromatographically. However, its

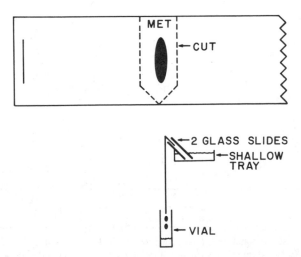

Figure 8-29. Elution of ^{35}S-methionine from a paper chromatogram. The upper diagram indicates the area of the chromatogram to be cut out and the lower figure depicts the sample elution apparatus.

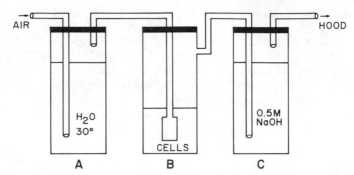

Figure 8-30. Schematic diagram of an apparatus that may be used to culture *Saccharomyces cerevisiae* in a closed atmosphere.

formation may be prevented if the paper chromatography is carried out in an inert atmosphere and the solution used for elution of methionine from the paper chromatogram contains 0.01 to 0.02M dithiothreitol or 2-mercaptoethanol (Reference 15). Storage solutions should also contain one of these agents. Methionine oxidized to the sulfoxide will not function as a precursor of protein synthesis and at high concentrations may even be inhibitory.

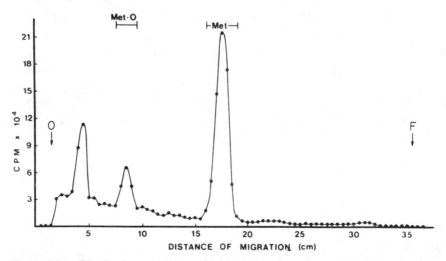

Figure 8-31. Distribution of radioactive materials after paper chromatography of a pancreatin digest of denatured yeast cells which had incorporated ^{35}S-sulfate from the medium. The chromatogram was run on Whatman 3MM paper and developed in *n*-butanol–acetic acid–water. Individual strips 0.5 cm long were counted in a liquid scintillation counter. Met, Met·O, O, and F indicate ^{35}S-methionine, ^{35}S-methionine sulfoxide, origin, and solvent front, respectively. [From R. Graham and W. M. Stanley, *Anal. Biochem.*, **47**:505 (1972).]

RADIOIMMUNOASSAY

Radioimmunoassay (RIA) procedures are capable of measuring trace amounts of any substance that can serve as an antigen or hapten. Resolution is limited principally by the specific activity of the radioactively labeled standard and by the specificity of the antibody preparation. At present the practical limit is on the order of 10^{-11} to 10^{-9} g antigen per sample. In the past many biologically important molecules such as drugs and hormones could be assayed only at low resolution using predominantly bioassay procedures. Today an ever-increasing number of these enzymatically inactive molecules are being reliably quantitated in samples of biological fluids such as blood, urine, and spinal fluid. Table 8-7 lists some molecules now being assayed routinely in this manner.

The basis of radioimmunoassays is diagrammed in Figure 8-32. An experimental sample of unlabeled antigen (open circles) is mixed with a constant, known amount of the same antigen, radioactively labeled (closed circles). To this mixture is added a limited amount of rabbit antibody prepared against the antigen being measured. It is critical to add sufficiently small amounts of antibody so that the system is in a condition of antigen excess even when unlabeled antigen is absent. Since the system is antibody limited, the amount of labeled antigen bound to the rabbit

Table 8-7. Compounds Quantitated by Radioimmunoassay

Adrenocorticotropic hormone
Aldosterone
Bradykinin
Calcitonin
Carcinoembryonic antigen
Cyclic AMP
Digitalis
Estradiol
Gastrin
Glucagon
Growth hormone
Human IgE
Insulin
Morphine
Parathyroid hormone
Prostaglandins
Secretin
Testosterone
Thyroglobulin
Vasopressin

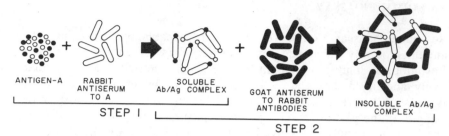

ANTIGEN-A   RABBIT   SOLUBLE
             ANTISERUM   Ab/Ag COMPLEX
             TO A
                                                          GOAT ANTISERUM
                                                          TO RABBIT   INSOLUBLE Ab/Ag
             STEP I   ANTIBODIES   COMPLEX

                                       STEP 2

Figure 8-32. Conceptual operation of the radioimmunoassay, RIA. The soluble Ab–Ag complex is both a product of the first step and a reactant in the second step of the assay.

antibodies is inversely proportional to the quantity of nonradioactive antigen present, as shown in Figure 8-33. This is basically an isotope dilution type of experiment. The final step of the procedure is separation of antigen–antibody complexes from free antigen, a prerequisite for measuring the amount of antigen bound. Although many methods are available the one most widely used experimentally is immunoprecipitation. This method of separation can be used successfully because (1) the antigenic determinants of an antibody molecule are not located very close to its antigen binding sites and (2) immunoprecipitation of an antibody molecule does not adversely affect its ability to bind antigen. In practice a second antibody preparation is added to precipitate the antigen–rabbit antibody complexes (Figure 8-32). A good candidate in this example would be goat antibody obtained following immunization of the animal with the γ-globulin fraction of rabbit serum. Some caution is needed in selecting the pair of antibody preparations to be used because some do not work well together, as is the case for donkey antibodies prepared against guinea pig γ-globulins.

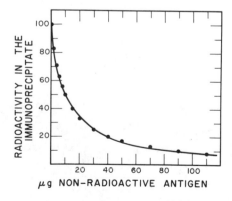

Figure 8-33. Precipitation of a standard radioactive antigen in the presence of increasing amounts of non-radioactive competitor. A curve such as this is required for standardization of radioimmunoassays.

^{125}I Labeling Procedures

Two principal methods exist for incorporation of ^{125}I or ^{131}I into purified antigens. Both involve oxidizing iodide to "active iodine," presumably a cationic form of iodine (I^+), which then reacts with ionized tyrosine residues of the antigen. The first method (12) employs lactoperoxidase and hydrogen peroxide (Figure 8-34) and the second method (14) uses

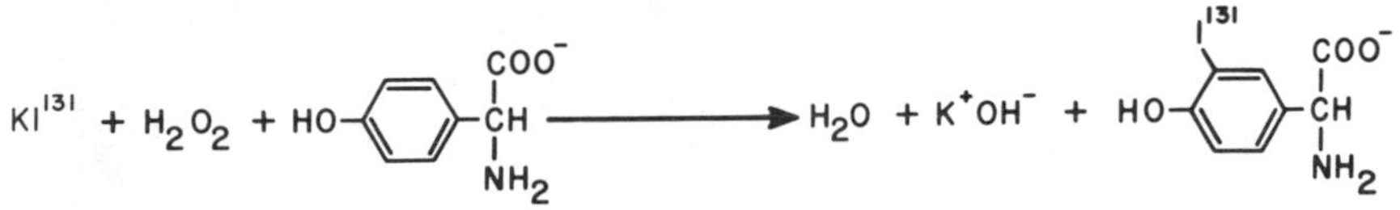

Figure 8-34. Synthesis of ^{131}I-iodotyrosine with the lactoperoxidase oxidation system.

hypochlorous acid generated by the slow decomposition of chloramine T in aqueous solution (Figure 8-35). These reactions are carried out in extremely small volumes ranging between 0.05 and 1.0 ml. Following incorporation of iodine into the antigen the reaction mixture is transferred to a molecular sieve column to separate unreacted iodine from that bound to protein. More convenient and gentle methods of protein iodination are beginning to develop and to be made available commercially. Two such methods may be found in references 16 and 17.

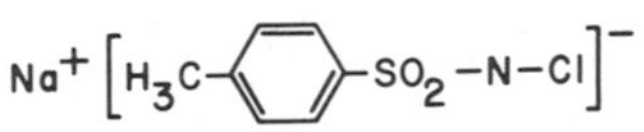

Figure 8-35. Structure of chloramine T.

Extreme care is required during these operations if loss of antigen immunoreactivity is to be avoided. Such losses may result from (1) alteration of the antigen resulting from replacement of the tyrosine proton by iodine, (2) chemical damage caused by oxidizing reagents and the reducing agents subsequently added to neutralize them, and (3) radiation damage to the protein. Covalent addition of a large atom such as iodine to the surface of a protein can potentially change its conformation. Addition of more than one molecule greatly increases the probability of such adverse conformational changes. Therefore, incorporation should be limited to one iodine molecule or less per molecule of protein. As shown in Table 8-8, values significantly below this level are common. Chemical damage is most often observed when the chloramine T method is used. It can be avoided to a great extent if chloramine T is added in small aliquots until incorporation has reached the desired level. Using such a "titration"

Table 8-8. Purification of Iodinated Hormones

Hormone	Iodination Procedure	I Atoms/Hormone Molecule	Methods of Separation of Iodohormone from Uniodinated Hormone	Reference
Adrenocorticotropin	Tracer iodination with chloramine T	0.01	Chromatography on carboxymethyl-cellulose column; salt gradient, ammonium acetate	R. J. Lefkowitz, J. Roth, W. Pricer, and I. Pastan, *Proc. Natl. Acad. Sci. US*, **65**:745 (1970)
Angiotensin	Standard chloramine T	?1.0	Chromatography on Dowex (AG) IX8 column in H_2O; chromatography on paper in pyridine, acetate ethanol; chromatography on Dowex 50X8 column in pyridine acetate gradient	S.-Y. Lin, H. Ellis, B. Weisenblum, and T. L. Goodfriend, *Biochem. Pharmacol.*, **19**:651 (1970)
Insulin	Tracer iodination with chloramine T	0.025–0.10	Chromatography on diethylamino-ethyl cellulose column; salt gradient, Tris–NaCl–urea	P. Freychet, J. Roth, and D. M. Neville, Jr., *Biochem. Biophys. Res. Commun*, **43**:400 (1971)
	Standard chloramine T	0.4–5.0	Starch gel electrophoresis	S. A. Berson and R. S. Yalow, *Science*, **152**:205 (1966)
Luteinizing hormone, releasing hormone	Lactoperoxidase	0.02	Polyacrylamide gel electrophoresis	Y. Miyachi, A. Chrambach, R. Mecklenburg, and M. B. Lipsett, *Endocrinology*, **92**:1725 (1973)
Oxytocin	Standard chloramine T (without metabisulfite)	0.01–0.02	Adsorption chromatography on Sephadex G-25 column in acetic acid	E. E. Thompson, P. Freychet, and J. Roth, *Endocrinology*, **91**:1199 (1972)
Vasopressin	Standard chloramine T (without metabisulfite)	0.02	Same as for oxytocin	J. Roth, S. M. Glick, L. A. Klein, and M. J. Peterson, *J. Clin. Endocrinol*, **26**:671 (1966)

From J. Roth, in *Methods in Enzymology*, Vol. 37 (B. W. O'Malley and J. G. Hardman, Eds.), Academic Press, New York, 1975, p. 223.

method decreases the concentration of chloramine T present at any one time and eliminates the need to add a reducing agent to neutralize excess oxidizing agent. Alternatively the lactoperoxidase procedure may be used; chemical damage is reduced but so is the efficiency of labeling. Radiation damage can be minimized by avoiding concentrated solutions of radioactivity for prolonged periods of time. The iodination reaction requires only a short time to reach completion, following which the reaction mixture is diluted during molecular sieve chromatography. Thereafter it is advisable to keep the solution dilute even if serum albumin must be added to stabilize the antigen. Care must be used, however, to select a source of albumin that does not in any way affect the immunochemical reactions.

All the problems discussed above require that the labeled antigen be checked for possible damage following the iodination procedure. In spite of all possible precautions, the presence of one mole of iodine per mole of protein may still have an effect. Therefore, it is most important to determine whether or not iodination alters the antigen with respect to its intended uses. Deleterious iodination effects may be detected as decreased enzyme specific activities or altered physicochemical properties of the antigen. The most sensitive probes, however, involve antigen binding to other molecules. For example, does the antigen bind as well to an affinity column or to its cognate antibody as it did prior to iodination? If the antigen is a hormone, does it bind as well to its receptor site and, once bound, does it still elicit the same physiological changes as before iodination? Much effort and careful consideration are required if the evaluation is to be at all rigorous.

Prior to any evaluation of the functional integrity of labeled molecules, they must be separated from unlabeled molecules that may still be present. This is especially important for hormones which can elicit physiological effects at minute concentrations. Separation is usually based on a physicochemical change in the molecule as a result of iodination. Figure 8-36 depicts one such separation using ion exchange chromatography and Table 8-8 lists a number of other methods that have also been used to successfully isolate various iodinated hormones.

Standardization of Radioimmunoassays

Several potentially serious conceptual and technical problems may arise during standardization and use of a radioimmunoassay. The first of these involves losses of antigen during assay. In most cases the protein concentrations found in an assay mixture are quite low, especially if purified preparations are used throughout. Under these conditions, losses

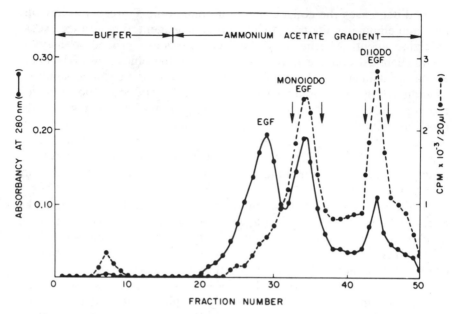

Figure 8-36. DEAE cellulose chromatography of iodinated and noniodinated epidermal growth factor. [From G. Carpenter, K. J. Lembach, M. M. Morrison, and S. J. Cohen, *J. Biol. Chem.*, **250**:4297 (1975).]

resulting from nonspecific absorption to glass and plastic vessels are substantial and must be eliminated. This may be done by adding serum to the reaction mixture (2% final concentration). Serum should be chosen that (1) does not contain any materials which would react with antibodies found in the assay and (2) does not affect the stability of the test antigens. The latter is exemplified by glucagon, which is inactivated by crystalline bovine serum albumin but not by horse serum.

The second problem is one of correctly selecting the appropriate quantities of antiserum. Both antibody preparations (see Figure 8-32) must be carefully titrated. Antibody against the antigen being tested must always be used at a concentration where there is antigen excess. The second antibody, prepared against γ-globulin analogous to the first antibody preparations, must always be used in moderately large excess. If the first antibody preparation is used at too high a concentration, all of the antigen-labeled standard and unlabeled experimental sample will bind to antibody, precluding competition. On the other hand, if the second preparation is used at too low a concentration, there will be incomplete precipitation of the first antibody, resulting in decreased assay resolution.

Since the titer of antiserum changes from animal to animal and within the same animal with time, it is advisable to use a single preparation for each set of experiments. When this is not possible each new preparation must be completely standardized. In the case of the secondary antibody preparation, only its ability to precipitate the primary antibodies must be quantitated. Primary antibody preparations, however, must be titrated for their ability to serve as both antibodies and antigens.

The amount of radioactivity in the final immunoprecipitate is determined using a scintillation counter if ^{14}C, ^{3}H, or ^{35}S is used as the labeling isotope. If, however, ^{125}I is used a second option is available. Since ^{125}I gives off γ-rays it is possible to use a γ-ray scintillation counter. This instrument operates in a manner similar to that of a standard scintillation counter, but uses a scintillation system (diagrammed in Figure 8-37) that is different from the one described in Chapter 3. The sample to be counted may either be suspended in water or left as a pellet. When lowered into the sample well of the counter, the sample is positioned close to a large silver iodide crystal containing trace amounts of thallium chloride. γ-Rays from the sample pass through the vial or test tube wall and into the crystal, where they interact with the silver iodide crystal, producing flashes of light which are detected by an adjacent photomultiplier tube. There are two advantages of using a gamma counter for this type of assay. First, the time required to accumulate statistically significant counting rates is usually quite small, thus increasing the number of samples that can be processed per unit time. Second, γ-ray counters are usually less complex electronically, do not require the use of scintillation vials or scintillation fluid, and are therefore considerably less expensive to purchase and operate.

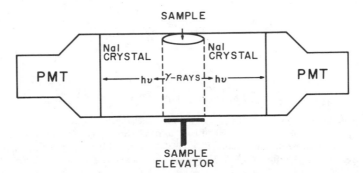

Figure 8-37. Schematic diagram describing the physical organization and operation of the γ-ray sensing elements of a γ-ray scintillation spectrometer. PMT indicates the photomultiplier tubes.

The data obtained from radioimmunoassay experiments may be plotted in a number of different ways. If a qualitative measurement of unknown antigen is desired, the amount of radioactive standard precipitated may be plotted as a function of the amount of unlabeled test sample added. A plot such as this is shown in Figure 8-38. In this example, no competition is observed when buffer is substituted for unlabeled experimental sample. Extract derived from a wild type organism containing lactate dehydrogenase, the antigen being measured, competes very well as evidenced by a dramatic decrease of radioactivity in the immunoprecipitate. Extract derived from a mutant strain devoid of lactate dehydrogenase activity also decreases the immunoprecipitable radioactivity, although less successfully, indicating that the mutant extract contains a protein that does not have lactate dehydrogenase activity but that does have determinant sites in common with purified lactate dehydrogenase. However, from these data it is difficult to compare the two competitive curves quantitatively. An alternative method of data presentation is shown in Figure 8-39.

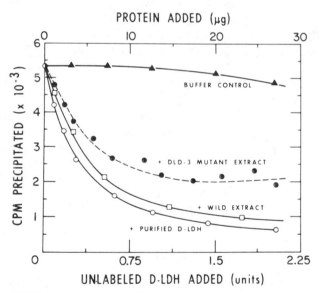

Figure 8-38. Inhibition of immune precipitation of ^{125}I-labeled D-lactate dehydrogenase by increasing amounts of unlabeled purified D-lactate dehydrogenase (○), a 0.1% Triton X-100 extract of wild type membrane vesicles (□), a 0.1% Triton X-100 extract of mutant membrane vesicles (●), and 0.1% Triton X-100 in 0.1 M potassium phosphate buffer, pH 7.1 (▲). The 0.1% Triton X-100 extracts of wild type and mutant vesicles contained the same amounts of protein. The inhibitory effect of the wild type extract was compared with that of a purified D-lactate dehydrogenase preparation containing the same number of units of enzyme activity. [From S. Short and H. R. Kaback, *J. Biol. Chem.*, **250**:4285 (1975).]

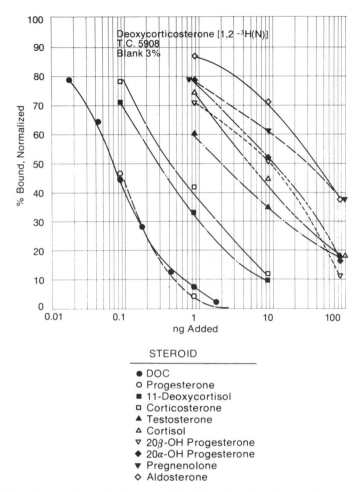

Figure 8-39. Competition of various steroids with radioactive deoxycorticosterone (DOC) for binding to DOC-specific antiserum. These data were generated by standard radioimmunoassay procedures. (Courtesy of New England Nuclear Corp., Worcester, Mass.)

Here the percentage of total radioactivity bound to the immunoprecipitate is plotted on semilogarithmic coordinates as a function of the amount of competitor added. The advantage of such a plot is that in many cases the central portion of the curve is a straight line, which aids in comparing the degrees of competition exhibited by standard and unknown samples. Also shown in this figure are the results observed when standard radioactive antigen (deoxycorticosterone, DOC) is challenged with various non-radioactive steroid analogues. Significant competition is observed with progesterone, 11-deoxycortisol, and corticosterone. The extent of competition is significant when it is considered that the hapten against which the antibody was produced is not likely to contain multiple determinants.

Although the latter method of analysis is better suited to quantitative interpretation, the data presented highlight perhaps the most serious conceptual problem involved with careful interpretation of radioimmunoassay results: what exactly is being measured. The merits of immunospecificity are to be lauded, but although an immunochemical reaction is indeed specific, it is unfortunately not absolutely specific. Antigenic determinants of different proteins may, and often do, share some similarities; since a single protein has more than one type of determinant the chances of finding such a similarity increases even more dramatically. In the case of an enzyme active site only one or a very few substrates bind to the active site well, and yet a variety of analogues may usually be found which bind to a limited extent (0.5–1% as well). The immense sensitivity of radioimmunoassay techniques, however, easily detects such slight binding. One must always ask, therefore, whether the observed competition is the result of a small amount of competitor whose identity is the same as the standard antigen or a somewhat larger quantity of material that has a similar antigenic determinant but is not the desired antigen at all. It must be remembered that in spite of its great flexibility and potential the immunochemical approach is indirect and requires a knowledge of the identity of the precipitated material—information that is often difficult if not impossible to obtain.

EXPERIMENTAL

Preparation of Avidin-immune Serum

8-1.   The following experiments will require (1) one 2.5 to 3.5 kg rabbit, (2) a highly purified preparation of avidin, (3) Freund's complete and incomplete adjuvants, (4) disposable Ouchterlony plates, and (5) ^{14}C-biotin, which can be obtained from Amersham Searle Corp.

8-2.   Four weeks prior to injection of antigen into the rabbit, bleed the animal and process the blood as described earlier in this chapter. Repeat this procedure twice at 7 day intervals. This allows 14 days between the last bleeding and antigen injection.

8-3.   During the 4 week period used to obtain control sera, determine the purity of a commercial avidin preparation using SDS–acrylamide gel electrophoresis as described in Chapter 5. Use only preparations exhibiting a single electrophoretic species in these experiments.

8-4.   Mix the following components in a 13×10 mm test tube.

> 1.0 ml 0.85% NaCl (w/v)
> 0.5 mg biotin
> 1.0 to 4.0 mg avidin

The amount of avidin needed may vary, but this range of concentrations has been used successfully.

8-5.   Mix the above components by gently vortexing the solution until all components are dissolved.

8-6.   Add 1.0 ml Freund's complete adjuvant and produce a stable emulsion either by vortexing the solution vigorously or by repeatedly drawing it into a syringe and expelling it forcibly through a 20 gage needle.

8-7.   Inject intramuscularly 1.0 ml of the stable emulsion produced in step 8-6 into one of the large hind thighs of the rabbit.

8-8.   Repeat step 8-7 14 days later but inject only 0.3 to 0.5 ml of the emulsion. Prepare fresh emulsion each time that injections are given and make all emulsions after the first one using Freund's incomplete adjuvant in place of the complete adjuvant.

8-9.   Seven days after the booster injection, bleed the rabbit and prepare cell-free serum as described in this chapter.

8-10. A crude assay for the presence of avidin-specific antibodies may be performed by adding increasing amounts (0.0, 0.1, 0.2, 0.3, or 0.4 ml) of serum to 30 μliters of an avidin solution containing 2 mg/ml. These components are mixed and allowed to incubate overnight at 4°C. If the serum contains antibodies a precipitate forms under these conditions. Very small amounts of precipitate may be detected only as an opalescent appearance of the test reaction mixture compared to that of the control from which avidin was omitted. A positive reaction to this test may be expected 3 to 5 weeks after the initial injection (step 8-7).

8-11. This procedure is applicable to most proteins which can be obtained in a highly purified form. Therefore, any of a variety of enzyme proteins may be used in place of avidin. Acid phosphatase is a very good choice for this purpose because it is easily assayed (see Chapter 10). It is advisable to prepare antisera against at least two unrelated proteins. These two types of sera will be used as part of the experiment as described below.

Quantitative Precipitation of an Antigen

8-12. Prepare a ^{14}C-biotin–avidin complex by dissolving 10 mg avidin and 7.5 μc ^{14}C-biotin in sufficient 0.85% saline solution to yield a final volume of 40 ml. Incubate this solution at 22°C for 5 minutes and then cool it to 4°C.

8-13. Number consecutively 16 15 ml Corex test tubes and place them in a rack.

8-14. To the first eight tubes add respectively the following quantities of the biotin–avidin solution prepared in step 8-12: 0.02, 0.05, 0.1, 0.15, 0.2, 0.3, 0.4, and 0.5 ml.

8-15. Add sufficient 0.85% saline solution to each of the test tubes to yield a final volume of 0.5 ml.

8-16. Repeat steps 8-14 and 8-15 with the second set of eight test tubes.

8-17. Place 0.3 to 0.5 ml of anti-avidin serum (step 8-9) in each of the first eight test tubes and the same quantity of control serum (step 8-2) in each of the second set of eight tubes. It is very important that all the test tubes contain exactly the same quantities of serum.

8-18. Mix the contents of each test tube gently with a vortex mixer and cover each tightly with Parafilm.

8-19. Incubate the test tubes for a convenient period greater than 16 hours at 4°C.

8-20. At the conclusion of the incubation period remove any immuno-precipitate that may have formed by centrifuging each tube for 20 to 30 minutes at $20,000 \times g$ (13,000 rpm in a Sorvall centrifuge or equivalent). It may be necessary to process only four tubes at one time if the precipitated material detaches easily from the tube wall upon standing.

8-21. Remove the supernatant solution using a Pasteur pipette. Take care to remove as much solution as possible. Wash the precipitate two or three times with 5 ml 0.85% saline solution. Some loss may occur at this point if great care is not exercised during the washing operations. The supernatant solutions derived from the washing operations may be discarded.

8-22. Place 0.2 ml of each supernatant solution (step 8-21) into 16 scintillation vials respectively and add 15 ml of an aqueous scintillation fluid to each.

8-23. Add 0.5 ml of a protein solubilizing agent such as Protosol to each of the 16 precipitates and vortex the test tube gently until the precipitated material is completely dissolved.

8-24. Transfer 0.3 ml of each of the solutions obtained in step 8-23 into scintillation vials and add 15 ml of an aqueous scintillation fluid to each.

8-25. Determine the amount of radioactivity in each of the 32 scintillation vials.

8-26. Plot the amounts of radioactivity observed in each of the 16 pellets and supernatant solutions as a function of the volume of ^{14}C-biotin–avidin complex added to that test tube. Such a plot is shown in Figure 8-15.

Double Diffusion of Avidin and Avidin-immune Serum in Ouchterlony Plates

8-27. Prepare an avidin solution by dissolving 4.0 mg avidin in 1.5 ml water. This is denoted as solution A.

8-28. Dilute 0.5 ml solution A with 0.5 ml water. This solution is denoted as solution B.

8-29. Place 0.045 ml avidin-immune serum (step 8-9) in the center well of a disposable micro-Ouchterlony plate such as that shown in Figure 8-18. Make certain that the wells are free of water before the solutions are placed in them and also that no bubbles are trapped at the bottom of the wells. This operation may be successfully performed using Lang–Levy pipettes.

8-30. Place 0.02 ml solution A in one of the peripheral wells and 0.02 ml solution B in a peripheral well adjacent to the one containing solution A.

8-31. Incubate the plates in a moist atmosphere at 4°C. Inspect the plates every 12 to 16 hours for several days. A precipitin band similar to that in Figure 8-21 should become visible during the second day of incubation.

8-32. Repeat this experiment using the second antigen and antibody combination generated in step 8-11.

8-33. Using various combinations of the two antigen and antibody preparations generate all the precipitin patterns shown in Figure 8-20.

References

1. B. D. Davis, R. Dulbecco, H. N. Eisen, H. S. Ginsberg, and W. B. Wood Jr., *Microbiology*, 2nd ed., Harper and Row, New York, 1973.

2. J. F. Miller, A. P. Mitchell, G. F. Davies, and R. B. Taylor, *Transplant. Rev.*, 1:3 (1969). Antigen-Sensitive Cells. Their Source and Differentiation.

3. J. Clausen, Immunochemical Techniques for the Identification and Estimation of Macromolecules, in *Laboratory Techniques in Biochemistry and Molecular Biology* (T. S. Work and E. Work, Eds.), American Elsevier, New York, 1969.

4. D. H. Campbell, J. S. Garvey, N. E. Gremer, and D. H. Sussdorf, *Methods in Immunology*, 2nd ed., Benjamin, New York, 1970.

5. D. M. Weir, Ed., *Handbook of Experimental Immunology*, F. A. Davis, Philadelphia, 1967.

6. D. J. Shapiro, J. M. Taylor, G. S. McKnight, R. Palacios, C. Gonzalez, M. L. Kiely, and R. T. Schimke, *J. Biol. Chem.*, 249:3665 (1974). Isolation of Hen Oviduct Ovalbumin and Rat Liver Albumin Polysomes by Indirect Immunoprecipitation.

7. N. H. Axelsen, J. Kroll, and B. Weeke, Eds., *A Manual of Quantitative Immunoelectrophoresis*, BioRad Laboratories, Richmond, Calif.

8. H. G. Minchin and T. Freeman, *Clin. Sci.*, 35:403 (1968). Quantitative Immunoelectrophoresis of Human Serum Proteins.

9. J. G. Feinberg, *Int. Arch. Allergy*, **11**:129 (1957). Identification, Discrimination and Quantification in Ouchterlony Gel Plates.

10. E. A. Kabat and M. M. Mayer, *Experimental Immunochemistry*, 2nd ed., Charles C. Thomas, Springfield, Ill., 1967.

11. R. D. Palmitter, *J. Biol. Chem.*, **248**:2095 (1973). Ovalbumin Messenger Ribonucleic Acid Translation.

12. J. I. Thorell and B.G. Johansson, *Biochem. Biophys. Acta*, **251**:363 (1971). Enzymatic Iodination of Polypeptides with ^{125}I to High Specific Activity.

13. J. Roth, Methods for Assessing Immunologic and Biologic Properties of Iodinated Peptide Hormones, in *Methods in Enzymology*, Vol. 37 (B. W. O'Malley and J. G. Hardman, Eds.), Academic Press, New York, 1975, p. 223.

14. W. M. Hunter, The Preparation of Radioiodinated Proteins of High Activity, Their Reaction With Antibody *in Vitro*: the Radioimmunoassay, in *Handbook of Experimental Immunology* (D. M. Weir, Ed.), F. A. Davis, Philadelphia, 1967.

15. R. Graham and W. M. Stanley, *Anal. Biochem.*, **47**:505 (1972). An Economical Procedure for the Preparation of L-(^{35}S)Methionine of High Specific Activity.

16. F. T. Wood, M. M. Wu and J. C. Gerhart, *Anal. Biochem.*, **69**:339 (1975). The Radioactive Labeling of Proteins with an Iodinated Amidination Reagent.

17. A. E. Bolton and W. M. Hunter, *Biochem. J.*, **133**:529 (1973). The Labeling of Proteins to High Specific Radioactivities by Conjugation to a ^{125}I-Containing Acylating Agent. Application to the Radioimmunoassay.

NOTE ADDED IN PROOF

A number of investigators have reported that the F_c portion of IgG binds tightly to the A protein of *Staphylococcus aureus* cell walls (*J. Immunol.* **104**:140, **105**:1116, 1970). Therefore, these organisms may be heat inactivated, fixed with formaldehyde, and used to precipitate antigen–antibody complexes (*Eur. J. Immunol.* **4**:29, 1974; *J. Immunol.* **115**:1617, 1975). Heat treated organisms may be substituted for the second antibody preparation normally used in radioimmunoassays (see pages 295, 296). They are also gaining popularity for direct immunoprecipitation (see pages 288–292), because with these methods the antigen–antibody complex is bound to a bacterium and as such can be more easily and thoroughly washed. This results in a significant lowering of nonspecific precipitation of antigens unrelated to those being studied.

Chapter 9

Centrifugation

The physical techniques most responsible for our current understanding of cellular makeup and operation are those involving the centrifuge. A wide variety of these instruments are available ranging in capacity from those handling 0.2 ml and less to those accommodating thousands of liters with relative ease. Some are crudely controlled with regard to speed and temperature whereas in others these parameters are regulated within limits of less than 5%. Technical applications are equally broad, encompassing the collection and separation of cells, organelles, and molecules. Such a range of uses precludes a thorough discussion of each, but a discussion of the basic operation and the considerations on which these variations are based is possible.

In its simplest form a centrifuge is composed of a metal rotor with holes in it to accommodate a vessel of liquid (see Figures 9-4 and 9-9) and a motor or other means of spinning the rotor at a selected speed. All the other parts found in today's modern centrifuges are merely accessories used to perform various useful tasks and maintain the environment within which the rotor operates. Our consideration of these techniques begins with very simple low and high capacity instruments and proceeds to the more complex machines used for refined analysis. The highly sophisticated techniques involved in quantitative analytical ultracentrifugation with the Model E ultracentrifuge are omitted here because they have been thoroughly and expertly reviewed elsewhere (1–3).

RELATIVE CENTRIFUGAL FORCE

Centrifugation is based on the fact that any object moving in a circle at a steady angular velocity is subject to an outward directed force, F. The magnitude of this force depends on the angular velocity in radians, ω, and the radius of rotation, r, in centimeters.

$$F = \omega^2 r \tag{1}$$

F is frequently expressed in terms of the earth's gravitational force and is then referred to as the relative centrifugal force, RCF, or more commonly as the "number times g."

$$\mathrm{RCF} = \frac{\omega^2 r}{980} \tag{2}$$

To be of use, however, these relationships must be expressed in terms of "revolutions per minute," rpm, the common way in which the operating speed of a centrifuge is expressed. Since rpm values may be converted to radians using the equation

$$\omega = \frac{\pi(\mathrm{rpm})}{30} \tag{3}$$

then

$$\mathrm{RCF} = \frac{\dfrac{(\pi\mathrm{rpm})^2}{30^2}(r)}{980}$$
$$= (1.119 \times 10^{-5})(\mathrm{rpm})^2 r \tag{4}$$

The considerations used to calculate the RCF exerted on a sample spinning in a centrifuge rotor require that the sample be located at a fixed distance, r, from the center of rotation. Owing to rotor design, r varies from top to bottom of the sample holder. The problem is illustrated in the following example.

Example. Calculate the RCF exerted at the top and bottom of a sample vessel spinning in a fixed angle rotor such as that shown in Figure 9-1. Assume that the rotor dimensions, r_{min} and r_{max}, are 4.8 and 8.0 cm, respectively, and that it is spinning at a speed of 12,000 rpm. By equation 4,

$$\mathrm{RCF}_{top} = (1.119 \times 10^{-5})(12,000)^2 4.8$$
$$= 7,734 \times g$$

and

$$\mathrm{RCF}_{bottom} = (1.119 \times 10^{-5})(12,000)^2 8.0$$
$$= 12,891 \times g$$

As can be seen here the centrifugal force exerted at the top and bottom of

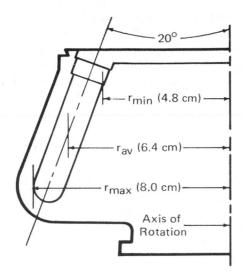

Figure 9-1. Cross-sectional diagram of an angle head rotor showing the distances from the axis of rotation to the top, middle, and bottom of the centrifuge tube. (Courtesy Spinco Division, Beckman Instruments, Inc., Palo Alto, Calif.)

the sample tube differs by nearly twofold. To account for this, RCF values may be expressed as an average RCF value (RCF_{ave}), which is the numerical average of the values exerted at the top and bottom of the same chamber. The important consideration, then, is to clearly define the value of r.

DESK TOP CLINICAL CENTRIFUGES

Clinical or desk top centrifuges are the simplest and least expensive centrifuges available. These instruments are most often used to compact or collect small amounts of substances that sediment rapidly (red blood cells, coarse or bulky precipitates, and yeast cells). The maximum speed of most desk model centrifuges is below 3000 rpm and all of them operate at ambient temperature. Although their speed and temperature of operation cannot be closely regulated, they can be used for a wide variety of applications that would otherwise needlessly employ larger and more sophisticated instruments.

HIGHSPEED CENTRIFUGES

Highspeed centrifuges are those operating up to speeds of 20,000 to 25,000 rpm. These machines, which account for the bulk of preparative applications, are usually equipped with refrigeration equipment to cool the rotor chamber. There are two major types of instruments in this category. The first is a relatively simple, high capacity, continuous flow centrifuge such as that shown in Figure 9-2. These instruments consist of a very fast electric motor or steam turbine drive unit, which is connected to a long tubular rotor via a cork clutch or V belt assembly. Their major application is the harvesting of yeast or bacteria from large cultures (5–500 liters). The culture is siphoned or pumped into the bottom of the spinning rotor. As the culture moves up, toward the top of the rotor, microorganisms are sedimented against the rotor walls whereas the clarified culture medium exits through an upper effluent port. After all of the medium has passed through the centrifuge the rotor is disassembled and the compacted cells are removed with a long spatula. In this manner 1 to 1.5 liters of culture per minute may be processed. Refrigerated coils may be added to this centrifuge but are really unnecessary if the culture is cooled by addition of ice prior to being subjected to centrifugation.

The second type of highspeed centrifuge is represented by the lower capacity refrigerated instrument shown in Figure 9-3. This centrifuge or a variation of it is used for most preparative techniques and can be found in nearly every biochemically oriented laboratory. Temperatures of the rotor chamber, although only coarsely controlled by a thermocouple in the bottom of the chamber, are easily maintained in the area of 0 to 4°C. Speed control is much more refined than in the clinical instruments discussed above, and this type of instrument also has a braking device to decrease the time required for rotor deceleration at the conclusion of centrifugation. In most cases, braking is accomplished by arranging for the drive motor to act as an electric generator; current generated in this way is then dissipated through a heavy duty resistor. A wide variety of rotors (Figure 9-4) are available and their potential is greatly increased by an even wider variety of adapters (Figure 9-5). These instruments are most often used to collect microorganisms, cellular debris, cells, large cellular organelles, ammonium sulfate precipitates, and immunoprecipitates. However, they cannot generate sufficient centrifugal force to effectively sediment viruses, small organelles such as ribosomes, or individual molecules.

THE ULTRACENTRIFUGE

Development of the ultracentrifuge with its ability to attain centrifugal forces in excess of $500,000 \times g$ (75,000 rpm, $r = 8$ cm) opened entirely new fields of investigation. It permitted the fractionation of subcellular

Figure 9-2. Sharples high capacity continuous flow centrifuge. Arrows indicate input and output ports of the cylindrical rotor. The model shown here is equipped with a high speed fractional horsepower electric motor. This may be replaced with a high pressure steam turbine drive. The centrifuge rotor, or bowl, as it is called, is shown in a cutaway view. (Courtesy of Sharples–Stokes Division, Pennwalt Corp., Warminster, Pa.)

Figure 9-3. Sorvall refrigerated highspeed centrifuge. Meters display the temperature and running speed; the two dials are used for selecting the speed and duration of centrifugation. (Courtesy of DuPont Instruments, Newtown, Conn.)

organelles previously observed only in electron micrographs; this in turn permitted assay of their enzymatic constituents, providing insights into structure–function relationships. Viruses could be isolated in pure form allowing careful definition of their composition. Molecules such as DNA, RNA, and protein could be carefully analyzed even to the point of resolving two types of DNA molecules differing only in the fact that one

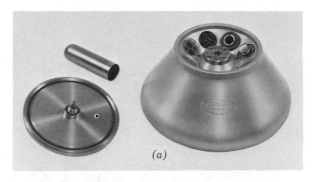

(a)

(b)

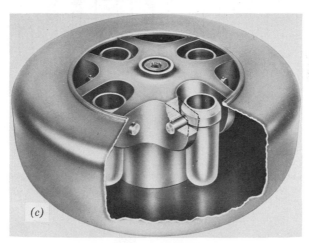

(c)

Figure 9-4. Various angle and swinging bucket rotors that may be used in a highspeed centrifuge. (*a*) The rotor accommodates eight 50 ml tubes or any of the adapters shown in the rotor. The concentric screws in the rotor cap are used to fasten the cap to the rotor (outer screw) and the rotor to the drive spindle (small inner screw). (*b*) The rotor accepts six 250 ml bottles. (*c*) The rotor is a swinging bucket model that accomodates four 50 ml tubes or appropriate adapters. The shield has been cut away to visualize the titanium yoke and tube holders. (Courtesy of DuPont Instruments, Newtown, Conn.)

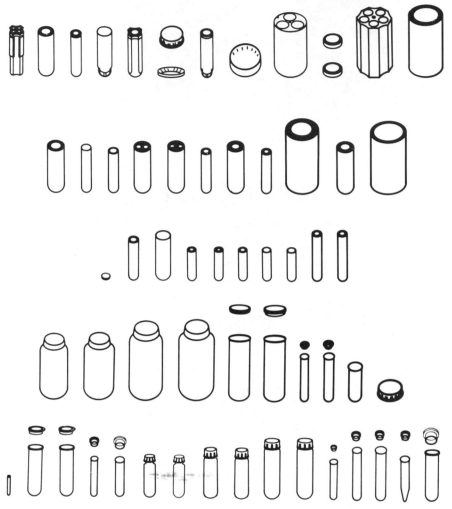

Figure 9-5. Tubes and adapters that can be accommodated by highspeed rotors. The volume of these vessels ranges from 1 to 500 ml. (Courtesy of DuPont Instruments, Newtown, Conn.)

of the two types contained ^{15}N in place of the naturally occurring ^{14}N isotope of nitrogen. Contemporary instruments such as that shown in Figure 9-6A consist of four principal parts: (1) drive and speed control, (2) temperature control, (3) vacuum system, and (4) rotors.

Drive and Speed Control

The drive assembly of most instruments is composed of a water-cooled electric motor connected to the rotor spindle by way of a precision gear box. Surprisingly, the drive shaft itself is only about $\frac{3}{16}$ in. in diameter. The small diameter of the shaft allows it to flex during rotation, thus accommodating a small degree of rotor imbalance without vibration or spindle damage. The speed of rotor rotation may be selected by means of a rheostat and monitored with a tachometer. However, neither the rheostat setting nor tachometer reading is precisely accurate. The best method of obtaining a highly accurate value is to determine the number of revolutions occurring over a period of 1 to 10 minutes. This may be done using a stopwatch and the instrument's odometer, which is usually calibrated in thousands of revolutions.

In addition to the speed control system, a second overspeed system was developed to prevent operation of a rotor above its maximum rated

Figure 9-6. Preparative ultracentrifuge (*a*) with the cover panels in place, and (*b*) from the front and (*c*) from the rear with the cover panels removed. *A*, armor plate surrounding the rotor chamber: *B*, drive and gear assembly: *C*, electronics bank housing the speed, temperature, and vacuum control and monitoring systems; *D*, vacuum pump; *E*, vacuum pump motor; *F*, refrigeration unit; *G*, diffusion pump; and *H*, drive oil reservoir. (Courtesy of Beckman Instruments, Palo Alto, Calif.)

Figure 9-6 (*Continued*)

speed. Such operation results in the rotor being torn apart or exploding. It is for this reason that the rotor chamber is always enclosed in heavy armor plate capable of containing any such explosion. The overspeed system consists of (1) a ring of alternating reflecting and nonreflecting surfaces attached to the bottom of the rotor (Figure 9-7), (2) a small but intense point source of light, and (3) a photocell. As shown in Figure 9-8, light from the point source is reflected from the reflecting portions of the overspeed ring and onto the photocell. The passing of reflecting and nonreflecting surfaces through the light beam as a consequence of rotor rotation chops the light and establishes a pulsing signal in the photocell output circuitry. The frequency of this signal, which is a function of the speed of rotation, is compared to a standard reference signal. If the

Figure 9-7. Rotor bottom showing the speed control ring consisting of alternating reflecting and nonreflecting surfaces. (Courtesy of Beckman Instruments, Palo Alto, Calif.)

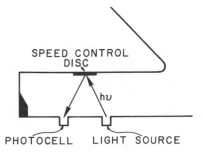

Figure 9-8. Diagram of the path followed by the light beam of an optical speed control system.

rotor-generated signal frequency surpasses that of the reference, the instrument is automatically shut down.

Temperature Control

The temperature monitoring system of an ultracentrifuge is also more sophisticated than that of the centrifuges described above. Temperature control in highspeed instruments involves placing a thermocouple in the rotor chamber, and monitoring only the rotor chamber temperature. In the case of an ultracentrifuge an infrared radiometric sensor placed beneath the rotor continuously monitors the rotor temperature directly, ensuring more accurate and responsive temperature control.

Vacuum System

A significant qualitative difference between highspeed and ultracentrifuges is the incorporation of a vacuum system into the latter. At speeds below 15,000 to 20,000 rpm only small amounts of heat are generated by friction between air and the spinning rotor. At greater speeds, however, air friction is significant and becomes severe above 40,000 rpm. To eliminate this source of heating, the rotor chamber is sealed and evacuated by two pumping systems operating in tandem. The first system is a mechanical vacuum pump similar to those in normal laboratory use, which can establish a vacuum down to 100 to 50 μ. Once the pressure in the chamber has decreased below 250 μ, a water-cooled diffusion pump is also brought into operation. Using both pumps it is possible to attain and hold vacuums of 1 to 2 μ. As would be expected, temperature control is significantly improved when the rotor chamber is evacuated.

Rotors

A large variety of rotors are available for use in modern ultracentrifuges, and fall into two classes: angle and swinging bucket. Both types are constructed from either aluminum alloys for low to moderate speeds or titanium for high speed operation. Angle rotors such as those in Figure 9-9 consist of a solid piece of metal with 6 to 12 holes machined at an angle between 20° and 45°. These rotors are most often used for applications involving total sedimentation or "pelleting" of a constituent. Their greatest advantage is a large capacity.

Swinging bucket rotors, on the other hand, consist of a rotor from which hangs three to six free moving buckets (Figure 9-10). These buckets hang vertically when the rotor is at rest and swing 90° to a horizontal position, under the influence of centrifugal force, as the rotor attains a speed of 200 to 800 rpm. This type of rotor was designed principally for incomplete sedimentation of a sample through some sort of gradient; details of these techniques are discussed below. Its major advantage is that gradient material may be placed in the centrifuge

Figure 9-9. Various configurations of angle rotors for the preparative ultracentrifuge. (Courtesy of Beckman Instruments, Palo Alto, Calif.)

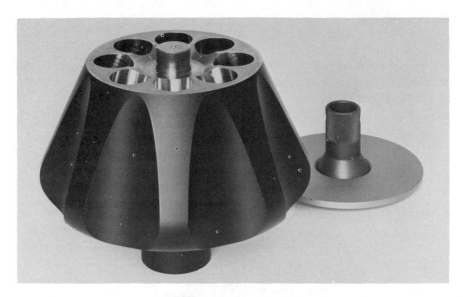

Figure 9-9 (*Continued*)

Figure 9-10. SW 40 swinging bucket rotor with the centrifuge tube holders or buckets in place. (Courtesy of Beckman Instruments, Palo Alto, Calif.)

tubes held vertically, but centrifugation occurs with the tubes held horizontally. In this position material sedimented to different areas of the tube appear as bands running across the tube rather than at an angle as in an angle rotor. Therefore, when a tube is removed from a swinging bucket rotor its contents do not reorient as they do in the case of an angle rotor.

SEDIMENTATION COEFFICIENTS

Early in this chapter it was stated that molecules or particles spinning around an axis are subjected to a centrifugal force, F. Under the influence of this force they sediment toward the bottom of the centrifuge tube at a velocity, v, described by the equation

$$v = \frac{dr}{dt} = \phi \frac{(\rho_p - \rho_m)}{f} \omega^2 r \tag{5}$$

where r is the distance (cm) from the axis of rotation to the sedimenting particle or molecule, ϕ is the volume of the particle (cm^3), ρ_p is the density of the particle (g/cm^3), ρ_m is the density of the medium (g/cm^3), f is the frictional coefficient (g/sec), and v is the radial velocity of sedimentation of the particle (cm/sec). A more common form of this equation is expressed in terms of the sedimentation coefficient, s, of the sedimenting particle, that is, its velocity of sedimentation per unit of force field, F.

$$s = \frac{dr}{dt} \cdot \frac{1}{\omega^2 r} \tag{6}$$

or

$$s = \phi \frac{(\rho_p - \rho_m)}{f} \tag{7}$$

The units of S are seconds, and since many biologically significant molecules possess sedimentation coefficients greater than 10^{-13} seconds (Figure 9-11), this quantity is defined as a Svedberg unit (S) in honor of Svedberg, the originator of this type of analysis. Therefore, ribosomal subunits or other particles possessing a sedimentation coefficient of 18×10^{-13} seconds are said to be 18S.

The frictional coefficient of a molecule (f in equation 7) depends on its size, shape, and the viscosity of the medium through which it is sedimenting. For a spherical molecule or particle of radius r_m, the

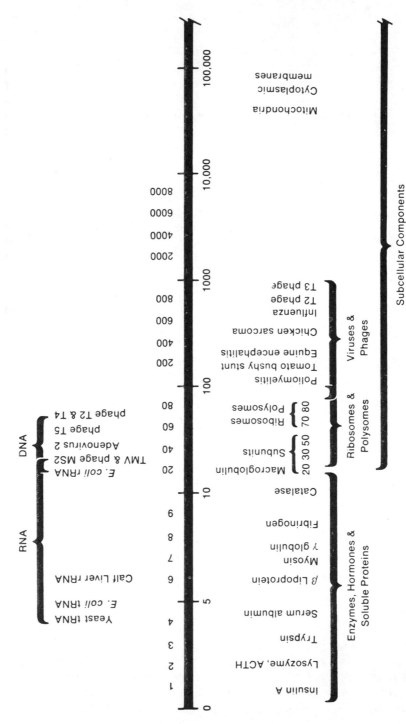

Figure 9-11. Sedimentation coefficients of various biological molecules, subcellular organelles, and organisms. (Courtesy of Beckman Instruments, Palo Alto, Calif.)

frictional coefficient may be calculated using the equation

$$f = 6\pi\eta r_m \tag{8}$$

where η is the viscosity of the medium in poises (g/cm · sec) and r_m is the molecule or particle radius (cm). From these considerations, it can be seen that the rate of sedimentation is governed by the size, shape, and density of the sedimenting particle or molecule, as well as by the viscosity and density of the medium through which it is moving.

Since there are a large variety of conditions under which a sedimentation coefficient may be determined, some means of normalization is necessary. Most often the sedimentation coefficient is corrected to the value that would be obtained in a medium with a density and viscosity of water at 20°C. The correction may be made mathematically using equation 7:

$$S_{20,w} = S_{t,m} \frac{\eta_{t,m}(\rho_p - \rho_{20,w})}{\eta_{20,w}(\rho_p - \rho_{t,m})} \tag{9}$$

where $s_{t,m}$ is the uncorrected sedimentation coefficient determined in medium m and temperature t, $\eta_{t,m}$ is the viscosity of the medium at the temperature of centrifugation, $\eta_{20,w}$ is the viscosity of water at 20°C, ρ_p is the density of the particle or molecule in solution (reciprocal of the partial specific volume), $\rho_{t,m}$ is the density of the medium at the temperature of centrifugation, and $\rho_{20,w}$ is the density of water at 20°C.

If equation 6 is integrated over the interval of t_0 to t_t, then

$$s = \frac{\ln r_t - \ln r_0}{\omega^2(t_t - t_0)} \tag{10}$$

where $t_t - t_0$ is the period of time required for the particle to move from position r_0 to position r_t. It is possible to use equation 10 to calculate the time required to sediment a particle or molecule with a known sedimentation coefficient:

$$t_t - t_0 = \frac{1}{s} \frac{(\ln r_t - \ln r_0)}{\omega^2} \tag{11}$$

If r_t and r_0 are set equal to the values for the radii at the top and bottom of the spinning cenu... fuge tube, then $T = t_t - t_0$ is the time required to bring about total sedimentation or pelleting of the sedimenting species. This value is sometimes referred to as the clearing time.

THE DENSITY GRADIENT

Thus far it has been assumed that sedimentation occurs through a homogeneous medium. However, in a preparative ultracentrifuge the smooth migration of particles through a homogeneous solution is disturbed by mechanical vibration, thermal gradients, and convection. These disturbances can be eliminated or largely alleviated by forming a gradient of some rapidly diffusing substance in the centrifuge tube. A wide variety of materials can be used, including sucrose, glycerol, cesium chloride, cesium sulfate, as well as more unusual compounds like ficoll and metrizamide. The gradient may be formed mechanically using a gradient maker such as that shown in Figure 9-12 (see Chapter 4 for an explanation of the functioning of this device) or by centrifugation as discussed below. The solution is most dense at the bottom of the tube and decreases in density up to the top of the tube. The gradient characteristics depend on the particular application, but two major types of techniques are commonly used, zone or sedimentation velocity centrifugation and isopycnic or sedimentation equilibrium centrifugation (Table 9-1).

Table 9-1. Characteristics of Sedimentation Velocity and Sedimentation Equilibrium Centrifugation

	Sedimentation Velocity	Sedimentation Equilibrium
Synonym	Zone centrifugation	Isopycnic, density equilibration
Gradient	Shallow, stabilizing—maximum gradient density below that of least dense sedimentating species	Steep—maximum gradient density greater than that of most dense sedimenting species
Centrifugation	Incomplete sedimentation, short time, lower speed	Complete sedimentation to equilibrium position, prolonged time, high speed

Sedimentation Velocity or Zone Centrifugation

Sedimentation velocity or zone centrifugation, in which the sedimenting species simply moves through a stabilizing gradient, is distinguished by use of a reasonably shallow gradient. This is necessary because, at maximum, the gradient density must be less than that of the least dense sedimenting species. During centrifugation sedimenting material moves through the gradient at a rate determined by its sedimentation coefficient. Therefore, it is important to terminate centrifugation before the first species reaches the bottom of the tube. This method works well for species that differ in size but not in density. As shown in Figure 9-13, two

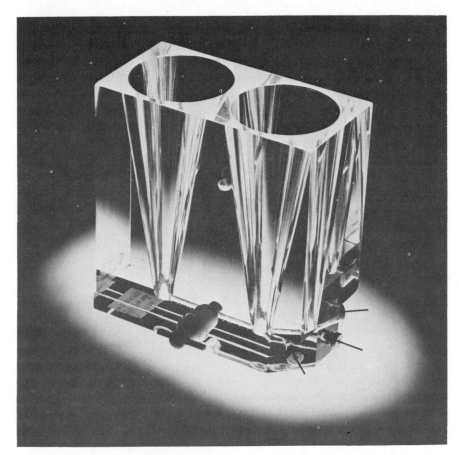

Figure 9-12. Gradient maker used to form linear density gradients over a broad range of volumes. A vibrator type of stirring motor is usually positioned above the unit. (Courtesy of Buchler Instruments, Fort Lee, N.J.)

proteins possessing nearly identical densities and differing in molecular weight by only threefold can be easily separated on such a gradient. On the other hand, Figure 9-14 demonstrates that subcellular organelles such as mitochondria, lysosomes, and peroxisomes, which have quite distinct densities but are similar in size, do not separate to any significant extent when subjected to this method of separation.

Sedimentation Equilibrium or Isopycnic Centrifugation

Isopycnic or sedimentation equilibrium centrifugation involves allowing the sedimenting species to move through the gradient until they reach a

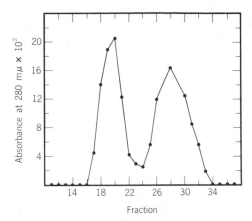

Figure 9-13. Separation of two proteins with molecular weights of 54,000 and 154,000 daltons using a 5 to 20% sucrose gradient. Centrifugation was performed for 9 hours at an average RCF of 67,968 × g. The smaller protein was observed in fractions 25–33, protein concentration was monitored by absorption at 280 nm.

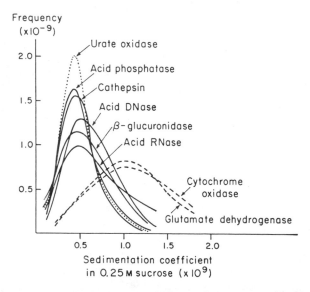

Figure 9-14. Frequency distribution of various marker enzymes for lysosomes (acid phosphatase, cathepsin, etc.), peroxisomes (urate oxidase) and mitochondria (cytochrome oxidase). Mitochondrial fraction from rat liver centrifuged in a linear gradient of 0.25 to 0.5M sucrose. [From H. Beaufay et al., *Biochem. J.*, **73**:628 (1959).]

point where their density and that of the gradient are identical. At this point no further sedimentation occurs because they are floating on a "cushion" of material that has a density greater than their own. This type of operation requires the gradient to be reasonably steep in order that at maximum, the gradient density is greater than the most dense sedimenting species. To permit all species to seek their equilibrium densities, centrifugation is continued for prolonged periods and at relatively higher speeds than would normally be used for sedimentation velocity techniques. This technique is used to separate particles similar in size but of differing densities. Since most proteins possess nearly the same density these methods are not usually employed for their separation. However, for situations where different densities are involved, isopycnic centrifugation is the method of choice. This is true for molecules, such as nucleic acids (Figure 9-15), as well as subcellular organelles, such as mitochondria, proplastids, and glyoxysomes (Figure 9-16A).

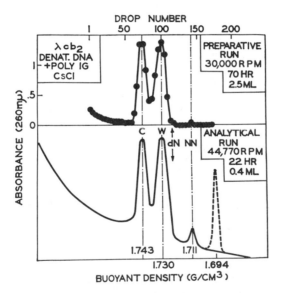

Figure 9-15. Poly I,G-effected separation of the complementary strands of λ phage DNA in preparative and analytical CsCl gradients. Upper trace represents the absorbance (260 nm) of the 4-drop (50 μl) fractions (total volume 2.5 ml) measured in a 20 μl microcuvette (2 mm light path). Lower trace represents the microdensitometric tracing of the photograph of the same undiluted material banded in the analytical ultracentrifuge (4°C, 3 mm cell) with added density marker DNA (*Cytophaga johnsonii*, 1.6945 g/cm³; dashed line). Peak C contains the DNA strands C, which preferentially bind poly I, G; the complementary strands W band under peak W. Symbols *dN* and *NN* indicate the positions (densities) of the denatured and native λcb₂ DNA, respectively. [From Z. Hradecna and W. Szybalski, *Virology*, **32**:633 (1967).]

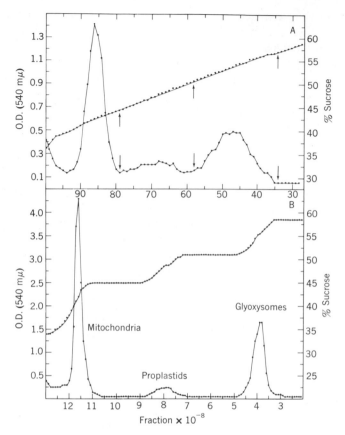

Figure 9-16. Protein distribution observed after isopycnic centrifugation of crude particles from castor bean endosperm on linear (*A*) and stepped (*B*) sucrose gradients. The upper curves of each panel show the measured sucrose concentration in each fraction and the lower curves, the protein. [From T. G. Cooper and H. Beevers, *J. Biol. Chem.*, **244**:3507 (1969).]

A helpful variation of isopycnic centrifugation, known as stepped gradients, may be used either to increase the capacity of a well defined gradient or to effect a greater separation of two species. Although these methods can be used to great advantage, they must be used with care and discrimination. Stepped gradients are produced by carefully layering solutions of different densities in the centrifuge tube. The system depicted in Figure 9-16*A* is used as an example of this technique and of the considerations involved in selecting "steps" of the appropriate size and density. The object here is to cleanly separate three organelles: mitochondria, $\rho = 1.19 \text{ g/cm}^3$; proplastids, $\rho = 1.23 \text{ g/cm}^3$; and glyoxysomes, $\rho = 1.25$ g/cm^3. As shown in the figure, these organelles are first separated on linear sucrose gradients. The desired sucrose concentrations for the "steps" are then identified on the linear gradient (arrows) as the highest densities observed for the sedimentation of each individual particle (57%, glyoxysomes; 50%, proplastids; and 44%, mitochondria). The stepped gradient is then constructed by placing a small amount of 60% sucrose to act as a "cushion" followed by larger amounts of 57%, 50%, and 44% sucrose solutions. A lower density solution, 33%, is used to separate the soluble, nonparticulate components of the sample from the first step. During centrifugation all the organelles except the mitochondria pass quickly through the low density step. Mitochondria, however, sediment through the 33% solution but then cannot penetrate the high density 44% step. In similar manner the proplastids and glyoxysomes are removed at the 50 and 57% steps, respectively. This method artificially sharpens the distribution of the isolated particles, permitting an increased capacity of the gradient. It must be emphasized, however, that any mitochondria successfully negotiating the 44% sucrose step continue to the next step, thus contaminating the proplastid function. Though it is important to select solution densities that minimize such contamination, it is necessary in each case to carefully ascertain the extent of cross-contamination present in the isolated fractions before the technique is used routinely. This may be done at low sensitivity with an electron microscope or at high sensitivity by assaying "marker" enzymes located specifically in each of these particles. Table 9-2 lists several activities that were monitored for this purpose in the above example.

Gradient Fractionation

Once a set of molecular species or subcellular organelles have been separated on a density gradient it is important to recover the separated components. One of two available methods may be used for this purpose. The first method involves puncturing the bottom of the centrifuge tube with a needle and allowing the contents to drip out as a result of gravity or to be drawn out with a peristaltic pump. Figure 9-17 depicts a commercial device that can be used for this purpose. If a peristaltic pump is employed care should be taken to keep the length of tubing used to a minimum. As the length of tubing increases, so does the degree of mixing, and this results in a serious decrease of resolution. The second method involves inserting a cover into the top of the centrifuge tube (see Figure 9-18). A small diameter glass or plastic tube is inserted through this cap to the bottom of the tube (alternatively, the bottom of the tube may be punctured with a needle). A solution with a density greater than any

Table 9-2. Distribution of Enzymes in the Subcellular Organelles from Castor Bean Endosperm

Fraction	Enzyme Activity (μmoles substrate utilized/g (fresh wt)/hr)					
	Citrate Synthetase	Isocitrate Lyase	Malate Synthetase	Malate Dehydrogenase	Succinate Dehydrogenase	Fumarase
9.5 K supernatant	2	66	44	22,150	0	20
Crude particulate[a]	155	342	450	16,700	87	334
Gradient supernatant	4 (1)	20 (8)	28 (3)	1,925 (11)	0	27
Mitochondria	218 (73)	8 (3)	46 (5)	10,700 (62)	65 (100)	394 (92)
Proplastids	2	9	27	311	0	4
Glyoxysomes	74 (25)	207 (85)	790 (89)	4,475 (26)	0	2

[a] After separation of the crude particulate fraction it was centrifuged on a stepped gradient and the three protein zones, as well as that part of the gradient above the mitochondrial zone ("gradient supernatant") were assayed separately. The figures in parentheses show the enzyme activity recovered in gradient supernatant, mitochondria, and glyoxysomes as a percentage of the total recovered in the four fractions from the gradient.

observed in the tube is then pumped into the bottom of the tube, forcing the contents out a hole in the cap. The exiting solution passes through a short tube and into waiting test tubes. The centrifuge tube cap should have a concave interior to minimize mixing and aid in the smooth passage of liquid from a large diameter to a small diameter tube.

Both these fractionation methods permit the contents of the centrifuge tube to drip sequentially into a set of test tubes. Ideally it would be desirable for each test tube or "fraction" to receive the same volume of solution. Unfortunately this is often not the case. Droplets obtained from the least dense region of the gradient are also the least viscous, and therefore exhibit a smaller amount of surface tension than do droplets obtained from denser regions. Since the size a droplet can attain before falling away from the tube is a function of its surface tension, the drops obtained from the top of the gradient are smaller than those obtained from the bottom. If quantitation of the gradient contents is desired, this problem must be eliminated or appropriate corrections made for it.

With the contents of the centrifuge tube distributed to a series of individual test tubes, it is possible to ascertain the distribution of desired components. In the case of subcellular organelles this may be done by locating specific "marker" enzyme activities. For example, succinate dehydrogenase and catalase are often used as "markers" for mitochondria and peroxisomes, respectively. Organelles and molecules lacking easily assayable enzyme activities are located either by their light absorption or by radioactive labeling.

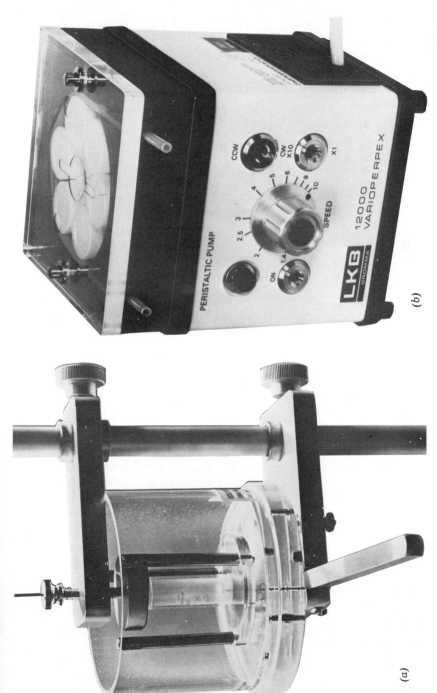

(a)

(b)

Figure 9-17. Sucrose gradient piercing unit (*a*) and three channel peristaltic pump (*b*). The piercing unit is equipped with a reservoir that may be filled with ice to keep the tube contents cold during fractionation. [Courtesy of (*a*) Buchler Instruments, Fort Lee, N.J.; (*b*) LKB Instruments, Rockville, Md.]

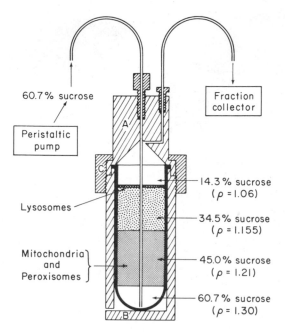

Figure 9-18. Collection of fractions from a discontinuous gradient. Parts *A* and *B* are made of lucite and are held tightly together by means of a screw cap, *C*. Airtight junctions are ensured by means of O rings. Tube contents are slowly driven upward into a collecting cone by injection of a dense sucrose solution. [From F. Leighton et al., *J. Cell Biol.*, **37**:482 (1968).]

Refractometric Determination of Concentration

In addition to component identification, a second important piece of information that can be obtained for each fraction is its average density. This is particularly important in the case of isopycnic gradient centrifugation because these methods separate particles or molecules on the basis of their density. Therefore, the density of the fraction containing a desired species is likely to provide a good estimate of the species' own density. The concentrations and therefore densities of most solutions used to establish gradients are proportional to their refractive indexes, a parameter that can be easily determined using a refractometer (Figure 9-19). Operation of this instrument is based on the fact that light travels through various media at different speeds. As a result of this, a beam of monochromatic light striking a thin layer of solution at an angle bends as shown in Figure 9-20. The degree of bending is quantitated as the angle of refraction (*r* in Figure 9-20). Snells' law, which describes the phenomenon, states that when a ray of light changes its direction on passing from

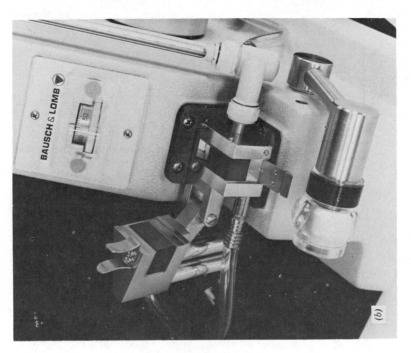

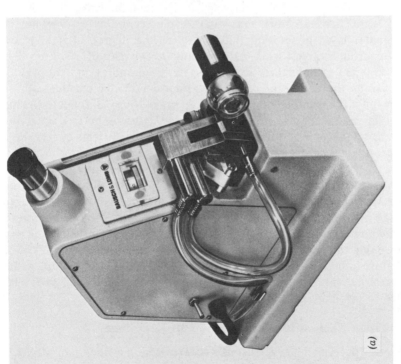

Figure 9-19. (a) Abbe 3L Refractometer. (b) Close-up of the prism and illuminator assembly. (Courtesy of Bausch and Lomb Co., Rochester, N.Y.)

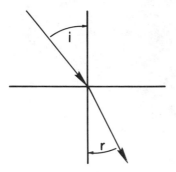

Figure 9-20. Light path through material that has a refractive index distinct from that of air. i denotes the angle of light beam incidence upon the material and r denotes the angle of refraction.

one medium to another, the ratio of the sine of the angle of incidence to that of the angle of refraction is a constant. This constant is called the index of refraction of the second medium with respect to the first. This may be formulated as

$$\eta = \frac{\sin i}{\sin r} = \frac{c_1}{c_2} \tag{12}$$

where η is the refractive index of the medium, i and r are the angles of incidence and refraction, respectively, c_1 is the light wave velocity in medium 1 (this is usually air), and c_2 is the velocity in the test medium. A refractometer is an instrument that measures the degree of light beam refraction brought about by a test solution. It consists of a source that projects a beam of light through the test material and an optical system that measures the degree of refraction. Figure 9-21 depicts the path of light through the sample and refracting prism onto a reflector. This reflector is pivoted, permitting the refracted light to be realigned for passage through a pair of compensating prisms. These prisms compensate for dispersion and chromatic aberration which arises from the use of a polychromatic light source. The corrected beam passes onto the eyepiece through a series of lenses and a two-way beam divider. Since the degree of correction needed to realign the beam is a function of sample refractive index, the pivot assembly has a scale attached to it. At the completion of beam alignment this scale may be illuminated by a second light source and read through the eye lens by means of a second lens system. Since temperature influences the observed refractive index of many samples, the entire sample area, including the refracting prism, is equipped with facilities to circulate coolant for temperature control.

Sedimentation Analysis in a Preparative Ultracentrifuge

Sedimentation coefficients are usually determined in a Beckman Model E analytical ultracentrifuge. This sophisticated instrument is equipped with

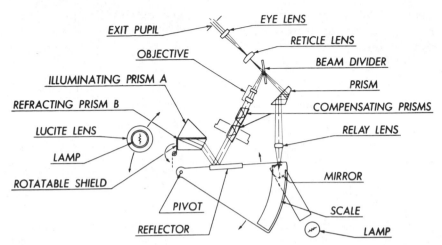

Figure 9-21.   Diagram of a light beam moving through the optical systems of an Abbe-3L Refractometer. The sample is placed between the illuminating (*A*) and refracting (*B*) prisms. (Courtesy of Bausch and Lomb Co., Rochester, N.Y.)

a system of optics which permits continuous visual monitoring of sedimenting material as it moves away from the axis of rotation; it can be used to obtain precise sedimentation coefficient values. However, many occasions arise when the analytical ultracentrifuge cannot be used, for example, if the desired species has not been purified to homogeneity. In these cases the preparative ultracentrifuge can be used to obtain a reasonably good approximation of sedimentation coefficient. The technique is predicated upon the observation by Martin and Ames (4) that the migration of most biological materials through an appropriately constructed density gradient is a linear function of the time of centrifugation (Figure 9-22). This means that the ratio of the distances traveled from the meniscus by two molecular species is always constant. Therefore, if a

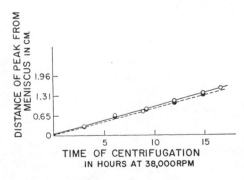

Figure 9-22.   Sedimentation of histidinal dehydrogenase (O) and imidazoleacetol phosphate transaminase (●) as a function of time. [From R. G. Martin and B. N. Ames, *J. Biol. Chem.*, **236**: 1372 (1961).]

species of unknown sedimentation coefficient is subjected to precisely the same conditions of centrifugation as a second species having a known sedimentation coefficient, the ratio, R, can be determined after any time of centrifugation

$$R = \frac{\text{distance traveled from meniscus by unknown}}{\text{distance traveled from meniscus by known}} \tag{13}$$

This is usually done with both species contained in one density gradient or in two identical gradients spinning in the same rotor. Since most macromolecules move at nearly constant rates,

$$R = \frac{s_{20,w} \text{ of unknown}}{s_{20,w} \text{ of known}} \tag{14}$$

A useful relationship that may be derived from a linear time dependence of sedimentation relates, the speed and time of centrifugation. These two parameters may be varied as long as the product of the force and its duration are constant:

$$r\omega_1^2 t_1 = r\omega_2^2 t_2 \tag{15}$$

where ω_1 and ω_2 are calculated from the two speeds of centrifugation and t_1 and t_2 are the durations of centrifugation at these speeds, respectively. The nature of the density gradient and the value of r, which depends upon the diameter of the rotor, must both be assumed constant.

Example. 40S ribosomal subunits sediment to the center of a 10 to 20% sucrose density gradient when the gradient is centrifuged for 140 minutes at 49,000 rpm. How fast must this gradient be centrifuged if the same result is to be observed in only 100 minutes? Since r must be assumed constant,

$$\omega_1^2 t_1 = \omega_2^2 t_2$$

$$t_1 = \frac{\left[\dfrac{(\pi)(\text{rpm}_2)}{30}\right]^2 t_2}{\left[\dfrac{(\pi)(\text{rpm}_1)}{30}\right]^2}$$

$$= \frac{\text{rpm}_2^2}{\text{rpm}_1^2} t_2$$

$$\text{rpm}_2 = \left(\frac{t_1 \text{rpm}_1^2}{t_2}\right)^{1/2}$$

$$= \left[\frac{(140)(49,000)^2}{100}\right]^{1/2}$$

$$= 57,978$$

A second use of this relationship involves moving a given species to a new position within a gradient. For example, if it is desired to move a given species 10% further into the gradient then the $\omega^2 t$ value should be increased by slightly less than 10%. It must be emphasized, however, that such a relationship is only approximate because as the species is moved to a new position within the gradient, the value of r changes. Also, if the nature of the gradient changes these simple mathematical relationships no longer hold. This more complex situation has been evaluated by McEwen (5).

Specific Design of a Density Gradient

Successful isolation of particular molecules and subcellular organelles by density gradient centrifugation requires careful consideration of the density gradient parameters. These parameters include (1) the material used to establish the gradient, (2) ionic strength, (3) viscosity, (4) osmotic characteristics, (5) gradient slope, (6) pH, and (7) the presence of stabilizing agents such as mercaptans, EDTA, enzyme substrates, and Mg^{2+}. The material most often used to establish a gradient is sucrose, which is cheap, available in high purity, and yields solutions with a density up to 1.28 g/cm³ (Figure 9-23). Glycerol is another compound that possesses quite similar characteristics but can be used only at densities less than 1.15 g/cm³. As shown in Figures 9-24 and 9-25, both these

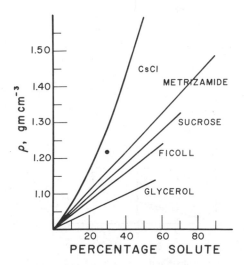

Figure 9-23. Densities yielded at 5°C from solutions of various materials used to form density gradients; ● indicates the density of a 30% solution of Ludox.

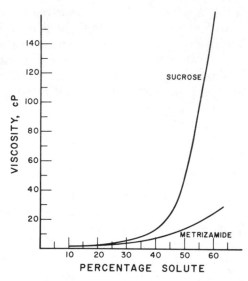

Figure 9-24. Viscosities of solutions containing various concentrations of sucrose and metrizamide. Glycerol behaves in a manner similar to that of sucrose.

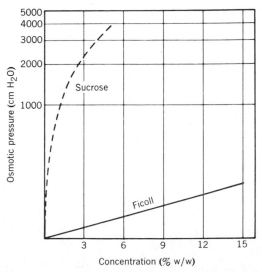

Figure 9-25. Osmotic pressure observed with solutions containing various concentrations of sucrose (dashed line) and Ficoll (solid line). (Courtesy of Pharmacia Fine Chemicals, Piscataway, N.J.)

compounds suffer two major disadvantages: they are very viscous at densities greater than 1.10 to 1.15 g/cm³ and exert very high osmotic effects even at very low concentrations. These liabilities prompted a search for substitute compounds that would yield a similar density range, but would also be inert and nonionized, and possess low viscosity and osmotic effects. Five such materials appear to fulfill these requirements: meglumine diatrizoate or Renografin, Urograffin, Ludox, Ficoll, and Metrizamide. The latter three compounds are used most often today. Ludox is a trade name for colloidal silica produced by Du Pont. A 40% solution of Ludox is often combined with polysaccharides such as dextran to form gradients that are used to isolate various types of whole cells such as red and white blood cells and unbudded yeast. Ficoll is the trade name for a high molecular weight polymer of sucrose and epichlorhydrin produced by Pharmacia Fine Chemicals. Metrizamide (Figure 9-26) is a relatively new addition to the list of gradient forming materials but has good potential because of its ability to readily generate densities up to 1.45 g/cm³. Gradients of metrizamide may be formed either with a gradient maker or by centrifuging a uniform solution of the compound. Use of this material does have one disadvantage, however, in that it moderately (up to 30%) quenches scintillation fluid, necessitating some correction when the materials separated on the gradient are assayed for their content of radioactivity.

Although these compounds are adequate for gradients involving proteins and whole cells, they are of little value when it comes to the separation of nucleic acids or organisms, such as viruses, that are composed predominantly of nucleic acids. This gap in technology was filled by the discovery by Meselson, Stahl, and Vinograd that cesium chloride could be used to establish gradients with densities ranging up to

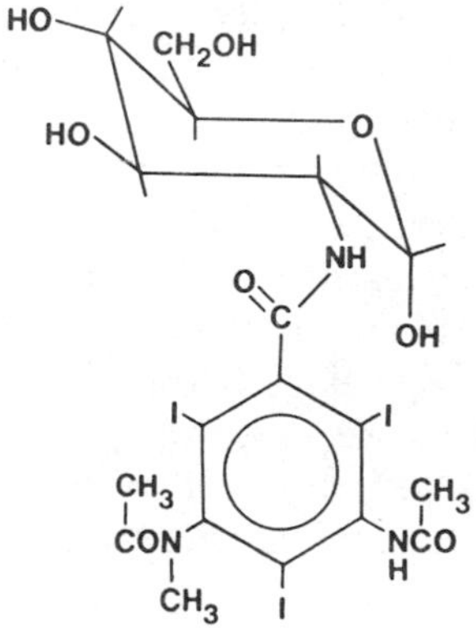

Figure 9-26. Structure of metrizamide.

1.70 g/cm³ (6, 7). Cesium chloride and more recently cesium sulfate linear gradients are almost always produced by centrifuging to equilibrium a uniform solution of the salt and sample to be resolved. During centrifugation the gradient forms by diffusion and the sedimenting species either sediment or "float" to a level within the gradient that has a density equivalent to their own. In addition to yielding high densities, cesium chloride gradients have the added advantage that DNA molecules sedimenting through them exhibit an almost linear relationship between their buoyant density and their guanosine plus cytosine contents (Figure 9-27). As can be seen from the figure, such linearity is not observed when the sulfate salt is used in place of cesium chloride. However, cesium sulfate has other characteristics in its favor. It forms twice as steep a gradient as the chloride salt and is therefore preferable for the separation of DNA molecules with widely different buoyant densities. DNA molecules are much more heavily hydrated in cesium sulfate than in cesium chloride solutions. This hydration reduces their buoyant density from about 1.7 g/cm³ in cesium chloride to about 1.4 g/cm³ in cesium sulfate, thus reducing the concentration of salt needed to form the gradients. At solution densities of 1.7 g/cm³, cesium chloride approaches its limit of solubility. The sedimentation of RNA molecules is almost always performed using cesium sulfate gradients because cesium chloride gradients cannot be produced at sufficient density. Although use of cesium sulfate is somewhat limited because some single stranded RNA molecules precipitate out, these solutions can be used for other single stranded RNA species as well as for all double stranded RNAs and RNA–DNA hybrid molecules.

A practical problem involved with the use of cesium salts for generating density gradients is their cost and their contamination by materials absorbing ultraviolet light. The presence of such contaminants precludes using the absorbance of nucleic acids at 260 nm as a means of assay. Optical purity salts may be obtained, but only at extremely high cost. However, the lower quality salts may be easily purified to optical quality using the following procedures. One to several pounds of cesium salt is dissolved in a minimum amount of distilled water and boiled for 5 minutes in the presence of finely divided, acid washed norit (charcoal). The hot solution is then filtered to remove the norit and again boiled until vigorous stirring can no longer prevent a crust from forming. At this point the solution is cooled to 0°C in an ice bath and the crystallized cesium salt recovered by filtration. The mother liquor may be repeatedly concentrated in this manner until 85–90% of the original weight of salt has been recovered. Acid-washed norit is prepared by suspending norit in 10

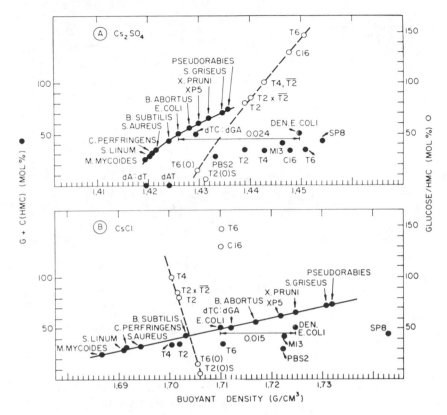

Figure 9-27. Buoyant densities (25°C) of native DNA's in Cs_2SO_4 (*A*) and CsCl (*B*) gradients as a function of base composition (●, solid lines), expressed as mole percent of G + C or G + HMC, or as a function of the glucose to HMC ratio (mole %), the latter for T-even phage DNA (○, dashed lines). Denatured *E. coli* DNA (Den. E COLI, ⊙) was prepared by heating DNA (10 μg/ml) for 10 minutes at 96 to 100°C in 0.02*M* sodium citrate (pH = 7.8) and rapidly chilling to 0°C. Based largely on the data of R. L. Erikson and W. Szybalski, *Virology*, **22**:111 (1964). [From W. Szybalski, in *Methods of Enzymology*, Vol. 12B, Academic Press, New York, 1968, p. 330.]

volumes of 1*N* HCl, boiling the mixture for 15 minutes, and washing the norit with water until the pH reaches neutrality. The washed norit is then dried. Since the preparation of norit is messy it is advisable to prepare a reasonable quantity at one time.

The battery of compounds described above can be used alone and in combination to produce a wide variety of density gradients; it is important to be aware of just how profoundly the composition of a gradient can

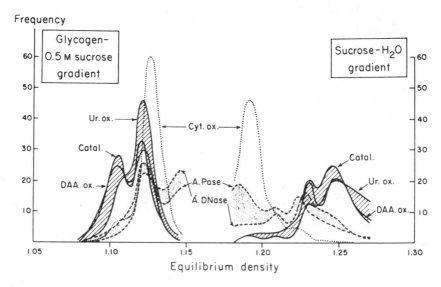

Figure 9-28.   Frequency distribution of equilibrium densities of mitochondria (cytochrome oxidase), peroxisomes (urate oxidase), and lysosomes (acid phosphatase). Mitochondrial fractions from rat liver equilibrated in a gradient of glycogen (initially linear from 0 to 30.6 g/100 ml) in $0.5M$ sucrose and in a linear sucrose gradient (59.7–117.0 g/100 ml). [From H. Beaufay et al., *Biochem. J.*, **92**:184 (1964).]

influence the final result. This is exemplified by Figure 9-28, which shows the density profiles of mitochondria (cytochrome oxidase), lysosomes (acid phosphatase and acid DNase), and peroxisomes (catalase, D amino acid oxidase, and urate oxidase) observed when mixtures of these organelles are subject to isopycnic centrifugation using sucrose or isoosmotic glycogen–sucrose to form the density gradient. In the case of the sucrose gradient the peroxisomes are significantly more dense than are the mitochondria. This relationship is reversed, however, when an isoosmotic glycogen–sucrose gradient is used in place of sucrose. Here a rather drastic alteration was needed to change the relative densities of the organelles. It must be emphasized that the same result can be accomplished more subtly. If a small amount of potassium chloride (in the range of millimolar quantities) is added to sucrose gradients that are used to separate mitochondria, proplastids, and glyoxysomes the distribution pattern shown in Figure 9-16*A* is no longer observed. Instead the glyoxysomes possess a density equivalent to that of the proplastids. If the ionic strength is increased further the density of the glyoxysomes can decrease to a value similar to that of mitochondria. These subtle influences must be carefully considered before arriving at the conclusion that two organelles or their marker enzymes are cosedimenting to the same buoyant density. Similar arguments, perhaps in somewhat different terms, can be made in the case of sedimenting molecules.

The final parameter to be manipulated during density gradient centrifugation is the sedimenting species themselves. Some caution was advised above concerning inadvertent modification of a species' sedimentation behavior. However, such changes can sometimes be used to advantage. For example, it is quite difficult to separate lysosomes (acid phosphatase) and peroxisomes (urate oxidase), as is shown in the control panel of Figure 9-29. Wattiaux et al. observed that lysosomes would

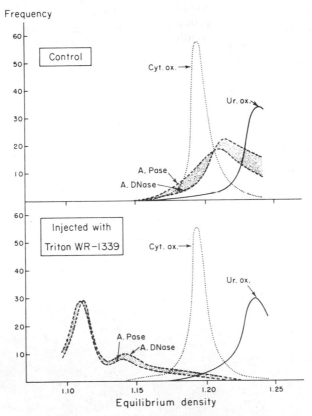

Figure 9-29. Effect of a previous injection of Triton WR-1339 on the equilibrium densities of particulate enzymes. Density equilibration of mitochondrial fractions from rat liver in an aqueous sucrose gradient. Upper panel: control. Lower panel: animal injected intravenously with 170 mg of Triton WR-1339 four days prior to sacrifice. [From Wattiaux et al., *Arch. Intern. Physiol. Biochem.*, **71**:140 (1963); *Ciba Foundation Symposium, Lysosomes*, Little, Brown, Boston, 1963, p. 176.]

accumulate nonionic detergents. Therefore, they injected Triton WR-1339 into their test animals 2 days before sacrifice. As shown in the lower panel of Figure 9-29, such treatment permitted the clean separation of these two types of organelles by changing the density of the lysosomes.

Large Scale Centrifugation in Zonal Rotors

One of the limitations of density gradient centrifugation is the amount of sample that can be accommodated. Even the largest swinging bucket rotors have a combined bucket capacity of only 100 ml. This precludes preparation of the quantities of material needed for many types of experiments. For example, it would be quite time-consuming to isolate the amount of glyoxysomes needed to purify citrate synthetase using the methods described above. To alleviate this problem Anderson and others have developed the zonal rotor. As shown in Figure 9-30, this rotor is essentially a somewhat flattened sphere whose interior is subdivided into four quadrants by means of a four-veined core. Figure 9-31 depicts a series of schematic diagrams which describe the operation of this rotor. Unlike conventional methods, the zonal rotor is empty at the beginning of an experiment. The speed is brought up to 3000 to 4000 rpm and then the gradient is "loaded" into the rotor. This is done by pumping the gradient solutions into a special movable seal and bearing assembly and then through holes drilled in the four veinings of the core and onto the perimeter of the rotor's interior. Unlike most gradients produced in a centrifuge, this gradient is established by pumping in the material of lowest density first. As the density of the solution increases, lighter density material already in the rotor is displaced toward the center of rotation. This operation is continued until the entire cavity is full. In

Figure 9-30. B–XV Zonal rotor showing the four-veined core. Arrows indicate ports at the ends and base of the veins. (Courtesy of Beckman Instruments, Palo Alto, Calif.)

addition to providing a means of conducting gradient solutions to the perimeter of the rotor's interior the veined core also prevents mixing and swirling of the rotor's contents owing to Coriolis forces generated during loading and unloading operations. Sample material is added to the top of the gradient (close to the center of rotation) through a second set of four small ports which are located at the base of each vein. Following these operations the loading seal and upper bearing assembly are removed, the rotor chamber is closed and evacuated, and the rotor speed is increased. At high speed the sedimenting species move radially until they reach a density similar to their own. At the conclusion of centrifugation the rotor speed is decreased to 3000 to 4000 rpm and contents are "unloaded" by pumping a very dense solution through the veins to the periphery of the rotor's interior. This addition forces the gradient containing the separated species out of the rotor through the central parts previously used for sample addition. The effluent solution is collected in a series of tubes in a manner similar to that described earlier.

Although the use of zonal rotors is somewhat more involved than conventional swinging bucket rotors it is important to emphasize that the end result is the same. An example of this is shown in Figure 9-32, which depicts the elution profile of a stepped sucrose gradient that was used to separate mitochondria, proplastids, and glyoxysomes. This profile is very similar to that shown in Figure 9-16A, but in that experiment the gradient volume was 54 ml. Here it is 1760 ml and accommodates the crude particulate fraction yielded from 1 lb of fresh tissue.

EXPERIMENTAL

Isolation of Mitochondria, Proplastids and Glyoxysomes on Linear and Stepped Sucrose Gradients

9-1. Soak a 500 ml volume of castor beans overnight in water at room temperature.

9-2. Germinate the soaked beans for 5 days at 30°C in moist vermiculite. If the vermiculite dries out, germination will be poor. Alternatively, if it is too moist the germinating seedlings may be overrun with a black mold. Sterilization of the vermiculite is useful if mold contamination becomes serious. CASTOR BEANS ARE EXTREMELY POISONOUS AND USUALLY FATAL IF TAKEN INTERNALLY.

9-3. At the conclusion of germination, recover the sprouted beans from beneath the vermiculite. Remove the embryo and cotyledons and discard them. Wash the endosperm tissue in distilled water and cool it in ice.

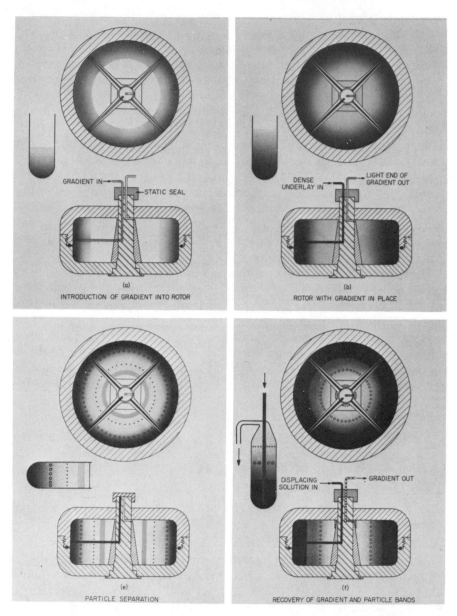

Figure 9-31. Schematic diagrams showing the operation of the B–XV zonal centrifuge rotor. Rotor shown at various stages of loading and unloading in top and side view. (*a*) Start of gradient introduction into rotor spinning at low speed, (*b*) completion of loading of gradient into rotor, (*c*) movement of sample layer into the rotor through the center line, (*d*) introduction of overlay into rotor to move the starting zone away from the core faces, (*e*)

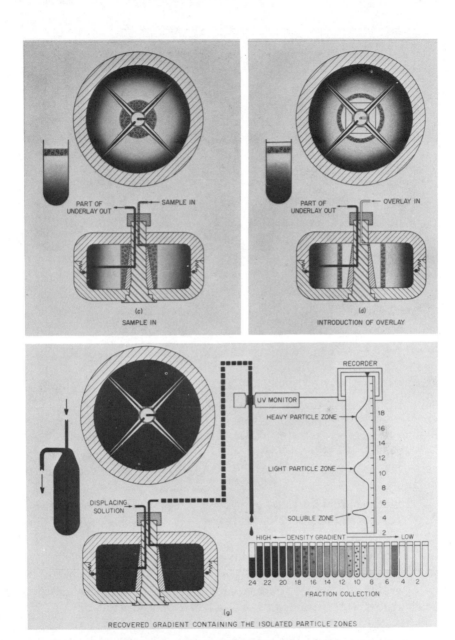

separation of particles at high speed, (f) displacement of separated zones out of rotor at low speed, and (g) completion of unloading and collection of sample tubes. Note that the swinging bucket equivalent of each step is also included. [From N. G. Anderson et al., *Anal. Biochem.*, **21**:235 (1967).]

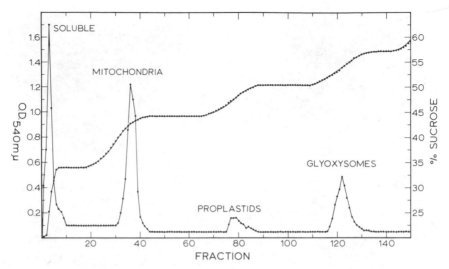

Figure 9-32. Separation of crude particulate fraction from castor bean endosperm using a B–XV zonal rotor.

9-4.   Prepare and chill 500 ml grinding medium consisting of the following components: $0.4M$ sucrose, $0.165M$ Tricine buffer adjusted to pH 7.5, $0.01M$ KCl, $0.01M$ MgCl$_2$, $0.01M$ EDTA adjusted to pH 7.5, and $0.01M$ dithiothreitol.

9-5.   Combine 60 g washed endosperm tissue with 90 ml chilled grinding medium and chop vigorously in an onion chopper for 6 minutes.

9-6.   Transfer the coarse brei to a cold mortar and grind until a smooth brei is obtained. The mortar should be kept in an ice bucket during the operation.

9-7.   Filter the brei through two layers of cheesecloth and centrifuge the filtrate for 10 minutes at $270 \times g$ to remove unbroken cells and large debris.

9-8.   Decant the supernatant obtained in step 9-7 into a clean, cold centrifuge tube and recentrifuge for 30 minutes at $10,800 \times g$.

9-9.   Carefully decant the supernatant yielded in step 9-8. Take care here because the pellet of sedimented organelles is very soft and may run. This supernatant fraction should be saved for subsequent enzyme assay.

9-10.   Gently resuspend the pellet from step 9-8 in 4 to 6 ml grinding medium.

9-11.   Sucrose solutions for both linear and stepped gradients may be prepared by adding appropriate amounts of sucrose to solutions of

0.01M EDTA adjusted to pH 7.5. Optimally the concentration of sucrose added should be determined with a refractometer rather than by weight. 100 ml each of the following solutions will be needed: 33, 44, 50, 57, and 60%.

9-12. For a linear gradient 16 ml 60% sucrose is placed in the bottom of a centrifuge tube (SW 25.2 rotor) and a linear 33 to 60% gradient is constructed on top of it.

9-13. Stepped gradients are constructed by carefully layering the following solutions in a centrifuge tube: 5 ml 60% sucrose, 10 ml 57%, 15 ml 50%, 15 ml 44%, and 7 ml 33%. Stepped gradients should be used as soon as they are prepared. If other centrifuge rotors are used the gradient volumes may be adjusted proportionally. The important consideration is that all the tubes be full and have exactly the same weight. Stepped and linear gradients should never be centrifuged together.

9-14. Layer 1 to 2 ml of the crude particles obtained in step 9-10 onto the top of each gradient.

9-15. Following the manufacturer's instructions, centrifuge the tubes at 2°C for 5 hours at 23,000 rpm in a Beckman SW 25.2 swinging bucket rotor. Other rotors may be used but appropriate adjustments of speeds and time will be necessary (see preceding text).

9-16. Fractionate the gradient as soon as possible after the conclusion of centrifugation by one of the following two methods. (a) If the preparation is to be used for determination of an enzyme profile it is collected in 10 drop fractions after puncture of the tube bottom. Profiles derived using this procedure are shown in Figure 9-16. (b) For the quantitative determination of enzymatic activities as in Table 9-2, an entire region of organelle protein is collected from stepped gradients in a single tube. These are then carefully stirred to ensure that a homogeneous solution of known volume is obtained.

9-17. Following are a set of assay conditions for marker enzymes of mitochondria (citrate synthetase, malic dehydrogenase, fumarase, and succinate dehydrogenase) and glyoxysomes (citrate synthetase, malate synthetase, and malic dehydrogenase). Some or all of these activities may be assayed across the density gradient. Their quantitative distribution is shown in Table 9-2.

Citrate Synthetase. The assay method is modified from that of Srere, Brazil, and Gonen. The reaction mixture contains, in a volume of 1.3 ml, $7.7 \times 10^{-2}M$ Tris (adjusted to pH 8.0), 1.5mM 5,5'-dithiobis-(2-nitrobenzoic acid), 7.7mM MgCl$_2$, 2.6mM oxalacetate, $1.8 \times 10^{-4}M$ acetyl-CoA, and from 2 to 25 μg of protein, depending on which fraction of gradient is being assayed. The assay is initiated by the addition of the acetyl-CoA and is followed at 412 nm. Calculations are made assuming a molar extinction coefficient of 1.3×10^7 cm^2/mole for the complex of dithiobis (nitrobenzoic acid) and CoA.

Malate Synthetase.   Assay is by a modification of the procedures of Hock and Beevers. The assay mixture contains, in a final volume of 1.3 ml, 7.7mM MgCl$_2$, $1.8 \times 10^{-4}M$ acetyl-CoA, $2.0 \times 10^{-2}M$ sodium glyoxylate, and 2 to 25 μg of protein. The reaction is initiated with glyoxylate and is followed at 412 nm.

Malate Dehydrogenase.   Assay is by the method of Ochoa. The reaction mixture contains, in a volume of 1.4 ml, $6.9 \times 10^{-2}M$ phosphate buffer (adjusted to pH 7.5), 3.4mM dithiothreitol, 6.9 mM MgCl$_2$, 2.3 mM oxalacetate, $2.5 \times 10^{-4}M$ NADH, and 1 to 5 μg protein. The assay is initiated by addition of oxalacetate and monitored at 340 nm.

Succinate Dehydrogenase.   Activity is determined by a modification of the procedure of Hiatt. The reaction mixture contains, in a final volume of 1.3 ml, $8.0 \times 10^{-2}M$ phosphate buffer (adjusted to pH 7.5), $7.9 \times 10^{-4}M$ phenazine methosulfate, $2.4 \times 10^{-2}M$ potassium cyanide, $1.6 \times 10^{-2}M$ succinate, $9.6 \times 10^{-5}M$ dichlorophenol indophenol, and 19 to 25 μg of protein. The reaction is initiated with succinate and is followed at 600 nm. The molar extinction coefficient for the dichlorophenol indophenol is 1.1×10^4 cm^2/mole.

Fumarase.   Fumarase is determined by measuring the conversion to malate to fumarate. This assay was used in the opposite direction by Racker. The mixture contains, in a volume of 1.3 ml, 8.0×10^{-2} phosphate buffer (adjusted to pH 7.5), 4.0mM dithiothreitol, 8.0mM sodium malate, and 2 to 25 μg of protein. The reaction is initiated with malate and is followed at 240 nm. The molar extinction coefficient of fumarate is 2.6×10^2 cm^2/mole.

9-18. An additional experiment that is most useful is the isolation of bacteriophage using cesium chloride gradients. Procedures for one such experiment involving the isolation of several varieties of phage are given in reference 8.

REFERENCES

1.   H. K. Schachman, in *Ultracentrifugation in Biochemistry*, Academic Press, New York, 1959.

2.   H. K. Schachman, Ultracentrifugation, Diffusion and Viscometry, in *Methods in Enzymology*, Vol. 4 (S. P. Colowick and N. O. Kaplan, Eds.), Academic Press, New York, 1957, p. 32.

3.   J. Sykes, Centrifugal Techniques for the Isolation and Characterization of Subcellular

Components from Bacteria, in *Methods in Microbiology*, Vol. 5B (J. R. Norris and D. W. Ribbons, Eds.), Academic Press, New York, 1971, p. 55.

4. R. G. Martin and B. N. Ames, *J. Biol. Chem.*, **236**:1372 (1961). A Method for Determining the Sedimentation Behavior of Enzymes: Application to Protein Mixtures.

5. C. R. McEwen, *Anal. Biochem.*, **20**:114 (1967). Tables for Estimating Sedimentation Through Linear Concentration Gradients of Sucrose Solution.

6. M. Meselson, F. W. Stahl, and J. Vinograd, *Proc. Natl. Acad. Sci. US*, **43**:581 (1957). Equilibrium Sedimentation of Macromolecules in Density Gradients.

7. M. Meselson and F. W. Stahl, *Proc. Natl. Acad. Sci. US*, **44**:671 (1958). The Replication of DNA in *Escherichia coli*.

8. T. G. Cooper, P. A. Whitney, and B. Magasanik, *J. Biol. Chem.*, **249**:6548 (1974). Reaction of *lac*-specific Ribonucleic Acid from *Escherichia coli* with *lac* Deoxyribonucleic Acid.

9. C. deDuve, J. Berthet, and H. Beaufay, Gradient Centrifugation of Cell Particles: Theory and Applications, in *Progress in Biophysics and Biophysical Chemistry*, Vol. 9 (J. A. V. Butler and B. Katz, Eds.), Pergamon Press, New York, 1959, p. 326.

10. C. deDuve, *J. Theoret. Biol.*, **6**:33 (1964). Principles of Tissue Fractionation.

11. C. deDuve, The Separation and Characterization of Subcellular Particles, *Harvey Lectures*, Series 59, Academic Press, New York, 1965, p. 49.

12. H. Pertoft, *Biophys. Biochim. Acta*, **126**:594 (1966). Gradient Centrifugation in Colloidal Silica–Polysaccharide Media.

13. H. Pertoft, *Exp. Cell Res.*, **46**:621 (1967). Separation of Blood Cells Using Colloidal Silica–Polysaccharide Gradients.

14. H. Pertoft, O. Back, and K. L. Kiessling, *Exp. Cell Res.*, **50**:355 (1968). Separation of Various Blood Cells in Colloidal Silica–Polyvinylpyrrolidone Gradients.

15. W. Szybalski, Use of Cesium Sulfate for Equilibrium Density Gradient Centrifugation, in *Methods in Enzymology*, Vol. 12B (L. Grossman and K. Moldave, Eds.), Academic Press, New York, 1968, p. 330.

16. C. deDuve and J. Berthet, *Intern. Rev. Cytol.*, **3**:225 (1954). The Use of Differential Centrifugation in the Study of Tissue Enzymes.

17. H. Beaufay, D. S. Bendall, P. Baudhuin, R. Wattiaux, and C. deDuve, *Biochem. J.*, **73**:628 (1959). Tissue Fractionation Studies. 13. Analysis of Mitochondrial Fractions from Rat Liver by Density-Gradient Centrifugation.

18. R. J. Britten and R. B. Roberts, *Science*, **131**:32 (1960). High Resolution Density Gradient Sedimentation Analysis.

19. R. Trautman, *Arch. Biochim. Biophys.*, **87**:289 (1960). Determination of Density Gradients in Isodensity Equilibrium Ultracentrifugation.

20. M. Ottesen and R. Weber, *Compt. Rend. Trav. Lab. Carlsberg*, **29**:417 (1955). Density-Gradient Centrifugation as a means of Separating Cytoplasmic Particles.

21. M. K. Brakke, *J. Am. Chem. Soc.*, **73**:1847 (1951). Density Gradient Centrifugation: A new separation technique.

22. M. K. Brakke, *Arch. Biochem. Biophys.*, **45**:275 (1953). Zonal Separations by Density-Gradient Centrifugation.

23. N. G. Anderson, *Exp. Cell Res.*, **9**:446 (1955). Studies on Isolated Cell Components. VIII. High resolution differential centrifugation.

24. N. G. Anderson, *Rev. Sci. Instrum.*, **26**:891 (1955). Mechanical Device for Producing Density Gradients in Liquids.

25. N. G. Anderson, Techniques for the Mass Isolation of Cellular Components, in *Physical Techniques in Biological Research*, Vol. 3: *Cells and Tissues* (G. Oster and A. W. Pollister, Eds.), Academic Press, New York, 1956.

26. N. G. Anderson, *Nature*, **181**:45 (1958). Rapid Sedimentation of Proteins through Starch.

27. A. P. Mathias and C. A. Wynther, *FEBS Letts.*, **33**:18 (1973). The Use of Metrizamide in the Fractionation of Nuclei from Brain and Liver Tissue by Zonal Centrifugation.

28. B. M. Mullock and R. H. Hinton, *Biochem. Soc. Trans.*, **1**:27 (1973). The Use of Metrizamide for the Isopycnic Gradient Fractionation of Unfixed Fibonucleoprotein Particles.

29. D. Rickwood, A. Hell, and G. D. Birnie, *FEBS Lett.*, **33**:221 (1973). Isopycnic Sedimentation of Chromatin in Metrizamide.

Chapter 10

Protein Purification

The thorough study of many biological systems has involved purification of one or more of the system components. Unfortunately, there are a large number of misconceptions about the manner in which to develop a successful purification scheme. One systematic way to develop such a scheme is discussed here using proteins as a model, but the same considerations, perhaps in different terms, would apply to any biologically significant molecule. The five basic steps of purification are as follows:

1. Development of suitable assay procedures.
2. Selection of the best source from which the molecule may be purified.
3. Solubilization of the desired molecule.
4. Stabilization of the molecule repeatedly at each stage of its purification.
5. Development of a series of isolation and concentration procedures.

Information concerning one or more of these steps may be available from the literature, but consideration of all the above steps may lead to greater efficiency or other improvements in the published procedures.

DEVELOPMENT OF AN ASSAY

A useful enzyme assay must meet four criteria: (1) absolute specificity, (2) high sensitivity, (3) high precision, and (4) convenience. Though some compromise of these criteria is usually necessary, it should be emphasized that any substantial compromise may severely limit general applicability of an assay.

The cardinal requirement of any assay is its specificity. Since most enzyme assays monitor disappearance of a substrate or appearance of a product, care must be exercised to ensure that only one enzyme activity is contributing to the monitored effect. Consider, for example, assay of phosphoenolpyruvate (PEP) carboxykinase, which catalyzes the reaction

$$PEP + CO_2 + GDP \rightleftharpoons OAA + GTP \tag{1}$$

Assay of this activity by measuring PEP disappearance or OAA production can be in serious error unless one can demonstrate the absence of enzymes such as PEP carboxylase,

$$PEP + HCO_3^- \longrightarrow OAA + P_i \tag{2}$$

pyruvate kinase,

$$PEP + ADP \longrightarrow pyruvate + ATP \tag{3}$$

or PEP carboxytransphosphorylase,

$$PEP + CO_2 + P_i \rightleftharpoons OAA + PP_i \tag{4}$$

These problems are especially acute at very early stages of purification where the necessary substrates of alternative reactions are likely to be present. The most convincing way of establishing the specificity of an assay is by studies of cofactor requirements and product identification under a variety of conditions. Since similar examples may be cited for almost any enzyme, one must be aware of all the possible reactions that may contribute to a given product accumulation or substrate utilization.

At the initial stages of any purification the specific activities of most enzymes are very low; therefore the assay must be highly sensitive. As purification progresses the absolute need for sensitivity decreases; however, the greater the amount of enzyme used for assay by insensitive methods, the less available for other purposes. A corollary of the need for high sensitivity is the need for high resolution. Subtle changes in activity may sometimes precede gross enzyme losses and corrective stabilization measures should be applied prior to these initial changes. Such observations require that these subtleties be visualized.

The accuracy and precision of an enzyme assay usually depend on the underlying chemical basis of techniques that are used. For example, if an assay involving reaction of an enzymatically produced aldehyde with phenylhydrazine to yield a phenylhydrazone product is carried out in buffer of the wrong pH, that is, a pH significantly above 6.0, the observed rate of phenylhydrazone formation will not accurately reflect the rate of enzymatically produced aldehyde, but rather the chemical rate of phenylhydrazone formation. The accuracy of an assay is a reflection of how closely the measured values are to the real values. Precision, on the other hand, is a reflection of the scatter observed in duplicate or triplicate assays of the same sample. For example, if an assay is performed identically 10 times and the observed experimental values range 5% above and below the mean value, the assay is of little value for detecting changes of much lower than 20%. Colorimetric assays are particularly susceptible to this problem.

Cost and convenience are the final considerations to be made. During purification the activity of the desired enzyme should be monitored after each step. The ease with which this can be done largely dictates how closely the activity is in fact monitored and how much the purification procedure is lengthened owing to resulting losses of time. These time losses may be significant because as the time required to purify a labile protein increases so does the probability of its inactivation.

SELECTION OF A SOURCE FROM WHICH A MACROMOLECULE MAY BE ISOLATED

An investigator may or may not have freedom to select the source from which a given molecule is to be purified. If the object is to obtain a given protein in quantity in order to study the protein itself or to use it in other studies, considerable freedom exists. If, alternatively, a specific component of a system being studied is to be purified, source selection is limited. In either case the following considerations may expedite the purification process. Purification is essentially a concentrating process in which the desired protein is concentrated, whereas other cellular proteins are not. Therefore, as the fraction of total cellular protein devoted to the specifically desired species increases, so does the ease of its purification. The key to selecting a good enzyme source is quantity. Tissues should be sought which contain large quantities of the desired protein and are themselves easily obtainable in large quantities. Prime candidates in this respect are the organs of large domesticated animals such as cows, pigs, and goats. An alternative source is microorganisms. Many of these organisms, such as yeast and bacteria, are easily grown or purchased in large quantities at reasonable cost. They may also be modified genetically and physiologically to increase the percentage of their total synthetic capacity devoted to synthesis of the desired protein. β-Galactosidase production, for example, may be increased genetically by using a strain of E. coli which carries genes of the lac operon on both its chromosome and an episome. Further increases are accrued by obtaining strains with appropriate mutations in the regulatory gene (i gene) for this operon.

Physiologically, growth of the organism in the presence of glycerol rather than glucose as sole carbon source increases production of β-galactosidase still further, since under these conditions of growth catabolite repression is minimized.

METHOD OF SOLUBILIZATION

Solubilization is required of any protein to be purified, because all the isolation procedures commonly used operate only in aqueous solution. In some cases this involves merely breaking open the cells containing the desired protein, whereas in others it may also involve removing the protein from a subcellular organelle. A wide range of solubilization methods exists, and the one chosen must take into account the characteristics of the cells to be broken and the nature of the proteins to be subsequently isolated. For example, if the protein is located in a cellular organelle such as the mitochondria and the object is to isolate the mitochondria first and then solubilize the protein from the purified organelles, very gentle methods of cell breakage would be advised to prevent damage to these organelles. If, alternatively, the object is merely to break open the cells and solubilize the protein without regard for its location, much harsher methods are possible. A second consideration that strongly influences the choice of solubilization procedures is the sample size. The methods used to break a few milligrams of cultured animal cells are likely to be of little use in processing several pounds of yeast or beef liver.

Osmotic Lysis

One of the most gentle procedures for cell breakage is osmotic lysis. This is performed by subjecting the cell to hypotonic conditions and a mild disruptive force such as drawing cells into a pipette. The gentle nature of this method limits its use to cell types that have only a cell membrane and exist as single cells. For cells with strong carbohydrate-containing cell walls such as bacteria, plants, and fungi, the method is useless. It can be used on a limited basis if the cell walls of these organisms are first removed enzymatically. For *E. coli*, lysozyme works effectively, as does glusulase (a mixture of sulfatase and glucuronidase) for some species of yeast. In addition to limited applicability, this method can be used only for small samples in dilute solution. As such, it is not used to a significant extent for large scale purifications of macromolecules, but may be used to isolate cellular organelles, that is, nuclei or small amounts of labile macromolecules such as messenger RNA. For applications involving nucleic acids a detergent may be included to assist breakage and protect the nucleic acids from enzymatic destruction.

Grinding

Grinding as a means of cell breakage covers a broad range of techniques, some gentle and others more harsh. The most gentle grinding is performed by hand, using a mortar and pestle. Slurries of minced tissue processed in this way are often used as a source of cellular organelles such as mitochondria, lysosomes, and microbodies (1). Hand driven Broeck and motor driven Elvehjem tissue homogenizers as shown in Figure 10-1 are also used. A mortar and pestle may be used to pulverize freeze dried or lyophilized tissues into fine powders, which are then suspended and extracted with appropriate aqueous solutions. This procedure works particularly well for dried baker's yeast purchased commercially. Grinding with a mortar and pestle may be made more efficient by adding an abrasive to the mixture. Fine alumina is often employed to break bacteria.

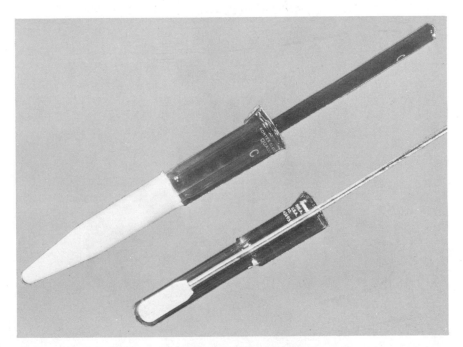

Figure 10-1. Broeck (upper) and Elvehjem (lower) homogenizers. These instruments are available in various sizes capable of accommodating sample volumes of 1 to 55 ml.

Fine glass beads (45–50 μ diameter) also work well, but may tend to bind certain proteins, limiting their applicability.

These methods are easily applicable for samples up to the size of rabbit livers or 25 g yeast. However, in excess of these quantities they are cumbersome and inefficient. For pulverizing dried yeast or other fungi in excess of 25 g a large variety of commercial mills are available which will handle quantities of material ranging from 100 g to many pounds with relative ease. When large quantities of yeast or other materials are pulverized with an electric mill, it is a good practice to include some dry ice in the mixture. This keeps the mill and dried material cold during milling, yet disappears as a gas before the material is suspended in

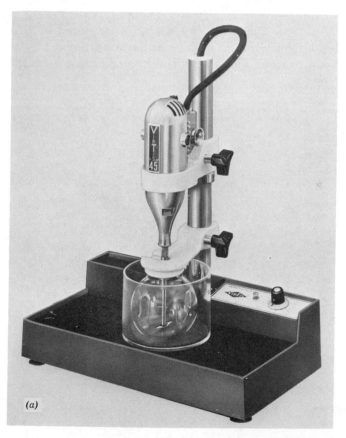

(a)

Figure 10-2. Electric blenders capable of handling sample volumes of 10 ml (*a*) to 4 liters (*b*). The instrument in (*a*) is equipped with a dish which may be filled with crushed ice in order to

solution. Care should be taken to work in an atmosphere of relatively low humidity. Otherwise, moisture will condense on the preparation and the presence of excess CO_2 will make whatever moisture that exists very acidic.

Blenders

Perhaps the most often used grinding method is a blender. Available in many sizes and variations, this instrument handles large and small samples with equal ease (Figure 10-2). It is, however, a relatively harsh method of breakage and care must be taken to keep the preparation cold

(b)

cool the sample during breakage. (Courtesy of VirTis Company, Gardiner, N.Y. and Waring Company, New Hartford, Conn.)

during grinding because the temperature will increase substantially (10°C or more) if grinding is continued more than 30 to 45 seconds. Although blenders are quite efficient for breaking plant and animal tissues they are ineffective for yeast and bacteria. This may be remedied if about one-third to one-half the volume of a heavy cell suspension is made up of fine glass beads. The cells are not broken by the blender blades but rather by the glass beads which collide with one another with cells caught between them.

Ultrasonic Waves

Sonic oscillators (Figure 10-3) provide a very efficient means of breaking cells or organelles. They are effective against bacteria and yeast at sufficiently long periods of application. The times may be shortened, however, by including glass beads in the cell suspension as described above. Sonicators are usually composed of two main units: an electronic generator which produces an ultrasonic signal of high intensity and a transducer which transmits these waves into the solution in contact with it. It is the shock and vibration set up by these sound waves that brings

Figure 10-3. Sonicator or sonic cell disruptor showing the sonic wave generator in the background and the titanium transducing element in the foreground. Needle tips capable of operating in small test tubes may be interchanged with the larger tip (arrow) shown here. (Courtesy of Heat Systems-Ultrasonics, Inc., Plainview, N.Y.)

about tissue destruction or moves the fine glass beads. Using various sized probes this instrument adequately accommodates samples from a few milliliters up to 1 liter. The major drawback is the large amount of heat generated at the transducer. Therefore, the preparation temperature must be carefully monitored and the duration of sonication held to a minimum, with intermittent cooling when necessary.

Presses

There are three major types of presses that may be used to break open microorganisms. They are the Hughes (2), French (Figure 10-4), and Eton (3) presses. All three of these instruments function by placing a cell suspension of small volume (5–50 ml) under 4000 to 8000 lb pressure and forcing it through a small opening. In the case of the French and Eton presses, the opening is a small hole of 1 mm or less. For the Hughes press it is the small distance between two flat plates. Breakage is brought about by shearing of the cells as they pass through the small orifice. This method is both gentle and thorough, but its application is restricted by its limited sample size range.

Removal of Proteins from Subcellular Components

It has become clear that many interesting and important proteins function only in association with subcellular components, that is, membranes, ribosomes, and nucleic acids. These observations have prompted development of methods to solubilize bound proteins as a prerequisite to their purification. For proteins that are loosely associated, suspension of the subcellular components in a solution of high ionic strength will usually suffice. Solutions of 0.5 to $5M$ KCl or NH_4Cl are often used for this purpose. However, these solutions are not effective for releasing proteins that serve a structural as well as catalytic function, for example, tightly bound membrane proteins. Here mild ionic or nonionic detergents such as deoxycholate, Brij, and the Triton compounds (polyoxyethylene ethers) must be used at low concentration, sometimes in conjunction with gentle sonication to aid in dismantling the subcellular structure.

STABILIZATION

The most difficult, time-consuming, and frustrating aspect of protein purification is stabilization of the protein in an active form. This difficulty derives from the fact that the entire stabilization process must be reconsidered at each step of purification. For example, the first step of phosphoenolpyruvate carboxylase purification from peanut cotyledons

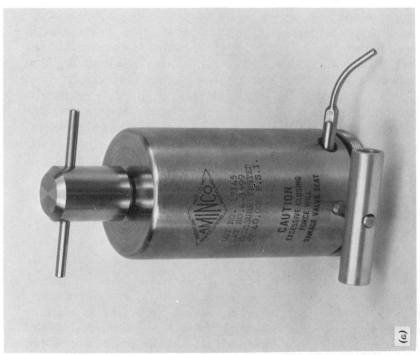

Figure 10-4. French pressure cell alone (*a*) and assembled into a hydraulic press (*b*). Diameter of the orifice through which cells pass may be adjusted using the valve located on the side of the pressure cell. This instrument accommodates sample volumes between 4 and 50 ml by lowering or raising the piston, respectively. The piston is raised to its maximum position in the cell that is assembled in the press and is completely lowered in panel (a). (Courtesy of American Instrument Company, Silver Spring, Md.)

involves grinding the tissue in an appropriate buffer and incubating the crude mixture overnight at 37°C. As one might expect, this treatment results in mass destruction of cellular protein, as indicated by a large curdy precipitate that forms during incubation. After the precipitate is removed, however, nearly all of the carboxylase activity is found in the supernatant solution. Unfortunately, the remarkable heat and protease resistance of the carboxylase is lost with the next step of purification. Explanation of these observations is based on the fact that proteins are not free molecules in solution. Quite the contrary, they exhibit strong hydrophobic and electrostatic attraction for one another. This results in the formation of large and complex protein aggregates. With each step of purification many proteins are lost from these aggregates, changing their composition and hence the procedures needed to stabilize them. Six major characteristics of a protein solution warrant consideration: (1) pH, (2) degree of oxidation, (3) heavy metal concentration, (4) medium polarity, (5) protease concentration, and (6) temperature.

pH

Regulation of solution pH, although easily accomplished using appropriate buffers, may involve a number of subtleties. For example, it is often assumed that pH values yielding the highest reaction rates are also those at which the enzyme is most stable; this, unfortunately, may not be true. There are a significant number of instances in which the pH optimum for assay and storage (stability) differ from one another by one or more pH units; thus it is advisable to determine the pH optima for enzyme assay and stability separately. Not only must the buffer have an appropriate pK_a, but also it must not adversely affect the protein. Phosphate and pyrophosphate buffers are notorious in this regard because they can act as competitive inhibitors of a large number of enzymes which catalyze reactions involving inorganic phosphate or organophosphate compounds as substrates or reaction products. Attention must be given to the concentration of buffer that is used. As a general rule, tissues with large vacuoles such as plants and fungi require higher capacities if pH is to be properly regulated. In some instances it is even necessary to neutralize the crude homogenates by dropwise additions of weak bases such as ammonium hydroxide. In these instances extreme care must be used so that the neutralization process doesn't denature the desired protein.

Degree of Oxidation

Most proteins commonly encountered contain a reasonable number of free sulfhydryl groups. One or more of these groups may participate in substrate binding and therefore are quite reactive. Upon oxidation sulfhydryl groups form intra- or intermolecular disulfide bonds, which usually result in loss of enzyme activity. A wide variety of compounds are available to prevent disulfide bond formation: 2-mercaptoethanol ($\rho = 1.12$ g/ml), cysteine, reduced glutathione, and thioglycolate. These compounds are added to protein solutions at concentration ranging from 10^{-4} to $5 \times 10^{-3} M$. In solution the following interchanges occur between enzyme and sulfhydryl reagent molecules:

$$\text{ENZ-S-S-ENZ} + \text{RSH} \rightleftharpoons \text{ENZ-SH} + \text{ENZ-S-S-R} \qquad (5)$$
$$\text{(inactive)} \qquad\qquad \text{(active)} \qquad \text{(inactive)}$$

$$\text{ENZ-S-S-R} + \text{RSH} \rightleftharpoons \text{ENZ-SH} + \text{R-S-S-R} \qquad (6)$$
$$\text{(inactive)} \qquad\qquad \text{(active)}$$

Since the equilibrium constants for these reactions are near unity, a large excess of the protective reagent is needed. This situation prompted identification of sulfhydryl reagents for which the above reactions are shifted far to the right. Dithiothreitol and its isomer dithioerythritol fulfill these criteria. As shown in Figure 10-5, dithiothreitol undergoes intramolecular disulfide bond formation to yield a sterically favored six-membered ring. Therefore, proteins are provided greater protection against oxidation by much smaller amounts of thiol when dithiothreitol is used as the protective agent.

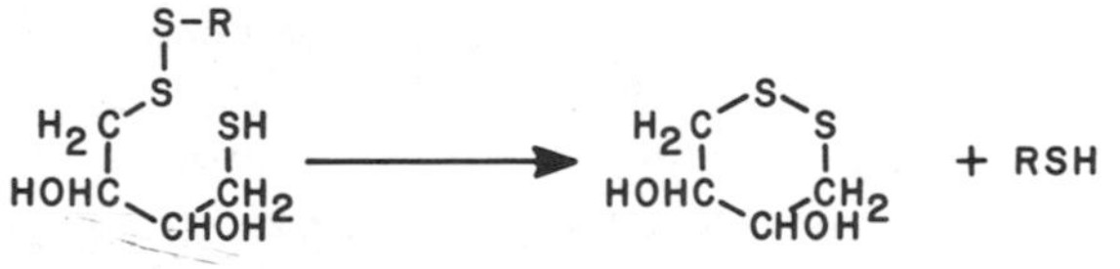

Figure 10-5. Reduction of dithiothreitol forming an intramolecular disulfide bond and a six-membered ring.

A second type of oxidizing agent is encountered in studies with some plant tissues. These are quinones, which are generated when cells are broken. It is the action of such compounds that is responsible for the browning of freshly cut apple and peach slices. The usual protective agent used against these strong oxidants is polyvinylpyrrolidone. However, even with this agent protein oxidation may remain a serious problem.

Heavy Metal Contamination

In addition to oxidation, sulfhydryl groups may react with heavy metal ions such as lead, iron, or copper. Principal sources of these heavy metals are the reagents used to make up the buffers, ion exchange resins used for protein separation, and the water in which solutions are prepared. Heavy metal contamination may be avoided by using water that has been glass distilled or passed through specially prepared mixed bed deionizing resins. Use of "house" distilled water without further purification is not recommended because in some cases it is produced by steam condensation in metal vessels, is stored in metal vessels, and arrives at the laboratory in metal pipes. Ion exchange resins may be purified of heavy metal contamination using the procedures outlined in Chapter 4. If, after these steps, trace amounts of heavy metals continue to be a problem, EDTA (ethylenediaminetetraacetic acid) may be included in the buffer solutions at a concentration of 1 to $3 \times 10^{-4} M$. This compound chelates most, if not all, deleterious metal ions.

Medium Polarity and Ionic Strength

Purification of proteins which reside in membranes or other subcellular components prompted consideration of solvent polarity. Although such problems were initially encountered with respect to membrane proteins, it is now clear that they apply in general for any protein. Proteins requiring a more hydrophobic environment may be successfully maintained in solutions whose polarity has been decreased using sucrose, glycerol, and in more drastic cases, dimethyl sulfoxide or dimethylformamide. Appropriate concentrations must usually be determined by trial and error but concentrations of 1 to 10% (v/v) are not uncommon. Use of these agents has permitted purification of many proteins that were at one time considered too labile for careful study. A few proteins, on the other hand, require a polar medium with a high ionic strength to maintain full activity. For these infrequent occasions, KCl, NaCl, NH_4Cl, or $(NH_4)_2SO_4$ may be used to raise the ionic strength of the solution. Such a requirement generally precludes use of ion exchange chromatography as a means of purification.

Protease or Nuclease Contamination

A problem that often occurs in the purification of proteins and nucleic acids is the presence of proteases and nucleases, respectively. These degradative proteins are not permitted to come in contact with their substrates in vivo except under carefully regulated circumstances, but during cell breakage they are liberated. Protease degradation of a protein being purified may be signaled by continued loss of enzyme activity regardless of the precautions taken to stabilize the desired protein, and it should be considered whenever such lability is encountered. There are several specific protease inhibitors, such as phenylmethylsulfonyl fluoride (PMSF), which can be used to eliminate the activity of some but not all proteases. However, they must be used with caution owing to the unexpected side effects. For example, PMSF and diisopropyl fluorophosphate (DIFP) inhibit not only proteases, but also a variety of other enzymes. Diethylpyrocarbonate was initially thought to be a useful means of inactivating RNase until it was claimed that it modified the RNA supposedly being protected from nuclease digestion. More recently it has been reported that this inhibitor merely increases the solubility of RNA in solution (4, 5, 15). Therefore, when inhibitors are used in a preparation, appropriate control experiments are needed to ensure that neither the catalytic nor the more delicate regulatory functions of the desired molecule are being altered.

Temperature

It is usually assumed that a protein is most stable at 0°C. However, it has been clearly shown that pyruvate carboxylase (6), isolated from avian livers, is cold sensitive and may be stabilized only at 25°C. A related problem is determination of the conditions under which a protein may be most safely stored. Some proteins survive best as concentrated solutions at 0°C, whereas others may require temperatures as low as −20 or −70°C to remain active. It is incorrect to assume, however, that colder conditions always yield greater stability, because freezing and thawing of some protein solutions is quite harmful. If this is observed, addition of glycerol or small amounts of dimethyl sulfoxide to the preparation before freezing may be of help. Storage conditions, like all the stability characteristics, must be determined by trial and error for each protein and may require updating at each stage of the purification.

ISOLATION AND CONCENTRATION

Only a relatively small number of major techniques for the isolation and concentration of proteins exist. These include (1) differential solubility, (2) ion exchange chromatography, (3) absorption chromatography, (4) molecular sieve techniques, (5) affinity chromatography, (6) electrophoresis, and (7) electrofocusing. Whether a technique will be successful for purification of a given protein is usually determined only by trial and error. However, a few general remarks are appropriate concerning construction of the scheme. The techniques are listed above in order of

increasing resolution and time consumption required for their performance (an exception is affinity chromatography, which has been handled
as a special case in Chapter 7). This is also the order of decreasing sample
capacity. Common sense advises use of the quick crude methods at early
stages of purification, and time-consuming high resolution techniques
only later. For example, electrophoretic methods capable of processing
milligram quantities of material are of little use for a starting sample of
several kilograms of soluble protein. It would be incorrect, however, to
suggest this as a hard-and-fast rule. At times the characteristics of the
molecules being purified lend themselves to unconventional but advantageous approaches. This is well illustrated by the methods used to purify
lactate dehydrogenase (LDH). Originally the protein was purified by
standard techniques and found to be a tetramer (134,000 daltons).
Conditions were established for dissociation of the protein into its
component subunits (33,500 daltons) and reassembly of isolated subunits
into a fully active tetrameric protein. This information was then used to
construct a more efficient purification scheme. Crude extracts were subjected to molecular sieve chromatography, and fractions containing
proteins with a molecular weight of about 135,000 were collected. The
collected fractions were subjected to conditions which separate LDH into
monomeric form and molecular sieving was again performed. This time
only the low molecular weight ($\sim 33,000$–$35,000$) proteins were collected.
A relatively pure enzyme preparation could be isolated from these
purified subunits. The art of protein purification is to recognize such
useful characteristics of a protein and use them effectively.

At times it is useful to repeat a given purification step, especially if one
or more other methods are used between the first and second applications
of the duplicated procedure. This approach is useful for the same reasons
offered in the discussion of stabilization. Behavior of a particular protein
during any given isolation method is dictated to some extent by other
proteins to which it is complexed. If the composition of these proteins is
changed, the behavior of the desired protein may change also. For
example, if ammonium sulfate is added to a relatively crude protein
extract derived from *Rhodospirillum rubrum*, the enzyme phosphoenolpyruvate (PEP) carboxykinase begins to precipitate as the concentration
reaches 45% of saturation, and total precipitation of the protein does not
occur until an ammonium sulfate concentration of 65% saturation is
attained. If such a precipitation procedure is performed several purification steps later, none of the PEP carboxykinase precipitates until the
ammonium sulfate concentration reaches 61%. This striking alteration in
behavior is the result of removing a significant number of proteins during
the intervening steps.

Differential Solubility

Proteins remain dissolved in solution because their charged surface residues interact with molecules of the solvent. If such interactions are prevented, the protein molecules interact principally with one another forming huge aggregates which precipitate out of solution. The ease or difficulty with which a protein is prevented from interacting with solvent is determined largely by the nature of its surface residues. However, details of such interactions and means of predicting the behavior of any given protein in this regard are not available. Specific behavior must be determined by trial and error. Five agents are commonly used to bring about protein precipitation: (*a*) inorganic salts, (*b*) pH or temperature variation, (*c*) organic solvents, (*d*) basic proteins, and (*e*) polyethylene glycol.

Salt Fractionation

The most often used method of protein precipitation is by addition of inorganic salts such as ammonium sulfate or potassium phosphate. In crude preparations of large volume, the salt is added in dry form. The speed at which salt is added to a protein solution is very important. For many proteins addition must be quite gradual; that is, a small amount is added and allowed to dissolve before making further additions. As the ammonium sulfate dissolves, a large amount of water is bound to each ammonium sulfate molecule. Therefore, as the number of ammonium sulfate molecules in solution increases, less water is available to interact with proteins that may be present. At some point there will no longer be sufficient water present to maintain solution of a set of proteins and they precipitate. "Salting proteins out of solution," as this is sometimes called, may be actually a dehydration process. However, other explanations have also been offered (14). If the rate of dehydration is too fast some proteins are denatured. On the other hand, if dehydration is too slow other proteins may denature. For any given protein the best rate of salt addition must be determined experimentally. If ammonium sulfate is added to an unbuffered or weakly buffered solution, care must be taken to monitor the pH. If adjustment is necessary, as it often is, a weak base such as dilute ammonium hydroxide or Tris base should be used.

A more gentle method of dehydration is the dropwise addition of a saturated ammonium sulfate solution whose pH has been previously adjusted. This is usually the method of choice for precipitation of purified or partially purified proteins in small volumes since the rate of protein dehydration can be carefully controlled. Unfortunately, this method is undesirable for large volumes, because ammonium sulfate addition greatly increases the sample volume. At 50% saturation, for example, the sample volume is doubled.

Finally, the quantity of ammonium sulfate added must be considered. A number of methods have been proposed to express the quantities of salt added. Figure 10-6 depicts an approach based upon the degree of saturation. At 25°C a saturated solution of ammonium sulfate is $4.1M$ in ammonium sulfate (767 g of salt/liter of water). The table indicates the amount of ammonium sulfate that must be added per liter of solution, either free of ammonium sulfate or at some initial concentration, to yield any of the final concentrations listed. Concentrations shown are in terms of percent saturation at 25°C. Since the solubility of ammonium sulfate does not decrease significantly as the temperature drops ($3.9M$ at 0°C), these values are useful regardless of the temperature. If saturated solutions of ammonium sulfate are used the amount to be added may be calculated using the equation

$$\frac{C_0 V_0 - C_i V_0}{C_i - C_a} = V_a$$

or

$$\frac{C_0 V_0 + V_a C_a}{V_0 + V_a} = C_i \tag{7}$$

where C_0 and C_i are the starting and final desired concentrations of ammonium sulfate in the experimental solution (expressed as percentage of saturation), respectively, C_a is the concentration of the ammonium sulfate solution to be added (expressed as percentage of saturation), V_0 is the initial volume of the protein containing solution, and V_a is the volume of ammonium sulfate solution to be added.

In practice precipitates are usually collected by centrifugation at intervals such as those shown in Table 10-1. In this case most of the enzyme activity is found in the 35 to 45% and the 45 to 55% precipitates. These data suggest that an initial fractionation of 0 to 35% and then a

Table 10-1. $(NH_4)_2SO_4$ Precipitation of Urea Carboxylase

Percentage $(NH_4)_2SO_4$ (saturated at 25°C, $4.1M$)	Activity Precipitated (arbitrary units)
0–25	0
25–35	112
35–45	6850
45–55	3020
55–65	27

Initial concentration of ammonium sulfate, % saturation	Final concentration of ammonium sulfate, % saturation																
	10	20	25	30	33	35	40	45	50	55	60	65	70	75	80	90	100
	Grams solid ammonium sulfate to be added to 1 l. of solution																
0	56	114	144	176	196	209	243	277	313	351	390	430	472	516	561	662	767
10		57	86	118	137	150	183	216	251	288	326	365	406	449	494	592	694
20			29	59	78	91	123	155	189	225	262	300	340	382	424	520	619
25				30	49	61	93	125	158	193	230	267	307	348	390	485	583
30					19	30	62	94	127	162	198	235	273	314	356	449	546
33						12	43	74	107	142	177	214	252	292	333	426	522
35							31	63	94	129	164	200	238	278	319	411	506
40								31	63	97	132	168	205	245	285	375	469
45									32	65	99	134	171	210	250	339	431
50										33	66	101	137	176	214	302	392
55											33	67	103	141	179	264	353
60												34	69	105	143	227	314
65													34	70	107	190	275
70														35	72	153	237
75															36	115	198
80																77	157
90																	79

Figure 10-6. Nomogram for determining the amount of ammonium sulfate needed per liter to yield various percentages of saturation. A saturated solution is $4.1M$ and $3.9M$ in ammonium sulfate at 25 and 0°C, respectively. The initial and final concentrations of ammonium sulfate are found on the vertical and horizontal scales, respectively. The point of intersection of two lines drawn from these points indicates the number of grams of ammonium sulfate that must be added to each liter of solution at the initial concentration to yield one of the final concentration. (Modified from *Methods of Enzymology*, Vol. 1, Academic Press, New York, 1968, p. 76.)

second fractionation between 35 and 55% might be used to shorten the procedure. However, before such a conclusion could be applied, it would be necessary to check its validity experimentally. After the desired amount of salt has been added it is advisable to allow the solution to stir slowly for a short time (10 to 60 minutes is commonly used) to make certain full equilibrium has been obtained. Following equilibration the precipitate is collected by centrifugation (at forces greater than 19,000 × g). A problem can arise at this point if too much glycerol or other such high density material has been added to stabilize the preparation. Normally the precipitate forms a compact pellet at the bottom of the tube during centrifugation, allowing the supernatant solution to be decanted easily. If the solution density is too high, a fluffy, easily dislodged pellet forms, making decantation almost impossible. The only remedy to this situation is to decrease the density of solution since increased times of centrifugation at higher speeds are usually ineffective.

A most useful modification of ammonium sulfate fractionation is called back-extraction. Ammonium sulfate fractionation of *E. coli* RNA polymerase is used to illustrate this method. This protein normally precipitates between 42 and 50% saturation. Instead of fractionating between 0 and 42%, followed by a second precipitation between 42 and 50% saturation, another scheme is used. The first fraction is taken between 0 and 33% saturation and discarded. The second fraction is taken from 33 to 50% saturation. Normally the collected precipitate would be dissolved in a minimal amount of buffer solution of suitably low ionic strength. However, for back-extraction the pellet is resuspended in a solution that is 42% saturated with ammonium sulfate. Some of the precipitated proteins dissolve but RNA polymerase remains precipitated and is collected by centrifugation after an appropriate equilibration period. The principle of this method is to precipitate more protein than is necessary and then extract the precipitate with appropriately chosen ammonium sulfate solutions. It is possible to carry out the entire ammonium sulfate fractionation using back-extraction. This is done by bringing the initial solution to 60 or 70% saturation and then extracting the pellet successively with solutions of decreasing ammonium sulfate concentrations. On most occasions, however, it is more advisable to use a combination of stepwise precipitation followed by back-extraction. The advantage of this procedure derives from the fact that proteins may be easily and nonspecifically carried out of solution in a precipitate, but dissolving from a precipitated form is more specific and fewer proteins are left contaminating the extracted precipitate.

There are two major advantages of ammonium sulfate fractionation. First it is usually possible to discard up to 75% of the crude protein at this step. This is a reasonable amount of purification (fourfold) for a method that is performed so quickly. Secondly, and perhaps more significantly, the protein solution can be greatly concentrated; the degree of concentration depends on how much solution is used to dissolve the precipitated proteins. The smaller volumes obtained increase the ease with which subsequent higher resolution methods are performed.

pH or Temperature Variation

The most difficult differential solubility techniques to perform are those involving variation of pH or temperature. Both of these procedures are quite drastic. As such, their application is generally limited. As shown in Chapter 4, proteins become negatively or positively charged as the pH is raised above or lowered below their isoelectric points. These charged forms are much more soluble than electrically neutral molecules. Therefore, changing the pH of a protein-containing solution may lead to precipitation of some of the proteins. The desired protein may precipitate and thus be separated from those left in solution, or conversely, it may remain in solution and the precipitated proteins may be discarded. The same problems are found here as for salt precipitation. How much pH variation is permissible and how rapidly the pH may be varied must be determined for each specific protein. The same is true of temperature variations. In the case of PEP carboxylase, discussed above, the temperature was increased only moderately (37°C) but maintained for 16 hours. In other applications the temperature may be increased to much higher levels (60–80°C), but remains elevated for only a short period. In this case the shorter the time required to go from one temperature to another, the better. This can be accomplished using thin walled, stainless steel vessels with large surface areas to increase the rate of heat exchange.

Organic Solvents

Organic solvents greatly decrease protein solubility. This decrease likely arises from two effects, a decrease in the dielectric constant of the medium and dehydration. By decreasing the dielectric constant of a solution the attractive forces between oppositely charged surface residues increase. This results in formation of huge aggregates which precipitate out of solution. Since organic solvents must also dissolve in and hence interact with water, protein dehydration occurs. In this regard organic solvents function in a manner similar to that described for salt precipitation. Four variables must be considered when using this method: (1) ionic strength of the medium from which precipitation occurs, (2) the organic solvent used,

(3) the temperature at which precipitation will be carried out, and (4) how the precipitation will be brought about mechanically.

At low ionic strength increasing amounts of organic solvents severely decrease protein solubility. Moderately increasing the ionic strength, however, tends to increase protein solubility. Although decreasing ionic strength decreases protein solubility, it may also affect protein stability. This is especially true of proteins composed of large numbers of subunits. The effects on any specific protein must, of course, be determined experimentally.

The organic solvents most often used to bring about precipitation include ethanol, methanol, and acetone. Dioxane and tetrahydrofuran are adequately soluble in water, but are usually contaminated with peroxides which are quite deleterious to most proteins. Hence these solvents are rarely used.

The temperature at which precipitation occurs is an important factor to be considered. Protein solubility in the presence of organic solvents decreases markedly with temperature. Therefore, at low temperatures (significantly below 0°C) less solvent is required to precipitate a given protein. The freezing point is also conveniently depressed by the presence of such solvents, making such an approach possible. However, the most useful dividend of low temperature (−20°C) precipitation is alleviation of the denaturing effects most organic solvents have on proteins. The underlying basis of this is not totally understood, but as a rule, the colder it is possible to maintain the preparation during precipitation, the greater yield of active proteins. This is most easily done by cooling the protein solution to 0°C and then adding solvents which have been cooled to much lower temperatures, that is, −30 to −60°C.

Mechanically there are three methods of performing an organic solvent precipitation. The first method is to grind the fresh tissue in a solvent such as acetone. This is usually done in a Waring Blendor because it is necessary to complete the entire operation as quickly as possible while maintaining a very low temperature. Extreme caution must be exercised here; only explosion proof, sealed grinders may be used. Standard inexpensive blenders have exposed motors which spark at the armature, creating extremely hazardous conditions. Under no circumstances should such an instrument be used. After the tissue is ground, the entire brei is filtered through coarse filter paper in a Büchner funnel. Again speed is the essence of success. As the amount of time acetone spends in contact with the protein solution increases, so does the extent of denaturation. The resulting cake of insoluble material is dried at low temperature and becomes a fine powder often referred to as an "acetone powder." This powder is suspended in an appropriate buffer solution for extraction.

Since many proteins are denatured by this procedure, a large degree of purification is gained if the desired species is not adversely affected and redissolves during extraction.

The second method is a modification of the first in which excess acetone (50% final concentration) is added to a crude homogenate produced by grinding tissue suspended in an aqueous buffer in a Waring Blendor. Here it is important to mix the solution thoroughly and quickly as the acetone is added. The most advantageous rate of addition must be determined experimentally, but rapid stirring prevents localized acetone concentrations from reaching denaturing levels.

The third method is useful only if the desired protein is quite resistant to organic solvent denaturation. Such resistance permits organic solvent fractionation to be performed in much the same manner as described for ammonium sulfate fractionation; the same considerations apply. As shown in Table 10-2, nearly all of the castor bean catalase in this preparation remains in solution at an ethanol concentration of 40% (v/v). Most proteins have already precipitated under these conditions and may be removed by centrifugation. Increasing the ethanol concentration to 60% precipitates catalase, permitting its recovery by centrifugation.

Table 10-2. Ethanol Precipitation of Catalase

Percentage Ethanol (v/v)	Activity Remaining in Solution (arbitrary units)
0–14	13,200
14–25	12,236
25–30	10,656
30–40	10,804
40–60	50

Basic Protein Precipitation

Polycations such as protamine or streptomycin may also be used for differential protein precipitation. These materials bind to negatively charged compounds and hence neutralize a large proportion of the charge they possess. Polycations can be used most advantageously if the desired protein is not precipitated by them. As shown in Table 10-3, PEP carboxykinase from *R. rubrum* remains soluble while three-quarters of the undesired proteins precipitate with protamine sulfate. On the other hand, the fact that polycations irreversibly remove many anionic proteins somewhat limits their use. When polycations are used they must be

Table 10-3. Protamine sulfate Precipitation of PEP Carboxykinase

Fraction	Protein (g)	Total Activity (arbitrary units)
Crude	1.35	3502
Protamine sulfate supernatant	0.33	3718

carefully neutralized before being added to the protein solution, since in unneutralized solutions the pH is 2 to 3.

Polycations are especially useful in the removal of nucleic acids. The large negative charge of such compounds makes complexation quite efficient. The advisability of removing nucleic acids early in a purification scheme arises from the fact that they bind significant amounts of proteins. This behavior greatly compromises the effectiveness of high resolution techniques because the proteins are held in huge aggregates by the nucleic acids and therefore cannot be separated on the basis of their individual characteristics.

The commercial availability of DNAse at reasonable cost has largely eliminated the need to use polycations for nucleic acid removal. A small amount of DNAse allowed to incubate in a crude homogenate for 30 to 60 minutes at 4°C will degrade the DNA to pieces that are sufficiently small as not to seriously compromise subsequent purification procedures. It should be emphasized, however, that such a procedure only clips the DNA into small pieces; it does not degrade it to the level of single nucleotide residues.

Polyethylene Glycol (PEG) Precipitation

Polyelectrolytes such as polyethylene glycol are finding increasing application as precipitating agents. A variety of factors influence the degree of specificity obtained with this method. Some of the more important factors are: the pH, ionic strength, and protein concentration of the solution as well as the molecular weight of the polyethylene glycol that is used. Many workers have used PEG 6000 with success, but a recent report claims that lower molecular weight species (PEG 400) yielded a higher specificity of precipitation and separations that were comparable to those obtained with gel chromatography (reference 16). Although some preliminary work is needed to optimize this mode of protein precipitation, it nicely complements the other precipitation methods that are available.

Dialysis and Concentration

Ammonium sulfate fractionation is often followed by ion exchange chromatography. However, ion exchange methods require that the sample be applied at low ionic strength. Unfortunately, when an ammonium sulfate pellet is dissolved, the ionic strength of the resulting solution is unacceptably high. A number of methods to lower the ionic strength have been developed. By far the easiest is to dilute the protein-containing solution with a very low ionic strength buffer. This is a useful method provided the sample is not excessively large or labile upon dilution. Its major consequence is addition of a rather large sample to the ion exchange resin. Most cellulosic ion exchange resins possess good flow rates, and sample size has minimal influence on resolution. Molecular sieve techniques, on the other hand, require very small volumes and therefore dilution is an unacceptable method for lowering ionic strength.

One of the oldest methods of salt removal is dialysis. This technique involves placing the protein solution in a bag made by tying knots (Figure 10-7) in the ends of cellulose dialysis tubing. The sealed bag is placed in a large volume of cold, buffered solution of low ionic strength which is being moderately stirred. Dialysis tubing has small pores in it (i.e., it is semipermeable) which permit passage of the small inorganic salt molecules but not the macromolecular proteins. As the small molecules diffuse out of the bag and into the dialysis buffer, the ionic strength of the protein solution decreases. This process continues until the salt concen-

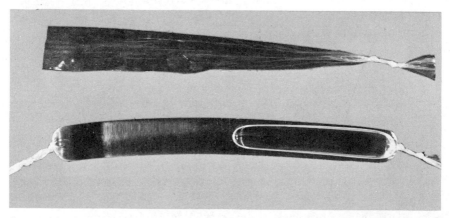

Figure 10-7. Dialysis tubing empty with only one end sealed and full of solution with both ends sealed. Note the filled tubing also contains an air bubble. Two knots rather than one are recommended for sealing the ends of the tubing.

tration inside and outside the bag are identical (5–6 hours is usually sufficient for equilibrium to be established). If at the end of the first dialysis the concentration of small molecules has not dropped sufficiently, the bag is merely transferred to fresh buffer. It should be emphasized that this procedure can remove not only salts but all sorts of small metabolites such as ATP and coenzymes.

Although dialysis would be expected to be a very gentle technique, many proteins are labile to it. Commercially obtained dialysis tubing is usually contaminated with heavy metals, proteases, and nucleases which could account for some of these observations. Therefore, before being used to dialyze protein or DNA solutions, dialysis tubing should be treated in the following manner. A convenient amount of tubing (10–20 ft) is placed randomly in a 4 liter beaker filled with a $0.5M$ solution of EDTA and boiled for $\frac{1}{2}$ hour. The solution is decanted, and this procedure is repeated eight more times using water in place of the EDTA solution. Following this treatment the tubing should be handled only with clean (protease- and DNAse-free) forceps or with disposable rubber gloves. It should never be handled by hand because fingers are a remarkably good source of both degradative enzymes.

Another means of salt removal is molecular sieve chromatography, a very efficient technique with total desalting possible in a matter of minutes for small samples. If large samples must be passed over a molecular sieve column, it is better to increase the column diameter than its length. This keeps the time needed to perform the chromatography at a minimum. Other considerations for this technique may be found in Chapter 6.

The most recent method of salt removal is fiber filter dialysis. This method, however, is most easily described after discussing the following protein concentration methods.

Concentration of biological solutions is often required in the course of purifying a given molecule. The solutions may range from crude culture filtrates to dilute purified enzyme solutions. Since some concentrating methods use procedures intimately related to dialysis, it is appropriate to treat the topic here. Protein solutions are often concentrated by ammonium sulfate precipitation, followed by dissolving the resulting pellet in a small amount of buffer. However, many occasions arise where the use of salt precipitation is undesirable or harmful; for these cases lyophilization or freeze drying is available. This method involves freezing the protein solution on the walls of a round bottomed flask and sublimating the liquid in a vacuum, as shown in Figure 10-8. The disadvantage of this method is that it concentrates both the protein and any inorganic salts that are present in the solution and may therefore be just as undesirable as

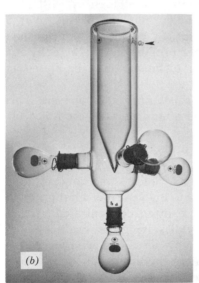

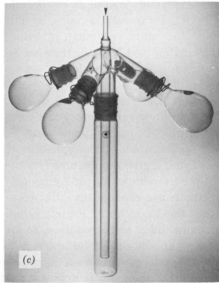

Figure 10-8. Equipment needed for lyophilization or freeze drying small samples. (*a*) The unit is cooled and evacuated by means of a self-contained refrigeration system and vacuum pump. Sample vessels are inserted into the rubber nozzles and the valve (little white knobs) opened for each nozzle that contains a vessel. (*b*, *c*) The units are evacuated by connecting them to a mechanical vacuum pump (connection indicated by arrows) and

ammonium sulfate precipitation. These problems have generated a series of methods that concentrate the proteins of a solution but not the small molecules that may also be present.

The first of these involves placing the solution to be concentrated in small diameter dialysis tubing. A small diameter is specified because this increases the effective surface area to volume ratio. After both ends of the tubing are sealed with the solution inside, the bag is covered with cold powdered polyethylene glycol 6000 and placed at 4°C. The dry powdered polyethylene glycol absorbs solution as it passes through the wall of the dialysis tubing. In this way the macromolecules inside the tubing are concentrated, but small molecules are lost through the tubing. Every 2 hours the bag is removed and the clumped, moist polyethylene glycol is removed from its surface. This procedure may be continued until the desired volume is reached. The polyethylene glycol may be used as it is provided by the manufacturer. It should not be dried because heating converts it to an unusable wax. This method is very convenient and is applicable for even the most labile proteins, but it is useful only for samples less than 50 ml and is somewhat time-consuming.

Entirely new approaches are based upon recent advances in membrane technology. It is now possible to produce membranes which have small holes or pores in them. A cross-section of such a membrane is shown in Figure 10-9. Remarkably, the pore size (Table 10-4) may be controlled to

Table 10-4. Molecular Weight Cutoff Points of Diaflo Ultra-filtration Membranes

Membrane Designation	Molecular Weight Cutoff	Approximate Average Pore Diameter (Å)
XM-300	300,000	140
XM-100	100,000	55
XM-50	50,000	30
PM-30	30,000	22
UM-20	20,000	18
PM-10	10,000	15
UM-2	1000	12
UM-05	500	10

cooled using a solution of acetone containing small pieces of dry ice. The unit in (b) is constructed to contain acetone and dry ice in the center, pointed chamber. The unit in (c) is constructed so that it can be inserted into a Dewar flask containing the acetone–dry ice bath. Where will sublimed water collect on each of these devices? (Courtesy of (a) VirTis Company, Gardiner, N.Y. and (b, c) Ace Glass, Vineland, N.J.)

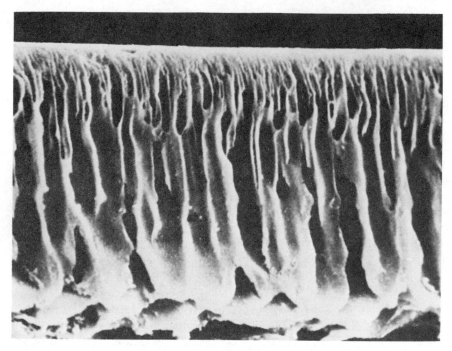

Figure 10-9. An electron micrograph of a cross-section through a Diaflo® membrane. Note the large pores at the exterior (bottom) side of the membrane and the much finer pored "skin" at the interior (top) side of the membrane. (Courtesy of Amicon Corporation, Lexington, Mass.)

permit passage of different sized molecules. A membrane disk is placed in a pressure cell and the sample solution is placed on top of it (Figure 10-10). The sample port is closed and pressure (10 to 80 psi) is exerted on the liquid surface using compressed nitrogen. The exerted pressure forces solvent and small molecules through the membrane (Figure 10-11). Large molecules are left behind in a more concentrated form. The solution is stirred during this operation to decrease clogging of the membrane pores with large molecules. Devices such as these are available in configurations capable of handling from 2 to 3 ml to several hundred liters (Figure 10-12). Concentration of most sample sizes in this manner is quite rapid and is very gentle so that little if any inactivation occurs. The availability of membranes of different pore sizes makes it possible to fractionate molecules of different sizes. The resolution of this method, however, is much less than is attainable using molecular sieve chromatography.

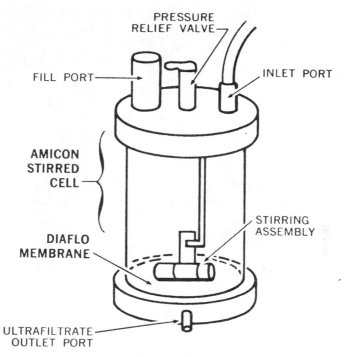

Figure 10-10. Schematic diagram of the Amicon pressure cell used to concentrate macromolecules in solution. (Courtesy of Amicon Corporation, Lexington, Mass.)

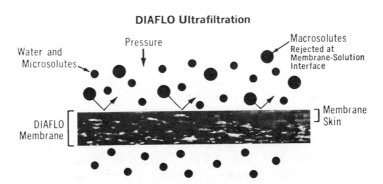

Figure 10-11. Schematic diagram demonstrating operation of Diaflo® membranes at the molecular level. (Courtesy of Amicon Corporation, Lexington, Mass.)

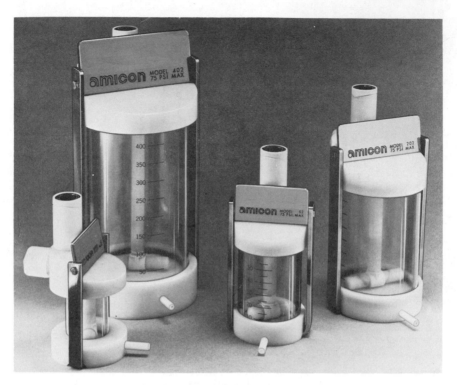

Figure 10-12. Amicon pressure cells capable of accommodating samples ranging from several to 500 ml. (Courtesy of Amicon Corporation, Lexington, Mass.)

The membranes described above are also manufactured in the shape of a hollow tube or fiber. A cross-section of one such fiber is shown in Figure 10-13. These fibers in large numbers form the basis of a new dialysis method called hollow fiber dialysis. Large numbers of these fibers are collected together in bundles in such a way that one solution flows through the central lumen of the fibers and a second solution surrounds the exterior of the fibers. Such a device is shown in Figure 10-14. The protein solution to be dialyzed is placed outside the fibers. This may be either in a static condition, using the beaker assembly shown in Figure 10-14, or it may be passed slowly over the fiber surfaces by means of a pump using a tubular assembly. Buffer solution of low ionic strength is then passed slowly through the lumen of the fiber. Small molecules from the protein solution may easily diffuse through membrane whereas large molecules do not. The mechanism underlying operation of these devices is very similar to that of simple dialysis tubing with the exception that the surface to volume ratio has been enormously increased. This permits the

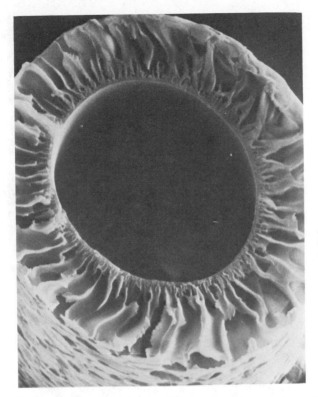

Figure 10-13. Electron micrograph of a cross-section through a membrane fiber used for hollow fiber dialysis. Note the similarity between this fiber and the membrane shown in Figure 10-9. (Courtesy of Amicon Corporation, Lexington, Mass.)

time required for dialysis to be decreased from 5 or more hours to half an hour for certain applications. The device shown in Figure 10-14 can also be used to concentrate proteins in solution. This is accomplished by evacuating the lumina of the fibers.

Ion Exchange Chromatography

Ion exchange chromatography has been discussed in detail in Chapter 4; in addition, there are a number of practical considerations that might be useful here. Previous discussions considered cellulosic ion exchange taking place in a column. Pouring a large volume of crude extract over a column of resin would require an unacceptably long period. There is an alternative method, however, called the batch procedure that can be used to advantage at times. Excess resin is suspended in the crude homogenate

Figure 10-14. Beaker assembly for hollow fiber dialysis. (Courtesy of Bio-Rad Laboratories, Richmond, Calif.)

and allowed to equilibrate for 20 to 40 minutes with intermittent stirring. Any proteins that bind to the resin do so during this time. Unbound proteins may then be removed along with most of the liquid by filtering the resin through coarse filter paper in a very large diameter Büchner funnel. The resin cake that is left can then be suspended in buffers of increasing ionic strength and refiltered. In this way the bound proteins can be divided into four to six crude fractions, for example, those eluting at the following salt concentrations: 0.01 to 0.08M, 0.08 to 0.16M, 0.16 to 0.24M, 0.24 to 0.32M, and 0.32 to 0.40M. Of course, if the desired protein was released in the 0.24 to 0.32M, for example, the procedure could be modified for only two elution steps: one at 0.01 to 0.20M and one at 0.20 to 0.35M. The value of this procedure is that it crudely divides the sample on the basis of charge very quickly. Therefore, any proteins not sticking to the resin are quickly lost. If the desired protein does not stick to the resin the technique is still quite useful because everything that does stick (usually about 50% of the soluble protein) is removed. The latter situation is called negative absorption, that is, a technique that is performed not to remove the desired protein from solution but rather to remove as many other proteins as possible. A second advantage of this method is the very short time required to perform it even on bulky crude homogenates, where it is of greatest value.

There are two problems that must be anticipated, however. First sufficient cellulosic resin must be added. This may be determined by removing small samples of the solution and determining the amount of resin needed to bind as much protein as will bind. Once this has been done the amount of resin needed per unit weight of soluble protein may be calculated. In practice, this allows one to perform a simple biuret protein determination and calculate the quantity of resin needed. It is also advisable to spend a small amount of time determining which of the cellulosic resins (i.e., DEAE, CM, cellulose phosphate) and which experimental conditions yield the best results on a small scale before proceeding to large scale operations.

The second potential problem concerns the ionic strength and magnesium concentration of the homogenate. Usually the buffer concentration used in homogenizing solutions ranges between 0.05 and 0.2M. If the solution has a salt concentration greater than 0.05M, this technique is likely to be seriously compromised. Therefore, a less concentrated buffer should be used initially, or if this is not possible, perhaps the homogenate can be appropriately diluted. It is usually a good practice to actually measure the ionic strength of the homogenate using conductance methods (these are discussed below) to ensure that proper conditions exist prior to adding the resin. If, of course, the desired protein binds tightly to the resin these considerations may not be needed. When phosphocellulose resin is used, magnesium addition to the buffers should be avoided because this cation destroys the capacity of the resin to function properly.

Another practical consideration concerns precautions to be taken in interpreting data obtained with ion exchange methods. Occasions arise when a given enzyme activity elutes from an ion exchange resin at two different salt concentrations. One could naïvely take this as evidence for two enzymes catalyzing the same reaction. However, such an interpretation is very hazardous in the absence of supporting evidence. One of the first things to be done if such a situation occurs is to rechromatograph each of the isolated peaks separately using the original conditions. If on rechromatography either of the isolated activities gives rise to a profile with two peaks (the original profile) it can be reasonably suspected that one protein is present and it is somehow being modified to give the second form. If, on the other hand, both isolated forms chromatograph as they did originally with respect to the salt concentration needed to release them from the resin, further analytical work is warranted to confirm the presence of two proteins carrying out the same reaction.

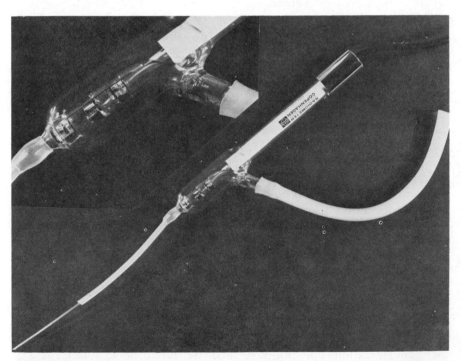

Figure 10-15. Electrode used for measuring the conductivity of a solution. The sampling end of the electrode is fitted with small bore rubber tubing and a glass capillary tube to aid in removing samples from large test tubes. The large rubber tubing fitted to the top of the electrode is connected to a reservoir of glass-distilled water and is used to thoroughly rinse the electrode between measurements. The close-up of the electrode shows the two platinum elements.

Conductance Measurement of Ionic Strength

Ionic solutions are capable of conducting an electrical current. This ability is called conductance, G, and is measured in mhos. The total conductance of a solution depends on the number and kinds of ions in solution and the characteristics of the electrodes. To eliminate the latter variable, measurements are made in terms of specific conductance. This is defined as conductance of a solution contained between two platinum electrodes exactly $1\,cm^2$ in area and $1\,cm$ apart. Such an electrode assembly is shown in Figure 10-15. Test solutions are drawn into the

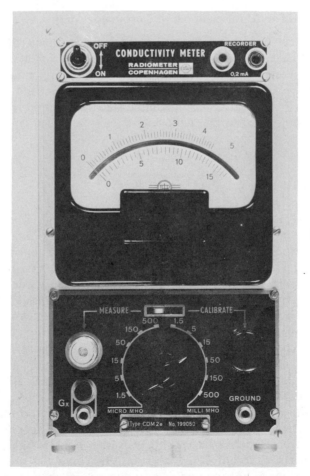

Figure 10-16. The face of a radiometer conductivity meter. The fitting marked measure is the socket used for attachment of the conductivity electrode. (Courtesy of Radiometer A/S, Copenhagen.)

electrode tube until both electrodes are totally submerged. The conductance may then be read from a meter which in effect measures the amount of current passing between the electrodes at a fixed potential difference. The electronic circuits are arranged such that conductance over large orders of magnitude may be measured (Figure 10-16). Since electrodes vary slightly from one set to another it is advisable to calibrate the instrument before routine use with a set of carefully prepared standard solutions of potassium chloride.

Electrophoresis and Molecular Sieve Chromatography

Earlier chapters have been devoted to both these topics, obviating the need for further discussion. Although preparative electrophoretic methods do exist they are not often found in purification schemes. Molecular sieve chromatography, on the other hand, is used extensively. However, owing to restrictions of sample size this method is usually one of the last techniques to be employed.

CRITERIA OF PURITY

Huge expenditures of time and resources are usually required to develop a successful purification scheme. Such investments would seemingly demand an absolute measure of preparation quality or purity. Unfortunately no single ultimate test exists; it is simply not possible to prove something is pure. The converse is easy to prove merely by demonstrating that under some set of conditions the sample may be divided into two parts that behave differently. The degree of purity depends ultimately on the resolution and type of method used. Preparations pure by low resolution techniques can easily be shown to be impure if the assay resolution is increased. If, for example, one uses a high resolution technique such as SDS electrophoresis, a degree of purity is established only to the extent that the preparation is homogeneous with respect to molecular weight. If an enzyme assay is used to detect contamination, resolution increases still farther, but again only for the specific enzymes assayed. The best criterion of purity is establishment of many criteria, each assaying a different characteristic. The following are additional indicators of purity, but are convincing only when considered together. (1) A pure preparation should exhibit a constant specific activity when chromatographed either using linear gradient ion exchange procedures or molecular sieve techniques. This may be shown by determining the amount of enzyme activity and protein in each fraction of the profiles. The ratio of the two should be the same across the whole profile. (2) A single band demonstrated on gel electrophoresis is taken as an indicator of purity. However, in reality it is only a demonstration that all species have a constant charge to mass ratio. If electrophoresis, on the other hand, is performed at a variety of pH values, confidence in the result may be increased. (3) Isoelectric focusing (7–9) may be used as another indication. This technique resolves only on the basis of isoelectric point. The cell contains 10,000 to 50,000 proteins, with isoelectric points distributed principally between pH 4 and 10. The overlap is clear: 7000 proteins per pH unit if the distribution is assumed to be uniform (which it is not; it clusters between pH 4 and 8). (4) The limitations of immunochemical purity were discussed in Chapter 8. (5) Techniques such as end group analysis and SDS acrylamide gel electrophoresis are of considerable value only if the purified material is composed of a single polypeptide chain or a group of identical subunits; this is not an extremely widespread characteristic of proteins. In short, the purity is only as good as the criteria used for its establishment. The more critical this requirement, the greater the number of techniques that must be used and the more skeptically the data obtained must be evaluated.

EXPERIMENTAL

Purification of Acid Phosphatase from Wheat Germ

10-1. The following solutions are required for assay of acid phosphatase.

Solution	Concentration (M)	Volume (ml)
(A) Sodium acetate buffer (pH 5.7)	1	100
(B) $MgCl_2$	0.1	50
(C) p-Nitrophenyl phosphate	0.05	10
(D) KOH	0.5	500

Reagent C should be stored frozen and protected from light. Fresh, unroasted wheat germ should be obtained from a biochemical supplier or other source.

10-2. The following solutions are required for purification of acid phosphatase.

Solution	Concentration (M)	Volume (ml)
(E) $(NH_4)_2SO_4$ (pH 5.5)	saturated at 4°C	1000
(F) Sodium EDTA (pH 5.7)	0.25	100
(G) $MnCl_2$	1	10

10-3. Place 1 qt methanol (tightly capped) in a -20°C freezer overnight.

10-4. Place at least 2 liters glass-distilled water at 4°C overnight. Figure 10-17 is a schematic flow chart of the operations involved in partial purification of acid phosphatase. Supernatant solutions enclosed in rectangles on the flow chart are to be sampled for later assay of enzyme activity and protein concentration.

10-5. Suspend 50 g wheat germ in 200 ml cold water with gentle stirring. Allow the mixture to stand 30 minutes with intermittent stirring.

10-6. Filter the coarse brei through cheesecloth. Squeeze as much liquid as possible from the cheesecloth.

10-7. Transfer the filtrate to centrifuge tubes. Be sure each pair of centrifuge tubes is carefully balanced (use a Harvard triple-beam balance) before they are placed in a centrifuge rotor.

10-8. Centrifuge the tubes in the cold (4°C) for 10 minutes at $10,000 \times g$ (9000 rpm in a Sorvall SS-34 rotor).

10-9. Carefully decant the supernatant from each tube into a 250 ml graduated cylinder and note the volume. This is supernatant I. Discard the pellets remaining in the tubes. A form similar to that shown in Figure 10-18 may aid in recording the volumes of solution yielded at each stage of the purification.

10-10. Remove 1.0 ml supernatant I for assay of protein and enzyme activity. Transfer the remainder of the solution to a 250 ml beaker situated in an ice bath, as shown in Figure 10-19.

10-11. While the extract is being gently stirred with a magnetic stirring bar add 2.0 ml 1M $MnCl_2$ (solution G in step 10-2) for each 100 ml supernatant I present.

10-12. Remove the precipitated material that forms using the procedures in steps 10-7 and 10-8.

10-13. Carefully decant the supernatant from each tube into a 250 ml graduated cylinder and note its volume. This is supernatant II. Discard the pellets remaining in the tubes.

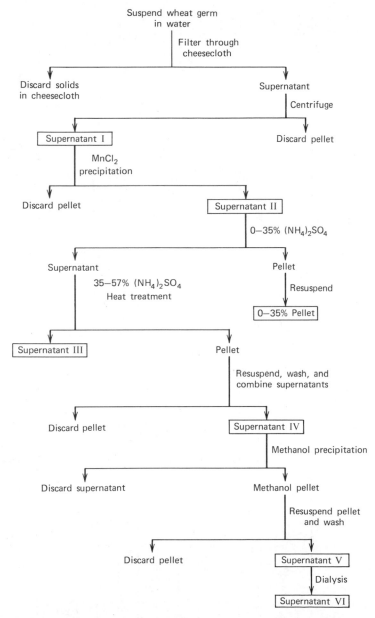

Figure 10-17. Schematic flow chart describing the major steps in the partial purification of acid phosphatase from wheat germ.

ENZYME: _____

DATE: _____

PROTEIN ASSAY: _____

ENZYME ASSAY: _____

PURIFICATION PROCEDURE	PROTEIN					ACTIVITY							
	VOL ml.	SAMPLE (ml)	OD	PROTEIN (mg)	MG/ ML	TOTAL PROTEIN	SAMPLE (ml)	CPM or ΔOD	CPM/ML or OD/ML	UNITS /ml	TOTAL UNITS	SPECIFIC ACTIVITY (U/mg prot)	RECOVERY	FOLD/ PURIFICATION

Figure 10-18. Purification form used to record the various data collected during purification of an enzyme.

Figure 10-19. Ice bath assembly used in ammonium sulfate precipitation of acid phosphatase.

10-14. Remove 1.0 ml supernatant II for assay of protein and enzyme activity. Transfer the remainder of the solution to a 600 ml beaker situated in an ice bath.

10-15. While stirring the solution gently with a magnetic stirrer, slowly add 54 ml of cold, saturated ammonium sulfate solution (solution E in step 10-2) for each 100 ml supernatant II solution. The final concentration of ammonium sulfate will be 35%. Ammonium sulfate addition may be conveniently accomplished using a 10 ml pipette and should require 5 to 10 minutes for completion. If the solution is stirred too rapidly some of the proteins will denature. This becomes apparent by the formation of an off-white foam at the surface of the solution. Such denaturation is to be avoided.

10-16. Following addition of ammonium sulfate, stir the solution for an additional 10 to 15 minutes.

10-17. Remove the precipitated protein using the procedures outlined in steps 10-7 and 10-8.

10-18. Decant the supernatant into a graduated cylinder and note its volume. Transfer the solution back into the cold 600 ml beaker for a second addition of ammonium sulfate.

10-19. Resuspend the pellets in 25 to 50 ml 0.05M sodium acetate buffer (prepared by diluting solution A). Resuspension may be easily accomplished using two or three wooden applicator sticks. When a smooth suspension is obtained remove any undissolved protein by centrifugation (steps 10-7 and 10-8). Decant the supernatant into a 100 ml graduated cylinder and note its volume. Then set this supernatant aside for later assay of protein and enzyme activity. Although little activity is expected in this fraction, it is advisable to preserve all fractions during a purification procedure until they have been demonstrated to contain negligible activity. The pellet of undissolved protein obtained here may be discarded.

10-20. While stirring the solution obtained in step 10-18 gently with a magnetic stirrer, slowly add 51 ml cold, saturated ammonium sulfate for each 100 ml solution volume (this is the volume determined in step 10-18). The final concentration of ammonium sulfate will be 57%.

10-21. Adjust a hot water bath to 70°C.

10-22. To perform heat precipitation of acid phosphatase place the beaker from step 10-20 in the hot water bath and stir the solution moderately with a thermometer. Allow the contents of the beaker to warm to 60°C and maintain that temperature for 2.0 to 2.25 minutes. After this incubation quickly plunge the beaker into an ice–water bath and continue stirring the contents until the temperature drops to 6 to 8°C.

10-23. Remove the precipitate by centrifugation using the procedures in steps 10-7 and 10-8.

10-24. Carefully decant the supernatant solution into a 250 ml graduated cylinder and note its volume. This is supernatant III and may be set aside (at 4°C) for later assay of protein and enzyme activity.

10-25. Resuspend the precipitate obtained in step 10-24 in 40 ml cold glass-distilled water ($\frac{1}{3}$ the volume of supernatant II) using wooden applicator sticks.

10-26. When an even suspension is obtained remove any undissolved protein by centrifugation (steps 10-7 and 10-8).

10-27. Carefully decant the supernatant into a 50 ml graduated cylinder and note its volume. Remove 1.0 ml for later assay. This is supernatant IV. The preparation may be stored frozen at this stage for several weeks without significant loss of activity.

10-28. Determine the protein concentration of supernatant IV and if necessary adjust it to 4 to 5 mg/ml by adding an appropriate amount of cold, distilled water. Determine the total volume of the resulting solution and add for each milliliter 0.09 ml 0.25M EDTA (solution F in step 10-2) and 0.05 ml saturated ammonium sulfate.

10-29. Add at moderate rates with moderate stirring 1.75 ml cold ($-20°C$) methanol for each milliliter of solution obtained in step 10-28. It is important that the methanol be as cold as possible when added to the protein-containing solution.

10-30. Remove the precipitated protein by centrifugation using the procedures in steps 10-7 and 10-8. Decant the supernatant from the precipitated protein and discard it.

10-31. Resuspend the precipitated protein in 10 ml cold water. Remove the undissolved protein by centrifugation and preserve the supernatant solution.

10-32. Resuspend the precipitated protein obtained from step 10-31 in a second 10 ml cold water. Remove the undissolved protein by centrifugation and combine this supernatant solution with that obtained in step 10-31. This is supernatant V. Note its volume and remove a 0.5 ml sample for assay purposes. The insoluble protein precipitate may be discarded.

10-33. Cut a convenient length (15 to 18 in.) of cellulose dialysis tubing ($\frac{1}{2}$ in. diameter). Wet the tubing with distilled water and tie two knots in one end of it, as shown in Figure 10-7.

10-34. Transfer supernatant V into the dialysis tubing bag by means of a Pasteur pipette. Be careful not to puncture the bag with the pipette.

10-35. Trap an air bubble in the top of the bag and tie two knots in the open end of the dialysis tubing.

10-36. Place the sealed dialysis bag in a 1 liter flask containing 1 liter of cold 5mM EDTA solution. Stir the contents of the flask slowly (using a large magnetic stirring bar) overnight at 4°C.

10-37. At the conclusion of dialysis carefully cut off the top of the dialysis bag and transfer its contents to a graduated vessel and note its volume. The preparation may be stored frozen in this condition and is designated as supernatant VI. A 0.5 ml sample should be removed for protein and enzyme assay.

10-38. At this point the preparation may be used in a number of different ways. It may be subjected to DEAE–cellulose chromatography at pH 7.4 using a linear salt gradient of between 0.005 and 0.1M Tris buffer adjusted to pH 7.4. Alternatively, it may be used at this stage of purity for kinetic studies as described beginning with step 10-42. Finally the preparation may be subjected to molecular sieve chromatography on an agarose gel such as Biogel 0.5M (0.5×10^6 dalton exclusion limit). If the last method is chosen the protein-containing solution must first be concentrated as described below.

10-39. To concentrate supernatant VI slowly add 4 g solid ammonium sulfate for each 10 ml solution. This is done in an ice bath as described in step 10-10 but on a smaller scale.

10-40. After the mixture is allowed to equilibrate for 10 to 15 minutes, collect the precipitated protein by centrifugation and dissolve it in 2 to 3 ml of a solution containing a final concentration of 0.2M ammonium sulfate and 1mM EDTA. Remove any undissolved protein by centrifugation and discard it.

10-41. The preparation may now be subjected to molecular sieve chromatography at room temperature using a 2.5 × 35 cm column of Biogel 0.5M agarose equilibrated with 0.2M ammonium sulfate and 1mM EDTA. Chapter 6 should be consulted for a detailed discussion of these procedures. A sample size of 1 to 2 ml, however, is recommended.

Warburg–Christian Protein Determination

Protein concentrations of the fractions eluted from either the DEAE cellulose or molecular sieve columns may be determined using the Warburg–Christian method (10). This method is based on the absorption of 280 nm light by tyrosine and tryptophan residues. Since the amounts of these residues vary greatly from one protein to another the method is of only qualitative usefulness when applied to protein mixtures. However, the ease, sensitivity, and short time required for the determination have made it the method of choice for monitoring protein concentration in column eluates. A second problem encountered with this assay is interference caused by contamination of the protein preparation with nucleic acids. This problem may be circumvented by taking advantage of the fact that nucleic acids absorb more strongly at 260 nm than at 280 nm and, conversely, proteins absorb much more strongly at 280 nm. Using crystalline yeast enolase and purified nucleic acid as standards, Christian and Warburg evaluated the errors caused by the presence of nucleic acids and constructed a correction table (see Table 10-5). The absorbance of each protein-containing sample is measured at 280 and 260 nm. The ratio of these two values is determined and used to select an appropriate correction factor from Table 10-5. The 280 nm absorbance is multiplied by this factor to yield the protein concentration in milligrams per milliliter:

$$(A_{280\,nm})(\text{correction factor}) = \text{mg/ml protein} \tag{8}$$

Solutions of high protein concentration must be diluted to yield an absorbance value below 1.0.

Establishment of Appropriate Assay Conditions for Acid Phosphatase

Before an assay is used in practice, experiments should be performed to determine precisely the conditions under which it is linear. This is particularly true of an assay that will be used in construction of a

Table 10-5. Correction Factors for the Warburg–Christian Method for Protein Determination

A_{280}/A_{260}	Correction Factor	Nucleic Acid (%)
1.75	1.12	0
1.63	1.08	0.25
1.52	1.05	0.50
1.40	1.02	0.75
1.36	0.99	1.00
1.30	0.97	1.25
1.25	0.94	1.50
1.16	0.90	2.00
1.09	0.85	2.50
1.03	0.81	3.00
0.98	0.78	3.50
0.94	0.74	4.00
0.87	0.68	5.00
0.85	0.66	5.50
0.82	0.63	6.00
0.80	0.61	6.50
0.78	0.59	7.00
0.77	0.57	7.50
0.75	0.55	8.00
0.73	0.51	9.00
0.71	0.48	10.00
0.67	0.42	12.00
0.64	0.38	14.00
0.62	0.32	17.00
0.60	0.29	20.00

purification table, where the range of enzyme specific activities (μmoles of product formed per minute per milligram of protein) is broad. Therefore, conditions for linearity of both time and protein must be established.

Time Curve

10-42. Carefully transfer the following reagents from step 10-1 into a 16×150 mm test tube:

Reagent	Volume (ml)
A	0.5
B	0.5
C	0.5
H_2O	3.3

Mix these components thoroughly by vortexing the tube at a moderate speed.

10-43. Place the tube in a 30°C water bath and allow it to equilibrate for 10 minutes.

10-44. Add 0.2 ml supernatant VI that has been appropriately diluted. The enzyme preparation must be diluted because at the high concentrations of acid phosphatase present in supernatant VI all of the substrate would be expended immediately. The proper degree of dilution must be determined by trial and error using reaction mixtures with a volume of 0.5 ml. The proper dilution should yield an absorbance of 0.3 to 0.4 for the assay mixture after 5 minutes incubation. If extremely high dilutions are needed it may be necessary to include serum albumin (1% final concentration) in the diluent to protect the enzyme against inactivation.

10-45. Mix the reaction components prepared in step 10-44 by moderate vortexing.

10-46. Immediately transfer a 0.5 ml sample from the large reaction mixture yielded at step 10-45 to a 13×100 mm test tube and add exactly 2.5 ml 0.5N KOH. The KOH will terminate the reaction and convert p-nitrophenol to its unprotonated colored form. Take the first sample as quickly as possible because it is the "zero" time point.

10-47. Remove 0.5 ml samples at 5, 10, 15, 20, 25, 30, and 35 minutes and process them as described in step 10-46.

10-48. Make certain the contents of each tube are mixed thoroughly by vortexing.

10-49. If the mixed solutions are cloudy owing to the presence of precipitated protein, centrifuge the tubes for 10 minutes at 4000 rpm in a desk top clinical centrifuge.

10-50. Determine the absorbance of each tube at 405 nm. A simple spectrophotometer such as the B & L Spectronic 20 is sufficient for these measurements.

10-51. Plot the data obtained as shown in Figure 10-20. Notice that the curve extrapolates through the "zero" time point rather than zero absorbance. The zero or blank value must be subtracted from each experimental value if these data are to be converted to enzyme units (μmoles p-nitrophenol produced/minute). It is more advisable to subtract the measured "zero" value from the experimental values than to merely use the "zero" time point to "blank" or zero the spectrophotometer. Although the latter method is more efficient the experimenter has no knowledge of the "zero" value when this method is used. Notice that the curve is no longer linear above absorbances of 1.2 to 1.4.

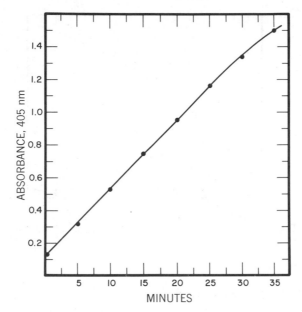

Figure 10-20. The effect of increasing times of incubation on the amount of *p*-nitrophenol produced by action of acid phosphatase.

Protein Curve

10-52. Carefully prepare eight 13×100 mm test tubes, each containing the following reagents.

Reagent	Volume (ml)
A	0.05
B	0.05

10-53. Add the following amounts of protein to each tube respectively: 0, 0.05, 0.1, 0.15, 0.2, 0.25, 0.3, and 0.35 ml. The protein solution should be supernatant VI, diluted as was described in step 10-44.

10-54. Add sufficient water to each tube to produce a final volume of 0.45 ml.

10-55. At 2 minute intervals initiate the reaction by adding 0.05 ml of reagent C to each tube, vortex it moderately, and place it in a 30°C water bath.

10-56. After incubating each tube for exactly 15 minutes terminate the reaction by adding 2.0 ml 0.5N KOH.

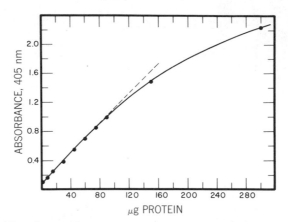

Figure 10-21. The effects of increasing protein on the amount of *p*-nitrophenol produced by action of acid phosphatase.

10-57. Repeat steps 10-48 to 10-50. Data from this type of experiment are shown in Figure 10-21. Note in this case it is not advisable to assay preparations of this purity at protein concentrations above 70 μg.

Determination of the Michaelis Constant of Acid Phosphatase for *p*-Nitrophenol Phosphate

10-58. Prepare a set of dilutions of *p*-nitrophenyl phosphate (Reagent C in step 10-1) such that when 0.1 ml of the diluted substrate is added to a 0.5 ml reaction mixture the following final concentrations are yielded: 0, 0.05, 0.06, 0.075, 0.10, 0.125, 0.25, 0.5, 1.0, and 5.0m*M*. Water may be used as diluent.

10-59. Add respectively 0.1 ml of the dilutions prepared in step 10-58 to each of 10 13 × 100 mm test tubes.

10-60. Add 0.05 ml reagent A, 0.05 ml reagent B, and 0.2 ml water to each of the 10 test tubes.

10-61. At 2 minute intervals, add 0.1 ml of an appropriate dilution of supernatant VI (step 10-44) to each tube. Mix the contents of the tube by moderately vortexing and place it in a 30°C water bath.

10-62. After each tube has incubated at 30°C for exactly 15 minutes, add 2.0 ml 0.5*N* KOH to terminate the reaction.

10-63. Repeat steps 10-48 to 10-50.

10-64. Repeat steps 10-58 to 10-63 but use 0.1 ml water and 0.1 ml 0.005*M* potassium dibasic phosphate (K$_2$HPO$_4$) in place of the 0.2 ml water used in step 10-60. This will yield a final phosphate concentration of 1mM in each reaction mixture.

10-65. Using an extinction coefficient for p-nitrophenol of 18.8×10^6 cm^2/mole, convert all the absorbance readings to μmoles p-nitrophenol formed.

10-66. Calculate the inverse of each of these values and the corresponding substrate (p-nitrophenyl phosphate) concentrations.

10-67. Plot these data as shown in Figure 10-22.

10-68. Determine the apparent K_m value for p-nitrophenyl phosphate in the presence and absence of inorganic phosphate.

10-69. V_{max} may also be determined from these data, but it must be kept in mind that this is the rate of cleavage for 15 minutes.

10-70. These data may be replotted using the method of Eadie and Hofstee (11, 12).

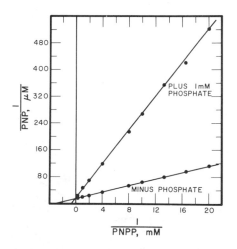

Figure 10-22. Lineweaver–Burke plot of the rate of p-nitrophenol production at increasing concentrations of substrate (p-nitrophenyl phosphate, PNPP). The same experiment was performed in the presence of 1mM inorganic phosphate.

Construction of a Purification Table

10-71. Assay each of the fractions preserved in steps 10-5 to 10-38 for protein concentration and enzyme activity. Use the Lowry method described in Chapter 2 for protein determinations. Use the enzyme assay procedures described in steps 10-52 to 10-57 to determine enzyme activity.

10-72. Calculate the specific activity (μmoles p-nitrophenol formed/minute/mg of protein) for each fraction assayed. A representative set of these data appears in Table 10-6.

Table 10-6. Partial Purification of Acid Phosphatase from Wheat Germ

Fraction	Total Volume (ml)	Total Protein (mg)	Total Enzyme Activity (nmole units/min)	Enzyme Specific Activity (units/mg)
Supernatant I	129	3354	171	0.051
Supernatant II	128	2432	160	0.066
0–35% Pellet	18	504	19	0.038
Supernatant III	324	810	72	0.089
Supernatant IV	66.5	798	124	0.155
Supernatant V	21.7	279.9	71	0.254
Supernatant VI	26.5	278	134	0.482

REFERENCES

1. T. G. Cooper and H. Beevers, *J. Biol. Chem.*, **244**:3507 (1969). Mitochondria and Glyoxysomes from Castor Bean Endosperm—Enzyme Constituents and Catalytic Capacity.

2. D. E. Hughes, *Brit. J. Exp. Pathol.*, **32**:97 (1951). A Press for Disrupting Bacteria and Other Microorganisms.

3. N. R. Eaton, *J. Bacteriol.*, **83**:1359 (1962). New Press for Disruption of Microorganisms.

4. N. J. Leonard, J. J. McDonald, R. E. L. Henderson, and M. E. Reichmann, *Biochemistry*, **10**:3335 (1971). Reaction of Diethyl Pyrocarbonate with Nucleic Acid Components. Adenosine.

5. I. Fedorcsak, L. Ehrenberg, and F. Solymosy, *Biochem. Biophys. Res. Commun.*, **65**:490 (1975). Diethyl Pyrocarbonate Does Not Degrade RNA.

6. J. J. Irias, M. R. Olmsted, and M. F. Utter, *Biochemistry*, **8**:5136 (1969). Pyruvate Carboxylase. Reversible Inactivation by Cold.

7. O. Vesterberg, Isoelectric Focusing of Proteins, in *Methods of Enzymology*, Vol. 22 (W. B. Jacoby, Ed.), Academic Press, New York, 1971, p. 389.

8. O. Vesterberg, Isoelectric Focusing and Separation of Proteins, in *Methods in Microbiology*, Vol. 5B (J. R. Norris and D. W. Robbins, Eds.), Academic Press, New York, 1971, p. 595.

9. H. Haglund, Isoelectric Focusing in pH Gradients—A Technique for Fractionation and Characterization of Ampholytes, in *Methods of Biochemical Analysis*, Vol. 19 (D. Glick, Ed.), Wiley, New York, 1971, p. 1.

10. E. Layne, Spectrophotometric and Turbidimetric Methods for Measuring Proteins, in *Methods in Enzymology*, Vol. 3 (S. P. Colowick and N. O. Kaplan, Eds.), Academic Press, New York, 1957, p. 447.

11. G. S. Eadie, F. Bernheim, and M. L. C. Bernheim, *J. Biol. Chem.*, **181**:449 (1949). Partial Purification and Properties of Animal and Plant Hydantoinases.

12. B. H. J. Hofstee, *Science*, **116**:329 (1952). On the Evaluation of the Constants V_m and K_m in Enzyme Reactions.

13. W. B. Jakoby, Ed., *Enzyme Purification and Related Techniques*, Vol. 22 of *Methods in Enzymology*, Academic Press, New York, 1971.

14. M. Dixon and E. C. Webb, Enzyme Fractionation by Salting-out: A Theoretical Note, in *Advances in Protein Chemistry*, Vol. 16 (C. B. Anfinsen et al., Eds.), Academic Press, New York, 1961, pp. 197–219.

15. S. L. Berger, *Anal. Biochem.*, **67**:428–437 (1975). Diethyl Pyrocarbonate: An Examination of its Properties in Buffered Solutions with a New Assay Technique.

16. W. Honig and M. R. Kula, *Anal. Biochem.*, **72**:502–512 (1976). Selectivity of Protein Precipitation with Polyethylene Glycol Fractions of Various Molecular Weights.

Appendix I

Concentration of Acids and Bases: Common Commercial Strengths

Acid or Base	Molecular Weight	Concentration		Specific Gravity
		(Moles/l)	% by Weight	
Acetic acid, glacial	60.1	17.4	99.5	1.05
Butyric acid	88.1	10.3	95	0.96
Formic acid	46.0	23.4	90	1.20
Hydriodic acid	127.9	7.6	57	1.70
Hydrobromic acid	80.9	8.9	48	1.50
Hydrochloric acid	36.5	11.6	36	1.18
Hydrofluoric acid	20.0	32.1	55	1.17
Lactic acid	90.1	11.3	85	1.20
Nitric acid	63.0	16.0	71	1.42
Perchloric acid	100.5	11.6	70	1.67
Phosphoric acid	80.0	18.1	85	1.70
Sulfuric acid	98.1	18.0	96	1.84
Ammonium hydroxide	35.0	14.8	28	0.89

International Scale (1936) of Refractive Indexes of Sucrose Solutions at 20°C

Index	Percent	Index	Percent	Index	Percent	Index	Percent
1.3330	0	1.3723	25	1.4200	50	1.4774	75
1.3344	1	1.3740	26	1.4221	51	1.4799	76
1.3359	2	1.3758	27	1.4242	52	1.4825	77
1.3373	3	1.3775	28	1.4264	53	1.4850	78
1.3388	4	1.3793	29	1.4285	54	1.4876	79
1.3403	5	1.3811	30	1.4307	55	1.4901	80
1.3418	6	1.3829	31	1.4329	56	1.4927	81
1.3433	7	1.3847	32	1.4351	57	1.4954	82
1.3448	8	1.3865	33	1.4373	58	1.4980	83
1.3463	9	1.3883	34	1.4396	59	1.5007	84
1.3478	10	1.3902	35	1.4418	60	1.5033	85
1.3494	11	1.3920	36	1.4441	61		
1.3509	12	1.3939	37	1.4464	62		
1.3525	13	1.3958	38	1.4486	63		
1.3541	14	1.3978	39	1.4509	64		
1.3557	15	1.3997	40	1.4532	65		
1.3573	16	1.4016	41	1.4555	66		
1.3589	17	1.4036	42	1.4579	67		
1.3605	18	1.4056	43	1.4603	68		
1.3622	19	1.4076	44	1.4627	69		
1.3638	20	1.4096	45	1.4651	70		
1.3655	21	1.4117	46	1.4676	71		
1.3672	22	1.4137	47	1.4700	72		
1.3689	23	1.4158	48	1.4725	73		
1.3706	24	1.4179	49	1.4749	74		

Density at 25°C of CsCl Solution as a Function of Refractive Index

Refractive Index (sodium D line, 25°C)	Density (g/cm³)
1.34400	1.09857
1.34880	1.15070
1.35340	1.20066
1.35800	1.25062
1.36260	1.30057
1.36720	1.35053
1.37180	1.40049
1.37640	1.45044
1.38100	1.50040
1.38560	1.55035
1.39020	1.60031
1.39480	1.65027
1.39940	1.70022
1.40400	1.75018
1.40860	1.80014
1.41320	1.85009
1.41780	1.90005

Periodic Table
of Elements

Periodic Table of the Elements

IA	IIA	IIIB	IVB	VB	VIB	VIIB	VIII			IB	IIB	IIIA	IVA	VA	VIA	VIIA	
1 H 1.0079																	2 He 4.0026
3 Li 6.939	4 Be 9.0122											5 B 10.81	6 C 12.01115	7 N 14.0067	8 O 15.9994	9 F 18.9984	10 Ne 20.183
11 Na 22.9898	12 Mg 24.312											13 Al 26.9815	14 Si 28.086	15 P 30.9738	16 S 32.06	17 Cl 35.453	18 Ar 39.948
19 K 39.102	20 Ca 40.08	21 Sc 44.956	22 Ti 47.90	23 V 50.942	24 Cr 51.996	25 Mn 54.9380	26 Fe 55.847	27 Co 58.933	28 Ni 58.71	29 Cu 63.54	30 Zn 65.37	31 Ga 69.72	32 Ge 72.59	33 As 74.92	34 Se 78.96	35 Br 79.909	36 Kr 83.80
37 Rb 85.47	38 Sr 87.62	39 Y 88.906	40 Zr 91.22	41 Nb 92.906	42 Mo 95.94	43 Tc 98	44 Ru 101.07	45 Rh 102.905	46 Pd 106.4	47 Ag 107.87	48 Cd 112.40	49 In 114.82	50 Sn 118.69	51 Sb 121.75	52 Te 127.60	53 I 126.904	54 Xe 131.30
55 Cs 132.905	56 Ba 137.34	57 La☆ 138.905	72 Hf 178.49	73 Ta 180.948	74 W 183.85	75 Re 186.2	76 Os 190.2	77 Ir 192.2	78 Pt 195.09	79 Au 196.967	80 Hg 200.59	81 Tl 204.37	82 Pb 207.19	83 Bi 208.980	84 Po 210	85 At 210	86 Rn 222
87 Fr 223	88 Ra 226	89 Ac★ 227	104 Ku 261	105 Ha 260	106 ( ) 260												

☆ Lanthanide Series	58 Ce 140.12	59 Pr 140.907	60 Nd 144.24	61 Pm 147	62 Sm 150.4	63 Eu 151.96	64 Gd 157.25	65 Tb 158.924	66 Dy 162.50	67 Ho 164.930	68 Er 167.26	69 Tm 168.934	70 Yb 173.04	71 Lu 174.97
★ Actinide Series	90 Th 232.038	91 Pa 231	92 U 238.03	93 Np 237	94 Pu 242	95 Am 243	96 Cm 247	97 Bk 247	98 Cf 249	99 Es 254	100 Fm 253	101 Md 256	102 No 254	103 Lr 257

From Cummins and Wartell, *Introduction to Chemistry.*

Index

Absorbance, 39–42
  complex solutions, 40–42
  spectra, 40, 41, 60
  spectrum catalase, 62, 63
Absorption coefficient, 39
Absorptivity constant, 39
Accuracy, 356, 357
Acetone powder, 375
Acid, Bronsted-Lowry, 1
Acidic resin, 139
Acid phosphatase
  assay, 165–167, 398–402
  kinetics, 402, 403
  purification, 391–398
Acrylamide
  recrystallization, 220
  structure of, 196
Acrylamide gel
  casting, 221, 223
  electrophoresis, 195–200
  pore size of, 197, 198
  removal from tube, 225
Activity coefficient, 4
O-Acylisourea, 244, 250
Adenosine isolation, 165
Adenylic acid, 56
Adjuvants, 265, 266, 268
ADP isolation, 165
Adsorption, 155
AE cellulose, 143
Affinity chromatography
  absorbent derivatives, 244–246, 248, 249
  batch process, 247
  coupling reaction pH, 245
  elution procedures, 247, 251–254
  ligand concentration, 243
  ligand coupling, 240–243
  ligand selection, 239, 240
  loading procedures, 247
  matrix, 238, 239
  mechanism of operation, 234, 235
  steps involved, 237

Affinity group linkage
  amino group, 245, 248
  imidazole group, 245, 249
  phenolic group, 245, 249
  sulfhydryl group, 245
Affinity matrix
  activation, 241–243
  calibration, 240
  nonaqueous derivatization, 246
  storage, 246
Agar, 173
Agaropectin, 173
Agarose-acrylamide gel, 210, 211
Agarose beads, 238
Alcian blue, 216
Aldopentoses, 56
Alkali error, 18
Alleles, 257
Allotype, 257
Aluminum hydroxide, 265
Amine activation, 241–243
Amino acids separation, 162–164
Amino acids titration, 28–30
ω-Aminoalkylation, 244, 248–250
Aminonapthosulfonic acid, 55
Ammonia, 58
Ammonium molybdate, 55
Ammonium persulfate, 196
Ammonium sulfate, 51, 370–374
  fluffy precipitate, 373
  nomogram, 372
Ammonium sulfate concentration calculation, 371
Ampholytes, 211
Amphoterism, 137
AMP isolation, 165
Amplification, 16, 81
Anamnestic response, 269
Angle of refraction, 334, 336
Angle rotor, 311, 321, 322
Angular velocity, 309–310
Anilinonaphthalene sulfonate, 214
Anion exchanger, definition, 136